THE NATURE OF EXISTENCE

THE NATURE OF EXISTENCE

JOHN MᶜTAGGART ELLIS MᶜTAGGART

VOLUME II

EDITED BY

C. D. BROAD

The right of the
University of Cambridge
to print and sell
all manner of books
was granted by
Henry VIII in 1534.
The University has printed
and published continuously
since 1584.

CAMBRIDGE UNIVERSITY PRESS
Cambridge
New York   New Rochelle   Melbourne   Sydney

Published by the Press Syndicate of the University of Cambridge
The Pitt Building, Trumpington Street, Cambridge CB2 1RP
32 East 57th Street, New York, NY 10022, USA
10 Stamford Road, Oakleigh, Melbourne 3166, Australia

First published 1927
Reprinted 1968
First paperback edition 1988

Printed in Great Britain by Redwood Burn Ltd, Trowbridge, Wilts.

British Library cataloguing in publication data
McTaggart, John McTaggart Ellis
The nature of existence.
1. Ontology
I. Title   II. Broad, C. D.
111'.1   BD311

Library of Congress cataloguing in publication data
McTaggart, John McTaggart Ellis, 1866-1925.
The nature of existence.
Includes bibliographical references and indexes.
1. Ontology.   2. Reality.   I. Broad, C. D. (Charlie
Dunbar), 1887–1971.   II. Title.
BD331.M313   1988   111   87-35748

ISBN 0 521 07426 6 hard covers
ISBN 0 521 35769 1 paperback

Typeset by Speedspools, Edinburgh

EDITOR'S PREFACE

THE late Dr McTaggart left behind him a paper of instructions in which he expressed the wish that I should undertake the publication of the concluding volume of his *Nature of Existence*, if he should die before the manuscript had been printed. This wish conferred an honour, whilst it imposed an obligation. I regret that the pressure of my academic work—heavy before, and greatly, suddenly, and permanently increased by McTaggart's lamented death—has prevented me from fulfilling my task earlier.

It was McTaggart's custom, before publishing a book, to make five successive complete drafts of it. Each draft, except the last, was submitted, when finished, to certain of his friends for criticisms and suggestions, and was exposed to the full force of his own unsparing judgment in respect of both literary form and logical rigour. The next draft would embody such additions and alterations as his own reflections or the comments of others seemed to make desirable. Naturally the changes in the later drafts were, as a rule, comparatively trifling. The position which had been reached at McTaggart's death, in the case of *The Nature of Existence*, was as follows. Drafts A and B were completed, and he had been busily engaged for some time in revising Draft B and writing Draft C. Draft C existed in typescript up to the end of Chap. XLVII, and in manuscript it extended to the end of what is now Section 567 of Chap. XLVIII. The book is therefore printed from Draft C up to the latter point, and thenceforward from Draft B. It seems unlikely that McTaggart would have made any very substantial modifications, if he had lived.

I have divided the book into numbered sections; have filled in all cross-references; and have constructed an *Analytical Table of Contents* and an *Index of Terms*, thus bringing it into line

with the first volume and with the rest of McTaggart's published
works. The only changes which I have made are verbal ones, and
they are few in number and slight in extent.

C. D. BROAD

Trinity College, Cambridge
October, 1927

ANALYTICAL TABLE OF CONTENTS

BOOK V. PRESENT EXPERIENCE AND ABSOLUTE REALITY

CHAPTER XXXII
THE INTRODUCTION OF THE EMPIRICAL

CHAPTER XXXIII. TIME

CHAPTER XXXIV. MATTER

CHAPTER XXXV. SENSA

CHAPTER XXXVI. SPIRIT

CHAPTER XXXVII. COGITATION

CHAPTER XXXVIII. IDEALISM

CHAPTER XXXIX
FURTHER CONSIDERATIONS ON SELVES

CHAPTER XL. VOLITION

CHAPTER XLII. DISSIMILARITY OF SELVES

BOOK VI. ERROR

CHAPTER XLIV. ERROR

CHAPTER XLVII
THE C SERIES—NATURE OF THE TERMS

CHAPTER XLVIII
THE C SERIES—NATURE OF THE RELATIONS

CHAPTER XLIX

THE RELATIONS OF THE THREE SERIES

CHAPTER LI
FURTHER CONSIDERATIONS ON TIME

CHAPTER LII
APPARENT MATTER AND APPARENT SENSA

CHAPTER LIII. APPARENT PERCEPTIONS

CHAPTER LIV. APPARENT JUDGMENTS

CHAPTER LV. APPARENT INFERENCE

CHAPTER LVI
OTHER APPARENT FORMS OF COGITATION

CHAPTER LVII. EMOTION AND VOLITION

CHAPTER LVIII. APPEARANCE AND REALITY

BOOK VII. PRACTICAL CONSEQUENCES
CHAPTER LIX
THE FUNDAMENTAL SENSE OF THE *B* SERIES

CHAPTER LX
THE FUNDAMENTAL SENSE OF THE *C* SERIES

CHAPTER LXI. THE FUTURITY OF THE WHOLE

CHAPTER LXII. IMMORTALITY

CHAPTER LXIII
PRE-EXISTENCE AND POST-EXISTENCE

CHAPTER LXIV. GOOD AND EVIL

CHAPTER LXVI
VALUE IN THE PRE-FINAL STAGES OF THE *C* SERIES

CHAPTER LXVII. TOTAL VALUE IN THE UNIVERSE

CHAPTER LXVIII. CONCLUSION

BOOK V

PRESENT EXPERIENCE AND ABSOLUTE REALITY

CHAPTER XXXII

THE INTRODUCTION OF THE EMPIRICAL

294. The remainder of this work will have a different object from that of the four preceding books, which were contained in the first volume. So far we have endeavoured to determine the general nature of the existent by arguments which required no empirical data except two—the fact that something did exist, and the fact that the existent was differentiated. But now we have to enquire what consequences of theoretical or practical interest can be drawn from the general nature of the existent, with respect to various parts of the existent which are empirically known to us. This enquiry will fall into three divisions.

295. In the first part, which will occupy the present Book, we shall have to consider various characteristics as to which our experience gives us, at the least, a *primâ facie* suggestion that they are possessed either by all that exists, or by some existent things. And two questions will arise about these characteristics. Starting from our conclusions as to the general nature of the existent, as reached in the earlier Books, we shall have to ask, firstly, which of these characteristics can really be possessed by what is existent, and which of them, in spite of the *primâ facie* appearance to the contrary, cannot be possessed by anything existent. And we must ask, secondly, of those which are found to be possible characteristics of the existent, whether any of them can be known to be actual characteristics of it.

296. The second part of the enquiry will be contained in Book VI. As a result of the first part we shall have reached the conclusion that some characteristics, both positive and negative, which appear to be possessed by the existent, are not really possessed by it. And this will be the case, not only with the characteristics which the existent is judged to possess, but also with the characteristics which it is perceived as possessing.

I shall speak of the nature which the existent really has, as its nature "In Absolute Reality," and of the nature which it appears

to us to have, as its nature "In Present Experience." I shall not
use either "present experience" or "appearance" in such a way
as to *exclude* absolute reality. For example, while an apparent
judgment is, as we shall see, something which appears to be a
judgment, but is not a judgment, an apparent perception is something which both appears to be a perception and is a perception.
The former will be distinguished from the latter by calling it
"mere appearance" or "only appearance."

We shall have to enquire how it can happen that the appearance should diverge from the reality. And we shall have to enquire also whether there are any uniform relations which can be
discovered between different variations of the appearance and
different variations of the reality. (For example, we shall find
reason in the next chapter to conclude that time is only an appearance. And we shall have to enquire whether the apparent relation
of earlier and later has itself any uniform relation to any real
relation between timeless realities. If this should be so, the
variations of the appearance in question will give us knowledge
about the relations of the reality of which it is the appearance,
and it may itself be called a *phaenomenon bene fundatum*[1].)

297. In the third place, there are various questions, which are
or appear to be of practical interest to us, of which it may be
possible to learn something by means of the results gained in the
earlier Books. These will be considered in Book VII. Of these
questions, some are usually stated in terms of various characteristics which we shall have found reason to suppose were only
apparently, and not really, possessed by the existent. With regard to these characteristics, we shall have to enquire what the
realities are which correspond to the appearances in question,
and we must then consider how the questions must be restated,
so as to apply to the existent, and how such questions should be
answered.

298. In these three Books the argument will necessarily be less
rigid than in the earlier enquiry. In our attempt to determine
the general nature of the existent, we aimed at absolute demonstration. Our results were either fallacious through some error in
the argument, or they were certain. We had occasion, at various

[1] Cp. Section 53.

points, to speak of probabilities, but this was only incidental. The assertion of those probabilities did not form steps in the main line of the argument, and did not affect the claim that the later stages had been absolutely demonstrated.

In our present enquiry it will be different. In the first part of it, indeed, some questions can be answered with absolute demonstration, but the answers will all be of a negative character. It will be possible to show that, having regard to the general nature of the existent as previously determined, certain characteristics, which we consider here for the first time, *cannot* be true of the existent, and it will be possible to show this in a way which, if not absolutely fallacious, can lead to nothing but a perfect certainty.

But it will not be possible to show with perfect certainty that any of those characteristics which we consider here for the first time *must* be true of the existent. The only manner in which such a proof could be effected would be to show that the general nature of the existent, as determined in the earlier enquiry, was such that nothing which we know or can imagine could have the general nature without having the characteristic in question. And this does not give more than a probability. For it is possible that that general nature could be possessed by something which had not that characteristic, but some other, which we have never experienced, and cannot now imagine. For example, we have reached the conclusion that nothing which exists can have simple parts. We may be able to show that nothing which we know or can imagine can be without simple parts except spirit. But this will not give us an absolute demonstration that everything that exists is spiritual, or indeed that anything that exists is spiritual. For there may be some characteristic, which we have never experienced or imagined, the possessor of which could be without simple parts although it was not spiritual. And it may be this characteristic which is found in part or all of the existent.

In the same way, when we pass to the second and third parts of our enquiry, all that can be shown, at any rate in most cases, is that a certain solution is possible, and that we know and can imagine no other. But here again, our inability to know or imagine another solution might be due only to the limitations of our experience. It is possible that some characteristic, which could only be known empirically, and which we have had no chance of knowing empirically, might be the key to an alternative possible solution, and that that solution might be the true one. In problems of this sort, therefore, our arguments may possibly attain a high degree of probability, but can never hope for certainty.

299. In these three Books we shall, as was said above, be dealing in part with facts which are only known empirically. And all empirical knowledge is either perception or knowledge based on perception. I do not perceive that my table is square, for I do not perceive the table at all. Nor do I perceive that Caesar was killed in the Senate House, or that all cows chew the cud. But my knowledge of all three propositions is based on my perception—*primâ facie* on my perception of sensa.

We defined perception in Section 44. Perception is knowledge; and it is distinguished from other knowledge by being knowledge by acquaintance, or awareness. It is distinguished, again, from other awareness by being awareness of substances, as opposed to that awareness of characteristics which tells us what a quality like yellow, or a relation like superiority, is in itself.

300. But, although perception is awareness of substances, we find that it always gives us knowledge *about* the characteristics of these substances. If it did not, we should have no knowledge about the characteristics of any particular substance, except the knowledge that it had those characteristics which we know *à priori* to belong to all substances. For all other knowledge about the characteristics of any particular substance is empirical, and, as we have just said, no empirical knowledge can be based on anything but perception. When I assert "this is a sensum of red," or "I am angry," it is clear that I am asserting a characteristic of a perceptum, and that the assertion can only be based on the perception. And when I assert that my table is square, or that Caesar was killed in the Senate House, or that all cows chew the cud, it is clear that to justify such assertions it is not sufficient to know certain sensa as substances. It is also necessary to know that those sensa have certain characteristics, which are such that their occurrence implies the truth of the propositions I am asserting.

301. A perception, however, cannot be knowledge *that* a substance has certain characteristics. For knowledge that anything is, or has, anything is a judgment, and not a perception. The best expression, I think, for the relation between the perception, the perceived substance, and the characteristics is to say that we perceive the substance *as having* the characteristics. The distinction between knowing that a substance has a characteristic and knowing the substance as having a characteristic is one which cannot, as far as I can see, be made clearer in words; but it is evident by introspection to anyone who contemplates the difference between his judgment "I am in pain," and the perception of himself on which that judgment is based.

We may go further than this. Not only is a substance perceived as having characteristics, but it may be perceived as having characteristics *having* themselves characteristics. For we frequently make such judgments as "I am intensely sleepy," or "the shade of red in *B* resembles the shade of red in *A* more closely than it does the shade of red in *C*." And it would be admitted that those judgments are in some cases well-founded. Now such judgments as those are assertions about the characteristics of characteristics. Intensity is asserted as a characteristic of the sleepiness which is a characteristic of myself. And the relations between the shades of red are characteristics of the shades, which are characteristics of the sensa. It is only by perception that I know that I have the characteristic of sleepiness, and it is only by perception that I can know how intense the sleepiness is. Again, it is only by perception that I know the three shades of red, and it is only by perception that I can know that the shade in *B* resembles that in *A* more closely than that in *C*.

302. It is commonly held that perception cannot be erroneous. By this is meant more than that, whenever there is a perception, there is a substance which is perceived. It is also meant that the substance perceived must really have any characteristic which it is perceived as having. In our subsequent judgments, based on the perceptions, there may be error. But in the perceptions themselves there can be none.

We shall, however, be forced to conclude that this is mistaken, and that perception can be erroneous. For the results of our enquiry in this Book into the nature of the existent will involve that the existent is other than it is perceived to be. Even if we provisionally admit the existence of judgments, we shall find that the differences between reality and appearance in respect of time, sensa, and the nature of spirit are such that *they* cannot be ascribed only to mistaken judgments about what is perceived, but involve errors of perception. And we shall also find reason to hold that there is really no cognition except perception, and that therefore all erroneous cognition, even if it appears as false judgments, is really erroneous perception. The difficulties involved in this will be discussed in Book VI.

It was necessary to say so much at this point about perception, in order to render intelligible our use of the term in Chapters XXXIII to XXXVI. But further considerations as to the nature of perception will be more conveniently discussed in Chapter XXXVII.

CHAPTER XXXIII

TIME

303. It will be convenient to begin our enquiry by asking whether anything existent can possess the characteristic of being in time. I shall endeavour to prove that it cannot.

It seems highly paradoxical to assert that time is unreal, and that all statements which involve its reality are erroneous. Such an assertion involves a departure from the natural position of mankind which is far greater than that involved in the assertion of the unreality of space or the unreality of matter. For in each man's experience there is a part—his own states as known to him by introspection—which does not even appear to be spatial or material. But we have no experience which does not appear to be temporal. Even our judgments that time is unreal appear to be themselves in time.

304. Yet in all ages and in all parts of the world the belief in the unreality of time has shown itself to be singularly persistent. In the philosophy and religion of the West—and still more, I suppose, in the philosophy and religion of the East—we find that the doctrine of the unreality of time continually recurs. Neither philosophy nor religion ever hold themselves apart from mysticism for any long period, and almost all mysticism denies the reality of time. In philosophy, time is treated as unreal by Spinoza, by Kant, and by Hegel. Among more modern thinkers, the same view is taken by Mr Bradley. Such a concurrence of opinion is highly significant, and is not the less significant because the doctrine takes such different forms, and is supported by such different arguments.

I believe that nothing that exists can be temporal, and that therefore time is unreal. But I believe it for reasons which are not put forward by any of the philosophers I have just mentioned.

305. Positions in time, as time appears to us *primâ facie*, are distinguished in two ways. Each position is Earlier than some and Later than some of the other positions. To constitute such a series there is required a transitive asymmetrical relation, and a collection of terms such that, of any two of them, either the first is in this relation to the second, or the second is in this relation to the first. We may take here either the relation of "earlier than" or the relation of "later than," both of which, of course, are transitive and asymmetrical. If we take the first, then the terms have to be such that, of any two of them, either the first is earlier than the second, or the second is earlier than the first.

In the second place, each position is either Past, Present, or Future. The distinctions of the former class are permanent, while those of the latter are not. If M is ever earlier than N, it is always earlier. But an event, which is now present, was future, and will be past.

306. Since distinctions of the first class are permanent, it might be thought that they were more objective, and more essential to the nature of time, than those of the second class. I believe, however, that this would be a mistake, and that the distinction of past, present, and future is as *essential* to time as the distinction of earlier and later, while in a certain sense it may, as we shall see[1], be regarded as more *fundamental* than the distinction of earlier and later. And it is because the distinctions of past, present, and future seem to me to be essential for time, that I regard time as unreal.

For the sake of brevity I shall give the name of the A series to that series of positions which runs from the far past through the near past to the present, and then from the present through the near future to the far future, or conversely. The series of positions which runs from earlier to later, or conversely, I shall call the B series. The contents of any position in time form an event. The varied simultaneous contents of a single position are, of course, a plurality of events. But, like any other substance, they form a group, and this group is a compound substance. And a compound substance consisting of simultaneous events may properly be spoken of as itself an event[2].

---

[1] p. 30.

[2] It is very usual to contemplate time by the help of a metaphor of spatial movement. But spatial movement in which direction? The movement of time consists in the fact that later and later terms pass into the present, or—which is the same fact expressed in another way—that presentness passes to later and later terms. If we take it the first way, we are taking the B series as sliding along a

307. The first question which we must consider is whether it is essential to the reality of time that its events should form an *A* series as well as a *B* series. [It is clear, to begin with, that, in present experience, we never *observe* events in time except as forming both these series.]We perceive events in time as being present, and those are the only events which we actually perceive. And all other events which, by memory or by inference, we believe to be real, we regard as present, past, or future. Thus the events of time as observed by us form an *A* series.

308. It might be said, however, that this is merely subjective. It might be the case that the distinction of positions in time into past, present, and future, is only a constant illusion of our minds, and that the real nature of time contains only the distinctions of the *B* series—the distinctions of earlier and later. In that case we should not perceive time as it really is, though we might be able to *think* of it as it really is.

This is not a very common view, but it requires careful consideration. I believe it to be untenable, because, as I said above, it seems to me that the *A* series is essential to the nature of time, and that any difficulty in the way of regarding the *A* series as real is equally a difficulty in the way of regarding time as real.

309. It would, I suppose, be universally admitted that time involves change. In ordinary language, indeed, we say that something can remain unchanged through time. But there could be no time if nothing changed. And if anything changes, then all other things change with it. For its change must change some of their relations to it, and so their relational qualities. The fall of

fixed *A* series. If we take it the second way, we are taking the *A* series as sliding along a fixed *B* series. In the first case time presents itself as a movement from future to past. In the second case it presents itself as a movement from earlier to later. And this explains why we say that events come out of the future, while we say that we ourselves move towards the future. For each man identifies himself especially with his present state, as against his future or his past, since it is the only one which he is directly perceiving. And this leads him to say that he is moving with the present towards later events. And as those events are now future, he says that he is moving towards the future.

Thus the question as to the movement of time is ambiguous. But if we ask what is the movement of either series, the question is not ambiguous. The movement of the *A* series along the *B* series is from earlier to later. The movement of the *B* series along the *A* series is from future to past.

a sand-castle on the English coast changes the nature of the Great Pyramid.

If, then, a *B* series without an *A* series can constitute time, change must be possible without an *A* series. Let us suppose that the distinctions of past, present, and future do not apply to reality. In that case, can change apply to reality?

310. What, on this supposition, could it be that changes? Can we say that, in a time which formed a *B* series but not an *A* series, the change consisted in the fact that the event ceased to be an event, while another event began to be an event? If this were the case, we should certainly have got a change.

But this is impossible. If *N* is ever earlier than *O* and later than *M*, it will always be, and has always been, earlier than *O* and later than *M*, since the relations of earlier and later are permanent. *N* will thus always be in a *B* series. And as, by our present hypothesis, a *B* series by itself constitutes time, *N* will always have a position in a time-series, and always has had one. That is, it always has been an event, and always will be one, and cannot begin or cease to be an event.

Or shall we say that one event *M* merges itself into another event *N*, while still preserving a certain identity by means of an unchanged element, so that it can be said, not merely that *M* has ceased and *N* begun, but that it is *M* which has become *N*? Still the same difficulty recurs. *M* and *N* may have a common element, but they are not the same event, or there would be no change. If, therefore, *M* changed into *N* at a certain moment, then at that moment, *M* would have ceased to be *M*, and *N* would have begun to be *N*. This involves that, at that moment, *M* would have ceased to be an event, and *N* would have begun to be an event. And we saw, in the last paragraph, that, on our present hypothesis, this is impossible.

Nor can such change be looked for in the different moments of absolute time, even if such moments should exist. For the same argument will apply here. Each such moment will have its own place in the *B* series, since each would be earlier or later than each of the others. And, as the *B* series depends on permanent relations, no moment could ever cease to be, nor could it become another moment.

311. Change, then, cannot arise from an event ceasing to be an event, nor from one event changing into another. In what other way can it arise? If the characteristics of an event change, then there is certainly change. But what characteristics of an event can change? It seems to me that there is only one class of such characteristics. And that class consists of the determinations of the event in question by the terms of the A series.

Take any event—the death of Queen Anne, for example—and consider what changes can take place in its characteristics. That it is a death, that it is the death of Anne Stuart, that it has such causes, that it has such effects—every characteristic of this sort never changes. "Before the stars saw one another plain," the event in question was the death of a Queen. At the last moment of time—if time has a last moment—it will still be the death of a Queen. And in every respect but one, it is equally devoid of change. But in one respect it does change. It was once an event in the far future. It became every moment an event in the nearer future. At last it was present. Then it became past, and will always remain past, though every moment it becomes further and further past [1].

Such characteristics as these are the only characteristics which can change. And, therefore, if there is any change, it must be looked for in the A series, and in the A series alone. If there is no real A series, there is no real change. The B series, therefore, is not by itself sufficient to constitute time, since time involves change.

312. The B series, however, cannot exist except as temporal, since earlier and later, which are the relations which connect its terms, are clearly time-relations. So it follows that there can be no B series when there is no A series, since without an A series there is no time.

313. We must now consider three objections which have been made to this position. The first is involved in the view of time which has been taken by Mr Russell, according to which past,

---

[1] The past, therefore, is always changing, if the A series is real at all, since at each moment a past event is further in the past than it was before. This result follows from the reality of the A series, and is independent of the truth of our view that all change depends exclusively on the A series. It is worth while to notice this, since most people combine the view that the A series is real with the view that the past cannot change—a combination which is inconsistent.

present, and future do not belong to time *per sê*, but only in relation to a knowing subject. An assertion that N is present means that it is simultaneous with that assertion, an assertion that it is past or future means that it is earlier or later than that assertion. Thus it is only past, present, or future, in relation to some assertion. If there were no consciousness, there would be events which were earlier and later than others, but nothing would be in any sense past, present, or future. And if there were events earlier than any consciousness, those events would never be future or present, though they could be past.

If N were ever present, past, or future in relation to some assertion V, it would always be so, since whatever is ever simultaneous to, earlier than, or later than, V, will always be so. What, then, is change? We find Mr Russell's views on this subject in his *Principles of Mathematics*, Section 442. "Change is the difference, in respect of truth or falsehood, between a proposition concerning an entity and the time T, and a proposition concerning the same entity and the time T', provided that these propositions differ only by the fact that T occurs in the one where T' occurs in the other." That is to say, there is change, on Mr Russell's view, if the proposition "at the time T my poker is hot" is true, and the proposition "at the time T' my poker is hot" is false.

314. I am unable to agree with Mr Russell. I should, indeed, admit that, when two such propositions were respectively true and false, there would be change. But then I maintain that there can be no time without an A series. If, with Mr Russell, we reject the A series, it seems to me that change goes with it, and that therefore time, for which change is essential, goes too. In other words, if the A series is rejected, no proposition of the type "at the time T my poker is hot" can ever be true, because there would be no time.

315. It will be noticed that Mr Russell looks for change, not in the events in the time-series, but in the entity to which those events happen, or of which they are states. If my poker, for example, is hot on a particular Monday, and never before or since, the event of the poker being hot does not change. But the poker changes, because there is a time when this event is happening to it, and a time when it is not happening to it.

But this makes no change in the qualities of the poker. It is always a quality of that poker that it is one which is hot on that particular Monday. And it is always a quality of that poker that it is one which is not hot at any other time. Both these qualities are true of it at any time—the time when it is hot and the time when it is cold. And therefore it seems to be erroneous to say that there is any change in the poker. The fact that it is hot at one point in a series and cold at other points cannot give change, if neither of these facts change—and neither of them does. Nor does any other fact about the poker change, unless its presentness, pastness, or futurity change.

316. Let us consider the case of another sort of series. The meridian of Greenwich passes through a series of degrees of latitude. And we can find two points in this series, S and S', such that the proposition "at S the meridian of Greenwich is within the United Kingdom" is true, while the proposition "at S' the meridian of Greenwich is within the United Kingdom" is false. But no one would say that this gave us change. Why should we say so in the case of the other series?

Of course there is a satisfactory answer to this question if we are correct in speaking of the other series as a time-series. For where there is time, there is change. But then the whole question is whether it is a time-series. My contention is that if we remove the A series from the *primâ facie* nature of time, we are left with a series which is not temporal, and which allows change no more than the series of latitudes does.

317. If, as I have maintained, there can be no change unless facts change, then there can be no change without an A series. For, as we saw with the death of Queen Anne, and also in the case of the poker, no fact about anything can change, unless it is a fact about its place in the A series. Whatever other qualities it has, it has always. But that which is future will not always be future, and that which was past was not always past.

It follows from what we have said that there can be no change unless some propositions are sometimes true and sometimes false. This is the case of propositions which deal with the place of anything in the A series—"the battle of Waterloo is in the past," "it is now raining." But it is not the case with any other propositions.

318. Mr Russell holds that such propositions are ambiguous, and that to make them definite we must substitute propositions which are always true or always false—"the battle of Waterloo is earlier than this judgment," "the fall of rain is simultaneous with this judgment." If he is right, all judgments are either always true, or always false. Then, I maintain, no facts change. And then, I maintain, there is no change at all.

I hold, as Mr Russell does, that there is no A series. (My reasons for this will be given below, pp. 18–23.) And, as I shall explain on p. 31, I regard the reality lying behind the appearance of the A series in a manner not completely unlike that which Mr Russell has adopted. The difference between us is that he thinks that, when the A series is rejected, change, time, and the B series can still be kept, while I maintain that its rejection involves the rejection of change, and, consequently, of time, and of the B series.

319. The second objection rests on the possibility of non-existent time-series—such, for example, as the adventures of Don Quixote. This series, it is said, does not form part of the A series. I cannot at this moment judge it to be either past, present, or future. Indeed, I know that it is none of the three. Yet, it is said, it is certainly a B series. The adventure of the galley-slaves, for example, is later than the adventure of the windmills. And a B series involves time. The conclusion drawn is that an A series is not essential to time.

320. I should reply to this objection as follows. Time only belongs to the existent. If any reality is in time, that involves that the reality in question exists. This, I think, would be universally admitted. It may be questioned whether all of what exists is in time, or even whether anything really existent is in time, but it would not be denied that, if anything is in time, it must exist.

Now what is existent in the adventures of Don Quixote? Nothing. For the story is imaginary. The states of Cervantes' mind when he invented the story, the states of my mind when I think of the story—these exist. But then these form part of an A series. Cervantes' invention of the story is in the past. My thought of the story is in the past, the present, and—I trust—the future.

321. But the adventures of Don Quixote may be believed by a child to be historical. And in reading them I may, by an effort of my imagination, contemplate them as if they really happened. In this case, the adventures are believed to be existent, or are contemplated as existent. But then they are believed to be in the A series, or are contemplated as being in the A series. The child who believes them to be historical will believe that they happened in the past. If I contemplate them as existent, I shall contemplate them as happening in the past. In the same way, if I believed the events described in Jefferies' *After London* to exist, or contemplated them as existent, I should believe them to exist in the future, or contemplate them as existing in the future. Whether we place the object of our belief or of our contemplation in the present, the past, or the future, will depend upon the characteristics of that object. But somewhere in the A series it will be placed.

Thus the answer to the objection is that, just as far as a thing is in time, it is in the A series. If it is really in time, it is really in the A series. If it is believed to be in time, it is believed to be in the A series. If it is contemplated as being in time, it is contemplated as being in the A series.

322. The third objection is based on the possibility that, if time were real at all, there might be in reality several real and independent time-series. The objection, if I understand it rightly, is that every time-series would be real, while the distinctions of past, present, and future would only have a meaning within each series, and would not, therefore, be taken as absolutely real. There would be, for example, many presents. Now, of course, many points of time can be present. In each time-series many points are present, but they must be present successively. And the presents of the different time-series would not be successive, since they are not in the same time[1]. And different presents, it would be said, cannot be real unless they are successive. So the different time-series, which are real, must be able to exist independently of the distinction between past, present, and future.

---

[1] Neither would they be simultaneous, since that equally involves being in the same time. They would stand in no time-relation to one another.

323. I cannot, however, regard this objection as valid. No doubt in such a case, no present would be *the* present—it would only be the present of a certain aspect of the universe. But then no time would be *the* time—it would only be the time of a certain aspect of the universe. It would be a real time-series, but I do not see that the present would be less real than the time.

I am not, of course, maintaining that there is no difficulty in the existence of several distinct A series. In the second part of this chapter I shall endeavour to show that the existence of *any* A series is impossible. What I assert here is that, if there could be an A series at all, and if there were any reason to suppose that there were several distinct B series, there would be no additional difficulty in supposing that there should be a distinct A series for each B series.

324. We conclude, then, that the distinctions of past, present, and future are essential to time, and that, if the distinctions are never true of reality, then no reality is in time. This view, whether true or false, has nothing surprising in it. It was pointed out above that we always perceive time as having these distinctions. And it has generally been held that their connection with time is a real characteristic of time, and not an illusion due to the way in which we perceive it. Most philosophers, whether they did or did not believe time to be true of reality, have regarded the distinctions of the A series as essential to time.

When the opposite view has been maintained it has generally been, I believe, because it was held (rightly, as I shall try to show) that the distinctions of past, present, and future cannot be true of reality, and that consequently, if the reality of time is to be saved, the distinction in question must be shown to be unessential to time. The presumption, it was held, was for the reality of time, and this would give us a reason for rejecting the A series as unessential to time. But, of course, this could only give a presumption. If the analysis of the nature of time has shown that, by removing the A series, time is destroyed, this line of argument is no longer open.

325. I now pass to the second part of my task. Having, as it seems to me, succeeded in proving that there can be no time without an *A* series, it remains to prove that an *A* series cannot exist, and that therefore time cannot exist. This would involve that time is not real at all, since it is admitted that the only way in which time can be real is by existing.

326. Past, present, and future are characteristics which we ascribe to events, and also to moments of time, if these are taken as separate realities. What do we mean by past, present, and future? In the first place, are they relations or qualities? It seems quite clear to me that they are not qualities but relations, though, of course, like other relations, they will generate relational qualities in each of their terms[1]. But even if this view should be wrong, and they should in reality be qualities and not relations, it will not affect the result which we shall reach. For the reasons for rejecting the reality of past, present, and future, which we are about to consider, would apply to qualities as much as to relations.

327. If, then, anything is to be rightly called past, present, or future, it must be because it is in relation to something else. And this something else to which it is in relation must be something outside the time-series. For the relations of the *A* series are changing relations, and no relations which are exclusively between members of the time-series can ever change. Two events are exactly in the same places in the time-series, relatively to one another, a million years before they take place, while each of them is taking place, and when they are a million years in the past. The same is true of the relation of moments to one another, if moments are taken as separate realities. And the same would be true of the relations of events to moments. The changing relation must be to something which is not in the time-series.

Past, present, and future, then, are relations in which events stand to something outside the time-series. Are these relations simple, or can they be defined? I think that they are clearly

---

[1] It is true, no doubt, that my anticipation of an experience *M*, the experience itself, and the memory of the experience, are three states which have different original qualities. But it is not the future *M*, the present *M*, and the past *M*, which have these three different qualities. The qualities are possessed by three different events—the anticipation of *M*, *M* itself, and the memory of *M*—each of which in its turn is future, present, and past. Thus this gives no support to the view that the changes of the *A* series are changes of original qualities.

simple and indefinable. But, on the other hand, I do not think
that they are isolated and independent. It does not seem that
we can know, for example, the meaning of pastness, if we do not
know the meaning of presentness or of futurity.

328. We must begin with the A series, rather than with past,
present, and future, as separate terms. And we must say that a
series is an A series when each of its terms has, to an entity X
outside the series, one, and only one, of three indefinable relations,
pastness, presentness, and futurity, which are such that all the
terms which have the relation of presentness to X fall between
all the terms which have the relation of pastness to X, on the
one hand, and all the terms which have the relation of futurity
to X, on the other hand.

We have come to the conclusion that an A series depends on
relations to a term outside the A series. This term, then, could
not itself be in time, and yet must be such that different relations
to it determine the other terms of those relations, as being past,
present, or future. To find such a term would not be easy, and
yet such a term must be found, if the A series is to be real. But
there is a more positive difficulty in the way of the reality of the
A series.

329. Past, present, and future are incompatible determinations.
Every event must be one or the other, but no event can be more
than one. If I say that any event is past, that implies that it
is neither present nor future, and so with the others. And this
exclusiveness is essential to change, and therefore to time. For
the only change we can get is from future to present, and from
present to past.

The characteristics, therefore, are incompatible. But every
event has them all[1]. If M is past, it has been present and future.
If it is future, it will be present and past. If it is present, it has
been future and will be past. Thus all the three characteristics
belong to each event. How is this consistent with their being
incompatible?

---

[1] If the time-series has a first term, that term will never be future, and if it
has a last term, that term will never be past. But the first term, in that case,
will be present and past, and the last term will be future and present. And the
possession of two incompatible characteristics raises the same difficulty as the
possession of three. Cp. p. 26.

330. It may seem that this can easily be explained. Indeed, it has been impossible to state the difficulty without almost giving the explanation, since our language has verb-forms for the past, present, and future, but no form that is common to all three. It is never true, the answer will run, that M *is* present, past, and future. It *is* present, *will be* past, and *has been* future. Or it *is* past, and *has been* future and present, or again *is* future, and *will be* present and past. The characteristics are only incompatible when they are simultaneous, and there is no contradiction to this in the fact that each term has all of them successively.

331. But what is meant by "has been" and "will be"? And what is meant by "is," when, as here, it is used with a temporal meaning, and not simply for predication? When we say that X has been Y, we are asserting X to be Y at a moment of past time. When we say that X will be Y, we are asserting X to be Y at a moment of future time. When we say that X is Y (in the temporal sense of "is"), we are asserting X to be Y at a moment of present time.

Thus our first statement about M—that it is present, will be past, and has been future—means that M is present at a moment of present time, past at some moment of future time, and future at some moment of past time. But every moment, like every event, is both past, present, and future. And so a similar difficulty arises. If M is present, there is no moment of past time at which it is past. But the moments of future time, in which it is past, are equally moments of past time, in which it cannot be past. Again, that M is future and will be present and past means that M is future at a moment of present time, and present and past at different moments of future time. In that case it cannot be present or past at any moments of past time. But all the moments of future time, in which M will be present or past, are equally moments of past time.

332. And thus again we get a contradiction, since the moments at which M has any one of the three determinations of the A series are also moments at which it cannot have that determination. If we try to avoid this by saying of these moments what had been previously said of M itself—that some moment, for example, is future, and will be present and past—then "is" and "will be" have the same meaning as before. Our statement, then, means that the moment in question is future at a present moment, and will be present and past at different moments of future time. This, of course, is the same difficulty over again. And so on infinitely.

Such an infinity is vicious. The attribution of the characteristics past, present, and future to the terms of any series leads to a contradiction, unless it is specified that they have them successively. This means, as we have seen, that they have them in relation to terms specified as past, present, and future. These again, to avoid a like contradiction, must in turn be specified as past, present, and future. And, since this continues infinitely, the first set of terms never escapes from contradiction at all[1].

The contradiction, it will be seen, would arise in the same way supposing that pastness, presentness, and futurity were original qualities, and not, as we have decided that they are, relations. For it would still be the case that they were characteristics which were incompatible with one another, and that whichever had one of them would also have the other. And it is from this that the contradiction arises.

333. The reality of the A series, then, leads to a contradiction, and must be rejected. And, since we have seen that change and time require the A series, the reality of change and time must be rejected. And so must the reality of the B series, since that requires time. Nothing is really present, past, or future. Nothing is really earlier or later than anything else or temporally simultaneous with it. Nothing really changes. And nothing is really in time. Whenever we perceive anything in time—which is the only way in which, in our present experience, we do perceive things—we are perceiving it more or less as it really is not[2]. The

---

[1] It may be worth while to point out that the vicious infinite has not arisen from the impossibility of *defining* past, present, and future, without using the terms in their own definitions. On the contrary, we have admitted these terms to be indefinable. It arises from the fact that the nature of the terms involves a contradiction, and that the attempt to remove the contradiction involves the employment of the terms, and the generation of a similar contradiction.

[2] Even on the hypothesis that judgments are real it would be necessary to regard ourselves as perceiving things in time, and so perceiving them erroneously. (Cp. Chap. xliv, p. 196.) And we shall see later that all cognition is perception, and that, therefore, all error is erroneous perception.

problems connected with this illusory perception will be considered in Book VI.

334. Dr Broad, in his admirable book *Scientific Thought,* has put forward a theory of time which he maintains would remove the difficulties which have led me to treat time as unreal[1]. It is difficult to do justice to so elaborate and careful a theory by means of extracts. I think, however, that the following passages will give a fair idea of Dr Broad's position. His theory, he tells us, "accepts the reality of the present and the past, but holds that the future is simply nothing at all. Nothing has happened to the present by becoming past except that fresh slices of existence have been added to the total history of the world. The past is thus as real as the present. On the other hand, the essence of a present event is, not that it precedes future events, but that there is quite literally *nothing* to which it has the relation of precedence. The sum total of existence is always increasing, and it is this which gives the time-series a sense as well as an order. A moment t is later than a moment t' if the sum total of existence at t includes the sum total of existence at t' together with something more."[2]

335. Again, he says that "judgments which profess to be about the future do not refer to any fact, whether positive or negative, at the time when they are made. They are therefore at that time neither true nor false. They will become true or false when there is a fact for them to refer to; and after this they will remain true or false, as the case may be, for ever and ever. If you choose to define the word *judgment* in such a way that nothing is to be called a judgment unless it be either true or false, you must not, of course, count judgments that profess to be about the future as judgments. If you accept the latter, you must say that the Law of Excluded Middle does not apply to all judgments. If you reject them, you may say that the Law of Excluded Middle applies to all genuine judgments; but you must add that judgments which profess to be about the future are not genuine judgments when they are made, but merely enjoy a courtesy title by anticipation, like the elder sons of the higher nobility during the lifetime

---

[1] *Op. cit.* p. 79. I have published my views on time, pretty nearly in their present shape, in *Mind* for 1908.

[2] *Op. cit.* p. 66.

of their fathers."¹ "I do not think that the laws of logic have anything to say against this kind of change; and, if they have, so much the worse for the laws of logic, for it is certainly a fact."²

336. My first objection to Dr Broad's theory is that, as he says, it would involve that "it will rain to-morrow" is neither true nor false, and that "England will be a republic in 1920," was not false in 1919. It seems to me quite certain that "it will rain to-morrow" is either true or false, and that "England will be a republic in 1920," was false in 1919. Even if Dr Broad's theory did enable him to meet my objections to the reality of time (which I shall try to show later on is not the case) I should still think that my theory should be accepted in preference to his. The view that time is unreal is, no doubt, very different from the *primâ facie* view of reality. And it involves that perception can be erroneous. But the *primâ facie* view of reality need not be true, and erroneous perception, as we shall see in Chapter XLIV, is not impossible. And, I submit, it is quite impossible that "it will rain to-morrow" is neither true nor false.

337. In the second place it is to be noted that Dr Broad's theory must be false if the past ever intrinsically determines the future. If X intrinsically determines a subsequent Y, then (at any rate as soon as X is present or past, and therefore, on Dr Broad's theory, real) it will be true that, since there is an X, there must be a subsequent Y. Then it is true that there is a subsequent Y. And if that Y is not itself present or past, then it is true that there will be a future Y, and so something is true about the future.

338. Now is it possible to hold that the past never does intrinsically determine the future? It seems to me that there is just as much reason to believe that the past determines the future as there is to believe that the earlier past determines the later past or the present.

We cannot, indeed, usually get a positive statement as simple as " the occurrence of X intrinsically determines the occurrence of a subsequent Y." But the intrinsic determination of the events can often be summed up in a statement of only moderate complexity. If the moon was visible in a certain direction last

¹ *Op. cit.* p. 73.   ² *Op. cit.* p. 83.

midnight, this intrinsically determines that, either it will be visible in a rather different direction next midnight, or the night will be cloudy, or the universe will have come to an end, or the relative motions of the earth and moon will have changed. Thus it is true that in the future one of four things will happen. And thus a proposition about the future is true.

And there are other intrinsic determinations which can be summed up in very simple negative statements. If Smith has already died childless, this intrinsically determines that no future event will be a marriage of one of Smith's grandchildren.

339. It seems, then, impossible to deny that the truth of some propositions about the future is implied in the truth of some propositions about the past, and that, therefore, some propositions about the future are true. And we may go further. If no propositions about the past implied propositions about the future, then no propositions about the past could imply propositions about the later past or the present.

If the proposition "the occurrence of X implies the occurrence of Y" is ever true, it is always true, while X is real, and, therefore, even according to Dr Broad's view of reality, it is always true while X is present and past. For it is dependent on the nature of X and the laws of implication. The latter are not changeable, and when an event has once happened, its nature remains unchangeable. Thus, if it were not true, in 1921, that the occurrence of any event in 1920 involved the occurrence of any event in 1922, then it could not be true in 1923, when both 1920 and 1922 are in the past. And this would apply to any two periods in time, as much as to 1920 and 1922.

340. There are, then, only two alternatives. Either propositions about the future are true, and Dr Broad's theory is wrong. Or else no proposition about any one period of time implies the truth of a proposition about any other period of time. From this it follows that no event at any point of time intrinsically determines any event at any other point of time, and that there is no causal determination except what is strictly simultaneous.

It is clear, from the rest of his book, that Dr Broad does not accept this last alternative, and it is difficult to conceive that anyone would do so, unless he were so complete a sceptic that he could have no theory as to the nature of time, or of anything else. For a person who accepted this alternative would not merely deny that complete causal determination could be proved, he would not merely deny that any causal determination could be proved, but he would assert that all causal determination, between non-simultaneous events, was proved to be impossible. But if this is not accepted, then some propositions about the future must be true[1].

341. In the third place, even if the two objections already considered should be disregarded, time would still, on Dr Broad's theory, involve the contradiction described above (p. 20). For although, if Dr Broad were right, no moment would have the three incompatible characteristics of past, present, and future, yet each of them (except the last moment of time, if there should be a last moment) would have the two incompatible characteristics of past and present. And this would be sufficient to produce the contradiction.

The words past and present clearly indicate different characteristics. And no one, I think, would suggest that they are simply compatible, in the way that the characteristics red and sweet are. If one man should say "strawberries are red," and another should reply "that is false, for they are sweet," the second man would be talking absolute nonsense. But if the first should say "you are eating my strawberries," and the second should reply "that is false, for I have already eaten them," the remark is admittedly not absolute nonsense, though its precise relation to the truth would depend on the truth about the reality of matter and time.

The terms can only be made compatible by a qualification. The proper statement of that qualification seems to me to be, as I have said (p. 21), that, when we say that M is present, we mean

[1] It might seem that the truth of propositions about the future would be as fatal to my theory as to Dr Broad's, since I am denying the reality of time. But, as will be explained later, although there is no time-series, there is a non-temporal series which is misperceived as a time-series. An assertion at one point of this series may be true of a fact at some other point in this series, which appears as a future point. And thus statements about the future might have phenomenal validity—they might have a one-to-one correspondence with true statements, and they might themselves be as true as any statements about the past could be. But Dr Broad's theory requires that they should have no truth whatever, while some statements about the past and present should be absolutely true.

that it is present at a moment of present time, and will be past at some moment of future time, and that, when we say that M is past we mean that it has been present at some moment of past time, and is past at a moment of present time. Dr Broad will, no doubt, claim to cut out "will be past at some moment of future time." But even then it would be true that, when we say M is past, we mean that it has been present at some moment of past time, and is past at a moment of present time, and that, when we say M is present, we mean that it is present at a moment of present time. As much as this Dr Broad can say, and as much as this he must say, if he admits that each event (except a possible last event) is both present and past.

Thus we distinguish the presentness and pastness of events by reference to past and present moments. But every moment which is past is also present. And if we attempt to remove this difficulty by saying that it *is* past and *has been* present, then we get an infinite vicious series, as pointed out on p. 22.

For these three reasons it seems to me that Dr Broad's theory of time is untenable, and that the reality of time must still be rejected.

342. It is sometimes maintained that we are so immediately certain of the reality of time, that the certainty exceeds any certainty which can possibly be produced by arguments to the contrary, and that such arguments, therefore, should be rejected as false, even if we can find no flaw in them.

343. It does not seem to me that there is any immediate certainty of the reality of time. It is true, no doubt, that we perceive things as in time, and that therefore the unreality of time involves the occurrence of erroneous perception. But, as I have said, I hope to prove later that there is no impossibility in erroneous perception. It may be worth while, however, to point out that any theory which treated time as objectively real could only do so by treating time, *as we observe it*, as being either unreal or merely subjective. It would thus have no more claim to support from our perceptions than the theories which deny the reality of time[1].

---

[1] By objectively real time, I mean a common time in which all existent things exist, so that they stand in temporal relations to each other. By subjectively real time, I mean one in which only the different states of a single self exist, so that it does not connect any self with anything outside it.

344. I perceive as present at one time whatever falls within the limits of one specious present. Whatever falls earlier or later than this, I do not perceive at all, though I judge it to be past or future. The time-series then, of which any part is perceived by me, is a time-series in which the future and the past are separated by a present which is a specious present.

Whatever is simultaneous with anything present, is itself present. If, therefore, the objective time-series, in which events really are, is the series which I immediately perceive, whatever is simultaneous with my specious present is present. But the specious present varies in length according to circumstances. And it is not impossible that there should be another conscious being existing besides myself, and that his specious present and mine may at the same time be of different lengths. Now the event M may be simultaneous both with X's perception Q, and with Y's perception R. At a certain moment Q may have ceased to be a part of X's specious present. M, therefore, will at that moment be past. But at the same moment R may still be a part of Y's specious present. And, therefore, M will be present at some moment at which it is past.

This is impossible. If, indeed, the A series was something purely subjective, there would be no difficulty. We could say that M was past for X and present for Y, just as we could say that it was pleasant for X and painful for Y. But we are now considering the hypothesis that time is objective. And, since the A series is essential to time, this involves that the A series is objective. And, if so, then at any moment M must be present, past, or future. It cannot be both present and past.

The present, therefore, through which events are really to pass, cannot be determined as being simultaneous with a specious present. If it has a duration, it must be a duration which is independently fixed. And it cannot be independently fixed so as to be identical with the duration of all specious presents, since all specious presents have not the same duration. And thus an event may be past or future when I am perceiving it as present, and may be present when I am remembering it as past or anticipating it as future. The duration of the objective present may be the thousandth part of a second. Or it may be a century, and the coronations of George IV and of Edward VII may form part of the same present. What reasons can we find in the immediate certainties of our experience to believe in the existence of such a present, which we certainly do not observe to be a present, and which has no relation to what we do observe as a present?

345. If we take refuge from these difficulties in the view, which has sometimes been held, that the present in the *A* series is not a finite duration, but a single point, separating future from past, we shall find other difficulties as serious. For then the objective time, in which events are, would be something entirely different from the time in which we experience them as being. The time in which we experience them has a present of varying finite duration, and is therefore divided into three durations—the past, the present, and the future. The objective time has only two durations, separated by a present which has nothing but the name in common with the present of experience, since it is not a duration but a point. What is there in our perception which gives us the least reason to believe in such a time as this?

346. And thus the denial of the reality of time turns out not to be so very paradoxical. It was called paradoxical because it required us to treat our experience of time as illusory. But now we see that our experience of time—centring as it does about the specious present—would be no less illusory if there were a real time in which the realities we experience existed. The specious present of our observations cannot correspond to the present of the events observed. And consequently the past and future of our observations could not correspond to the past and future of the events observed. On either hypothesis—whether we take time as real or as unreal—everything is observed as in a specious present, but nothing, not even the observations themselves, can ever really *be* in a specious present. For if time is unreal, nothing can be in any present at all, and, if time is real, the present in which things are will not be a specious present. I do not see, therefore, that we treat experience as much more illusory when we say that nothing is ever present at all, than when we say that everything passes through some present which is entirely different from the only present we experience.

347. It must further be noted that the results at which we have arrived do not give us any reason to suppose that *all* the

elements in our experience of time are illusory. We have come to
the conclusion that there is no real A series, and that therefore
there is no real B series, and no real time-series. But it does not
follow that when we have experience of a time-series we are not
observing a real series. It is possible that, whenever we have an
illusory experience of a time-series, we are observing a real
series, and that all that is illusory is the appearance that it is a
time-series. Such a series as this—a series which is not a time-
series, but under certain conditions appears to us to be one—may
be called a C series.

And we shall see later[1] that there are good reasons for supposing that such a C series does actually exist, in every case in
which there is the appearance of a time-series. For when we
consider how an illusion of time can come about, it is very
difficult to suppose, either that all the elements in the experience are illusory, or that the element of the serial nature is so.
And it is by no means so difficult to account for the facts if we
suppose that there is an existent C series. In this case the
illusion consists only in our applying the A series to it, and in
the consequent appearance of the C series as a B series, the
relation, whatever it may be, which holds between the terms of
the C series, appearing as a relation of earlier and later.

348. The C series, then, can be real, while the A and B series
are merely apparent. But when we consider how our experience
is built up, we must class C and A together as primary, while
B is only secondary. The real C series and the appearance of the
A series must be given, separately and independently, in order
to have the experience of time. For, as we have seen, they are
both essential to it, and neither can be derived from the other.
The B series, on the other hand, can be derived from the other
two. For if there is a C series, where the terms are connected by
permanent relations, and if the terms of this series appear also
to form an A series, it will follow that the terms of the C series
will also appear as a B series, those which are placed first, in the
direction from past to future, appearing as earlier than those
whose places are further in the direction of the future.

349. And thus, if there is a C series, it will follow that our
experience of the time-series will not be entirely erroneous.

[1] Chap. XLV, p. 213.

Through the deceptive form of time, we shall grasp some of the true relations of what really exists. If we say that the events M and N are simultaneous, we say that they occupy the same position in the time-series. And there will be some truth in this, for the realities, which we perceive as the events M and N, do really occupy the same position in a series, though it is not a temporal series.

Again, if we assert that the events M, N, O are all at different times, and are in that order, we assert that they occupy different positions in the time-series, and that the position of N is between the positions of M and O. And it will be true that the realities which we see as these events will be in a series, though not in a temporal series, and that they will be in different positions in it, and that the position of the reality which we perceive as the event N will be between the positions of the realities which we perceive as the events M and O.

350. If this view is adopted, the result will so far resemble the views of Hegel rather than those of Kant. For Hegel regarded the order of the time-series as a reflection, though a distorted reflection, of something in the real nature of the timeless reality, while Kant does not seem to have contemplated the possibility that anything in the nature of the noumenon should correspond to the time-order which appears in the phenomenon.

351. Thus the C series will not be altogether unlike the time-series as conceived by Mr Russell. The C series will include as terms everything which appears to us as an event in time, and the C series will contain the realities in the same order as the events are ranged in by the relations of earlier and later. And the time-series, according to Mr Russell, does not involve the objective reality of the A series.

But there remain important differences. Mr Russell's series is a time-series, and the C series is not temporal. And although Mr Russell's time-series (which is identical with our B series) has a one-to-one correspondence with the C series, still the two series are very different. The terms of the B series are events, and the terms of the C series are not. And the relation which unites the terms of the B series is the relation of earlier and later, which is not the case with the C series. (We shall consider what is the relation of the terms of the C series in Chapter XLVIII.)

CHAPTER XXXIV

MATTER

352. The universe appears, *primâ facie*, to contain substances of two very different kinds—Matter and Spirit. The existence, however, both of matter and of spirit, has been denied by different schools of philosophy. And we must now enquire what light can be thrown on this question by the help of the results reached in the last three Books. In this chapter we shall consider the existence of matter.

353. In settling what shall be called by the name of matter, thought has started from the denotation, rather than from the connotation. It was generally accepted that matter was a term which was to be applicable to rocks, to gases, to human bodies, to tables, and so on, provided that these things had more or less the characteristics which they appear *primâ facie* to have. And there is, I think, a general agreement that such matter can be defined by means of the characteristics commonly known as the Primary Qualities of matter—size, shape, position, mobility, and impenetrability. What the exact definition should be would be a more difficult question. We should have to enquire first whether any of these characteristics were implied in any of the others—in which case those which were implied could be left out as superfluous. And, even among independent elements, it might be asked whether any of them are excessive—whether, for example, a substance, though it did not possess impenetrability, would be material if it possessed size, shape, and position. But for our purpose it is not necessary to reach an ideal minimum definition. It is sufficient if we can say—as the ordinary use of the word entitles us to say—that everything which is matter has all these five qualities[1].

354. The question of the definition of matter is not affected by any consideration as to what are sometimes called the Secondary Qualities of matter. These are colour, hardness, smell, taste, and

---

[1] We shall have to consider later, however, how far mobility is essential to matter, p. 42.

sound[1]. These characteristics are held by some philosophers not to be qualities of matter at all, but to be effects produced by matter on an observing subject. Others, however, hold that they are really qualities of matter[2]. But in any case they would not enter into the definition. For it would be generally admitted that a substance which had the primary qualities would be material, independently of its possession of the secondary qualities; and, on the other hand, that, if it were possible for a substance to have the secondary qualities without the primary qualities, such a substance would not be material.

355. Taking matter, then, as something which possesses the primary qualities, we have to ask whether it is possible that any existent substance can be matter. It is clear that, if this is the case, there must be other characteristics belonging to all matter, besides those we have mentioned. For we have seen that nothing that exists can have simple parts. And the absence of simple parts is not a primary or secondary quality. Nor is it implied in any such quality. Indeed it has generally been held that matter consists of simple parts.

There might be no difficulty in the assertion that every part of matter was again divided into parts, if that assertion were taken by itself. But we saw in the last two Books that the absence of simple parts in any substance produced a contradiction unless the parts of that substance were determined by determining correspondence. If matter does exist, then, its parts must be determined by determining correspondence. Is this possible?

356. Matter has dimensions in space and in time, but, according at least to the ordinary view, it has no other dimensions.

---

[1] Hardness is, of course, to be distinguished from impenetrability. Hardness is a quality which varies in degree, and which is possessed in very different degrees by granite and by butter. Impenetrability admits of no degrees, and is possessed by butter as much as by granite, since each of them excludes all other matter from the place where it is.

[2] It is curious that the name of secondary qualities of matter is generally given to these qualities by philosophers who hold that they are effects produced in the observing subject, and not really qualities of matter. Locke, indeed, who was the first person to use the name, applied it, not to the effects on the observing subject, but to the powers of the objects to produce these effects in us by their primary qualities. But this is not now the common usage.

Let us consider whether it can, in these dimensions, be divided into parts of parts to infinity by determining correspondence.

To begin with space. If matter is to be infinitely divisible in space, all matter must be divided into a set of spatial parts which are primary parts, each of which has a sufficient description, by correspondence to which sufficient descriptions of the secondary parts of all grades are determined. What sort of sufficient description could such primary parts have?

Anything which is in space has qualities of two sorts. It has qualities which are strictly spatial—size, shape, and position. But besides these it may have other qualities—for example, impenetrability, colour, hardness, sound, smell, and so on—which can belong to things which have spatial qualities, and which perhaps can only belong to things which have spatial qualities, but which are not themselves spatial qualities in the sense in which size, shape, and position are. In the course of our enquiry into the possibility of infinite division in space, I shall call the non-spatial qualities of spatial objects by the abbreviated name of non-spatial qualities.

Can the required sufficient descriptions of the primary parts be composed of non-spatial qualities? It seems clear that they cannot, because no sort of correspondence between such qualities of primary parts could determine sufficient descriptions of secondary parts to infinity.

Take, for example, colour. Let the primary parts be sufficiently described, one as blue, one as red, and so on. What would the correspondence be between a determinant primary part and a determinate secondary part? Could it be that the colour of the determinate should resemble that of the determinant? But this is impossible, because then we should have, for example, a primary part which was blue, while the secondary part of it, which corresponded to the red primary part, would be red. And it is obvious that a thing cannot be blue if a part of it is red.

Nor could this be avoided by saying that the part of the blue which corresponded to the red primary part might be a resultant of blue and red, the part which corresponded to the yellow primary part a resultant of blue and yellow, and so on. For then the whole would not be blue, or any other colour. And the hypothesis requires that it should be blue.

Nor, again, could the difficulty be avoided by saying that the primary part as a whole might have the general quality of being some sort of blue, while each of its parts was a more definite shade of blue. For no substance can be blue in general without being some definite shade of blue. The only meaning that could be given to the phrase "*B* is blue in general, but not any definite shade of blue," would be that each of its parts had some definite shade of blue, which was a different shade for each part. In this case the sufficient description of the primary part would depend on the sufficient descriptions of its secondary parts. This could not give us determining correspondence, which requires that the sufficient descriptions of each secondary part should depend on the sufficient descriptions of two primary parts—the primary part of which it is a part, and the primary part to which it corresponds.

Or could it be that, while the primary parts were sufficiently described by one sort of non-spatial qualities, the secondary parts were described by other sorts? Could, for example, the primary parts be sufficiently described by their colour, while the secondary parts of the first grade which corresponded to them should be sufficiently described by their taste? In that case the part of the blue primary which corresponded to the red primary might be blue and sweet, the part which corresponded to the yellow primary might be blue and sour, and so on.

But this is impossible. In the first place it would involve that matter should possess an infinite number of sorts of qualities, analogous to colour, taste, and so on—one sort for each of the infinite series of grades of secondary parts. And there is not the least reason to suppose that matter does possess any greater number of such sorts of qualities than the very limited number which are empirically known to us. And, in the second place, no one kind of determining correspondence could determine a given taste by a given colour, a given sound by a given taste, and so on infinitely. There would have to be a separate law of correspondence for each of the infinite number of grades of parts which had to be determined. And in that case the determining correspondence would not remove the contradiction[1].

[1] For then the nature of the primary parts might include sufficient descriptions of the members of the sets of parts, but it would not imply them without including them. (Cp. Sections 192–194.)

But, further, even if it were possible to give sufficient descriptions of spatial parts of parts of matter to infinity by means of their non-spatial qualities, this would not be sufficient. For all these parts, being spatial, must have spatial qualities, and, if a contradiction is to be avoided, it will be necessary that these qualities also should be determined by determining correspondence.

This will be necessary for two reasons. In the first place, if the spatial qualities are not determined by determining correspondence they must be independently fixed. And the concurrence of such independently fixed qualities with the non-spatial qualities which are determined by determining correspondence will be a concurrence which is ultimate and undetermined. Of these ultimate concurrences there will be an infinite number, since the number of the parts will be infinite.

We saw in Section 190 that it was impossible to accept the view that there could be an infinite number of such ultimate concurrences between qualities. And, therefore, the special qualities of the parts of matter must be determined by determining correspondence.

In the second place, it is clear that the spatial qualities of the members of a set of parts imply the spatial qualities of the whole of which they are a set of parts. If we know the shape and the size of each one of a set of parts of A, and their position relatively to each other, we know the size and shape of A. And if, in addition to this, we know the position of each part of the set in relation to anything external, B, we know the position of A in relation to B. On the other hand, the size, shape, and position of the whole implies that it has parts which have size, shape, and position— for otherwise it could not be divided into parts in respect of its spatial dimensions. And if it does not also imply what the size, shape, and position of these parts are, it presupposes them. We shall thus have an infinite series of terms, in which the subsequent terms imply the precedent, while the precedent presuppose the subsequent. And, as was shown in Section 191, such a series will involve a contradiction, since every term in it will have a presupposition, and yet will have no total ultimate presupposition. The only way to avoid this is for the spatial qualities of any precedent term to imply the spatial qualities of all subsequent terms.

And this, when the series is infinite, can only be done by determining correspondence[1].

357. Is it, then, possible to determine the spatial qualities of the spatial parts of matter by means of determining correspondence? If so, there would have to be one or more primary wholes of matter, such that any whole, A, had a set of parts, B and C, whose size, shape, and relative positions were given as ultimate facts. (These primary parts, of course, might also be differentiated from each other by non-spatial qualities.) Then the law of correspondence would have to be that B and C had each a set of parts whose members corresponded, in shape and in position relatively to the other members, with the members of every set of parts of A. This would imply the shapes and positions of parts within parts of B and C to infinity, and the size of each part would follow from the shapes and positions of the parts in combination with the sizes of B and C themselves.

But, if we look further, we shall see that it is not possible to have such determining correspondence in respect of the spatial qualities of matter, unless we can establish it also in respect of the non-spatial qualities, which we have already seen to be impossible.

358. Space is sometimes held to be relative, and sometimes to be absolute, but in neither case could there be determining correspondence by spatial qualities, unless each part was differentiated by non-spatial qualities. If we take space to be relative, then all the spatial qualities of matter are relational qualities which arise from the relationship of one piece of matter with another. And there cannot be such relationships unless the pieces of matter are otherwise differentiated from each other. M and N cannot be differentiated from each other merely by the fact that M's relation to N is different from N's relation to M. Nor

---

[1] We saw in Section 225 that an argument like the argument in this paragraph would not apply in *all* cases of the further qualities of an infinite series of parts of parts already determined by determining correspondence. But the reason it would not apply, as we then saw, is that it cannot be shown that in all such cases the fixing of the presuppositions in the subsequent terms of the series would imply the fixing of the presuppositions in the precedent terms; and therefore the total ultimate presuppositions need not disappear. But in the case of spatial qualities, as was shown in the text, the fixing of the presuppositions of the subsequent terms *would* imply the fixing of the presuppositions of the precedent terms.

could they be differentiated by their different relations to a third substance, if the third substance had no other ground of differentiation than that it stood in different relations to M and N[1].

359. What is the position if space is taken as absolute? The usual theory of absolute space is that it is made up of indivisible points. In that case it is clear that matter cannot be infinitely divisible in space at all, since the matter which occupies each indivisible point is itself spatially indivisible.

But perhaps another theory of absolute space is possible. The units of such a space might be, not indivisible points, but areas, each of which, as an ultimate fact, possessed a certain size and shape, and stood in certain relations to all the other areas. Might not the primary parts of matter be such as occupied, each of them, one of these areas. And then would it not be possible to determine, within each primary part, an infinite series of parts within parts by means of the law of correspondence suggested on p. 37.

This, however, cannot be done unless each part in the infinite series of sets of parts is also differentiated by its non-spatial qualities. Let us suppose that B and C are differentiated by their spatial qualities in the way mentioned on p. 37, and that they are also differentiated by their non-spatial qualities—B, for example, being blue, and C red—but that the differentiation by non-spatial qualities stops there, so that all the parts of B are homogeneously blue, and are not differentiated by any other non-spatial qualities. Then the theory requires that, within B, there are $B!B$ and $B!C$ which correspond in shape and relative position to B and C respectively. And these parts are not differentiated from each other by any non-spatial qualities.

But can there be two such parts, differentiated from each other only by their relations to B and C? It seems to me clear that, if they are not otherwise differentiated, there can be no parts which answer to the descriptions $B!B$ and $B!C$. If they were otherwise differentiated—if, for example, one was violet and one was indigo, or one was hard and the other soft—then they could

---

[1] It is, of course, possible that two terms, which do not differ in their original qualities, may be differentiated by relational qualities arising out of their relationships to some other terms. But then those other terms must themselves be differentiated by means of qualities other than those arising from their relationships to other terms. (Cp. Section 104.)

answer to the descriptions *B ! B* and *B ! C* respectively, and those descriptions would afford additional sufficient descriptions of them. But, by the hypothesis, all *B* is homogeneous in respect of its non-spatial qualities. And in that case it seems evident that there are no parts to which the descriptions *B ! B*, *B ! C*, can apply, and consequently no differentiation.

But, it may be asked, must not this conclusion be fallacious? If the descriptions of parts as *B ! B* and *B ! C* would give sufficient descriptions of those parts, how can we say that there are no parts to which the descriptions can apply, unless those parts are otherwise differentiated? Will not those descriptions themselves mark out such parts?

This seems, at first sight, a serious objection. And no doubt the view I have put forward seems paradoxical. But I believe that when we look closer into the nature of space, we shall see that the objection is not valid, and we shall incidentally see why the view I am advocating appears paradoxical in spite of its truth.

360. I submit that it belongs to the nature of space that nothing spatial can be discriminated from anything else, in respect of its spatial qualities, except by means of descriptions of its parts. A description of the whole which does not describe it by means of descriptions of its parts will not discriminate it from other spaces. There are two ways in which a spatial whole can be described by means of descriptions of its parts, and it is always in one of these two ways that we do discriminate spatial wholes.

The first of these ways is to describe it by pointing out some quality which is possessed by all its parts, and which is possessed by nothing in spatial contact with the whole. Thus we can discriminate a particular surface by the fact that all its parts are blue, while everything which touches it is red. It is also the case when we mark off a section of a homogeneous blue line by measuring it against a non-graduated stick. For then every part of that section has the quality of being in contact with that stick—a quality not possessed by any other part of the blue line. This, of course, would not discriminate the section in question, unless the stick were discriminated. But that is discriminated by the fact that every part of it has qualities, in respect of colour and hardness, which are not possessed by anything that is in contact with it.

The second way is that each part, F, G, and H of some set of parts of E, should have its own discrimination from all spatial things in contact with it, so that we can discriminate E as the substance which has a set of parts consisting of F, G, and H. In this way we might discriminate the representation of England on a map. We might say that it was as much of the map as contained within itself the representation of Northumberland, the representation of Cumberland, and so on for all the other counties, the representation of each county being discriminated by the fact that it had, through all its parts, some colour which belonged to no part of the map in contact with it[1]. And in this way we discriminate a section of a blue line when we measure it against a graduated stick. For then we say that the section is composed of the parts F, G, and H; that each of them is in contact with a part of the stick; and that the parts of the stick are discriminated, by various qualities, from the parts of the stick on each side of it, and from the air that surrounds it on its other sides.

The reason why the view which I am maintaining appears paradoxical, is, I believe, that we instinctively suppose that this second method of discrimination must be applicable in the case of $B!B$ and $B!C$. We assume that B must be divided into so many parts which can be discriminated from each other in this way that $B!B$ and $B!C$ will not cut across any of them, but that each of them will fall entirely in $B!B$ or $B!C$. Then $B!B$ and $B!C$ could be discriminated as the parts of B which contained, one of them the parts F, G, H, etc., the other the parts I, J, K, etc.

But it is clear that the second method of discrimination cannot be applied to the discrimination of parts within parts to infinity. For the discrimination of any whole, when determined in this

---

[1] If we tried to give in this way a *sufficient description* of the representation of England, we should require sufficient descriptions of the representations of counties. And if two of these, which did not touch one another, were of the same colour, we should have to add other qualities to make up the sufficient description. But we are speaking in the text only of the discrimination of the representation of England from the rest of the map. And for this the conditions in the text are sufficient.

way, depends on the discrimination of its parts. And therefore, when the discrimination is to infinity, no space could be discriminated except by starting from the last term of a series which has no last term, and therefore no space could be discriminated at all.

If, therefore, on that theory of absolute space which we are now considering, spaces are to be discriminated into parts of parts to infinity, it follows that—at any rate after some finite number of grades—all discrimination must be effected in the first way—that is, for each spatial part, L, there must be some quality which is shared by all the parts of L, and is not shared by anything in spatial contact with L. And this quality must be non-spatial. For, as we have seen, all spatial differences require such a discrimination, and cannot provide it.

On this theory of absolute space, then, spatial division to infinity by spatial qualities is only possible if each part is also differentiated by non-spatial qualities. And we have seen (p. 35) that this is also the case on any theory of relative space, while the more usual theory of absolute space renders infinite division impossible (p. 38). It follows that, on any theory of space, spatial division to infinity by spatial qualities is possible only if each part is also differentiated by non-spatial qualities.

If the parts of parts to infinity are to be differentiated by non-spatial qualities, it is necessary that those non-spatial qualities should be determined by determining correspondence. For, if they were not, there would have to be an infinite number of ultimate coincidences between, on the one hand, the determination of parts by spatial qualities by means of determining correspondence, and, on the other hand, the differentiation of those parts by non-spatial qualities. And we saw in Section 190 that an infinite number of such coincidences must be rejected. The non-spatial qualities, then, must be determined to infinity by determining correspondence. And we saw (p. 35) that this was impossible.

Our conclusion is, then, that matter cannot be divided into parts of parts to infinity in respect of its spatial dimensions. For, if so, there would have to be determining correspondence based either on non-spatial qualities or on spatial qualities. And we

have seen (1) that it could not be based on non-spatial qualities,
(2) that, if it could, it would be necessary to base it also on spatial
qualities, (3) that the possibility of basing it on spatial qualities
depends on its being independently based on non-spatial qualities,
which, as we have just said, is impossible.

361. Can matter be divided into parts of parts to infinity in
respect of its temporal dimension? It seems clear that what was
said of the nature of space, whether taken as absolute or relative,
on pp. 33–42, is true also of time. And, therefore, by arguments
similar to those which we have used in the case of space, we should
be led to the conclusions (1) that the necessary determining
correspondence could not be based on non-temporal qualities
(using the words temporal and non-temporal in senses analogous
to those in which we have used the words spatial and non-spatial),
(2) that, if it could, it would be necessary also to base it on
temporal qualities, (3) that the possibility of basing it on temporal qualities depends on its being independently based on non-
temporal qualities, which, as we have just said, is impossible.

362. But the question here is not so simple as with spatial
qualities. It is quite certain that, if we are to use matter in the
ordinary sense of the word, it must have spatial qualities. Nothing
which had not size, shape, and position would be called matter.
But it is by no means so certain that anything which had size,
shape, and position, would not be called matter, if it was shown
not to be really mobile, but only apparently mobile. At any rate,
it would be very much like what we call matter.

Now we have seen, in the last chapter, that nothing is really
in time, but that whatever appears to constitute a time-series
does really constitute a C series. The fact that matter could not
be divided into parts of parts to infinity in respect of a temporal
dimension is therefore irrelevant. For, if matter were real, it would
not have a temporal dimension but would have a dimension in
the C series. And might it not be possible that this C series might
have such a nature as would admit of its being divided into parts
of parts to infinity?

But this is not possible. The time-series consists of terms,
joined by the relation of earlier and later, which terms are different in their non-temporal qualities. (If they were not different

in their non-temporal qualities, there would be no change, and therefore no time.) The C series, which appears as the time-series, must therefore consist of terms different in their non-temporal qualities, joined by some relation which is not that of earlier and later. In the case of a C series which was a dimension of matter the non-temporal series could only differ, either in respect of spatial qualities, or in respect of the non-spatial qualities which matter possesses. And we saw, when we were discussing space, that these qualities will give no ground for the differentiation of matter into parts of parts to infinity. Therefore there can be no differentiation of matter, in respect of the C series, into parts of parts to infinity.

Thus matter cannot be divided into parts of parts to infinity either in respect of its spatial dimensions, or of that dimension which appears as temporal. And matter, as usually defined, and as we have defined it, has no other dimensions. It cannot therefore be divided into parts of parts to infinity. And therefore it cannot exist.

363. But, it may be objected, there are two alternatives which we have not considered. We have seen that the qualities of matter which are implied in its being matter will not provide for its division to infinity. But, in the first place, that which is matter might have other qualities, which might provide sufficient descriptions for the infinite series of parts in space, or in apparent time, or in both. And, in the second place, matter might consist of a number of units which were materially simple and indivisible—that is, were not divisible in the dimensions of space or of apparent time—but which, in addition to their material qualities, had other qualities such as to determine, by determining correspondence, sufficient descriptions of an infinite series of parts within parts. A substance, of which either of those hypotheses was true, would certainly not be the matter whose existence has hitherto been asserted by anyone. It will be more convenient to postpone the consideration of the possibility of such a substance to Chapter XXXVIII (p. 116). We shall then find reason to reject it. But at any rate it cannot affect our conclusion that matter, with the nature ordinarily attributed to it, cannot exist.

It is to be noted that our conclusion that matter does not exist depends on our conclusion that none of the qualities which matter, if it existed, would have, can determine sufficient descriptions of parts of parts to infinity. It would therefore follow that if anything else, which was asserted to exist, had those qualities, and no others—or no others which would determine such descriptions—then that thing also could not exist. This result will be important when, in the next chapter, we deal with the question of the reality of sensa.

364. I conceive, then, that I have proved, on the basis of the results reached in earlier parts of this work, that matter *cannot* exist. But it may be worth while to consider, whether, apart from those results, we have any reason to believe that matter *does* exist.

We judge matter to exist, but we do not perceive it as existing[1]. And therefore, if we accept the conclusion that a particular piece of matter exists, such a conclusion can only be justified as an inference from something which we perceive.

I do not say that we *arrive* at our conclusion by such an inference. In almost every case in which a man arrives at such a conclusion as "there is a tree in the field," he does not start from the existence of certain sensa as premises, and then explicitly infer from them that there is a tree. But the question before us is not how the belief can be reached, but how it can be justified. Now there seems no way of justifying it, except an appeal to the sensa. And this is, in fact, the justification which we all use, if we feel a necessity for justifying such a belief. If, for example, *B* should deny that there was a tree in that field, *A*'s natural reply would be "there is, for I see it." If we translate this from colloquial into exact language, it takes the form that *A* had perceived certain sensa, and now argues that he could not have done so, unless there was a tree. And if he subsequently comes to believe that the inference is not valid—if, for example,

---

[1] As has already been said, we shall later reach the conclusion that whatever appears as judgment is really perception, and that consequently we do perceive certain existent things as matter, though erroneously. But in these sections I am, as was said above, abstracting from the results reached in this work. And, *primâ facie*, we judge matter to exist, and do not perceive it as existent.

he is convinced that the sensa in question were experienced in a dream—he abandons his belief in the existence of the tree.

If the belief in the existence of particular pieces of matter must, if justified at all, be justified by inference, the same will be the case about a belief in the existence of matter in general. For matter, as we have said, is never perceived. And it will scarcely be asserted that the proposition "some matter exists" is self-evident *à priori*.

365. Sensa on the other hand are, *primâ facie*, perceived as existent, and not only judged to exist. They furnish, therefore, a basis on which judgments of existence can rest. And we need not object to the propositions that sensa must have causes, and that it is highly improbable that each percipient is the sole cause of the sensa he perceives. But, granted that I am justified in inferring the existence of something outside myself which is the cause or part-cause of my sensa, and assuming that the sensa have the qualities which they appear to have, what is my justification for asserting that that cause is of the nature of matter? Why am I entitled to exclude such conclusions as those of Berkeley, of Leibniz, and of Hegel, all of whom assigned to the sensa of each percipient a cause outside himself, and all of whom denied the existence of matter?

I know of only two answers which are given to this question. It is said, in the first place, that the cause of the sensa must have those qualities which we have taken as constituting the nature of matter, because the sensa have those qualities. Now this involves the principle that cause and effect must resemble one another. If we are not entitled to affirm this, we are not justified in arguing from the possession of characteristics by sensa to the possession of the same characteristics by their causes.

It is certainly true that every cause must resemble its effect in certain respects. They must both exist, both be substances (in the sense in which we are using that term), both be subject to general laws. But these similarities are shared by all causes and all effects, and do not support the view that a cause has to resemble its effect in any special manner—in any manner in which it does not resemble other things.

There are, no doubt, cases in which such a resemblance does exist. In the first place there are cases where the cause and the effect have a common element. Sugar and fruit are part of the cause of jam (not the whole cause, for that includes the jam-maker), and the same matter which is the sugar and fruit is the matter which is the jam. And in cases where there is no common element, there may yet be a special resemblance. The motion of an engine in a particular direction at a particular speed is the cause of a carriage which is coupled to it moving in the same direction at the same speed. The happiness of A is the cause of the happiness of the sympathetic B.

But often there is no special resemblance. The happiness of A causes the misery of the envious C. A blow from a stone causes a bruise. An east wind causes a bad temper. My volition causes a movement of my body. The ambition of Napoleon causes bullet holes in the walls of Hougoumont. What special resemblances are to be found here?

And even in cases where there is some special resemblance, it would be a mistake to argue that the cause must resemble the effect in all particulars. We cannot infer that sugar and currants would stick to our fingers because currant jam does so, that the breaking of a coupling would stop the motion of the engine because it would stop the motion of the carriage, or that A's happiness is a mark of a sympathetic nature because that is the case with the happiness it causes in B.

The principle, then, of the special resemblance of cause and effect must be rejected as invalid. And, even if it were valid, its application as a proof of the existence of matter would involve fatal inconsistencies.

366. In the first place, there is the case of dreams. When, in waking life, I have certain visual sensa, I am justified, it is said, in concluding that the cause of the sensa must be a material hen's egg, with qualities resembling the qualities of the sensa. But it would be universally admitted that if I had certain visual sensa in a dream, I should be mistaken in concluding that their cause must be a material roc's egg. Yet the sensa in the dream are just as real as the others; it is just as necessary that they should have a cause; and the roc's egg would resemble them in the same way that the hen's egg resembles the others. Why may I make the inference in one case, and not in the other?

And, if we confine ourselves to waking life, there are still inconsistencies. When I see a hot poker, I perceive sensa of form and sensa of colour. Now the ordinary theory of matter makes the matter the cause of the sensa of colour, as well as of the sensa of form. Yet, while it asserts the matter to be straight, it denies that it is red. It is thus admitted that the external causes of sensa do not always resemble them. Why should we suppose that they must do so in the case of the straightness?

The distinction between primary and secondary qualities renders the theory of the existence of matter less tenable than it would otherwise be. In the first place, there is the inconsistency, which we have just noticed, of asserting that we can argue from some of our sensa to causes which resemble them, and that we cannot do so from other sensa. And, in the second place, on this theory, matter, while it is really extended, is destitute both of colour and of hardness, since these are secondary qualities. Now extension is only known to us by sight and touch. When it is known by sight, it is invariably conjoined with colour. When it is known by touch, it is invariably conjoined with hardness. We cannot even imagine to ourselves a sensum which has extension without having either colour or hardness. How then can we imagine matter which has neither colour nor hardness?

That which is unimaginable can, no doubt, exist. But the argument for the existence of matter, which we are at present considering, has now reached a climax of inconsistency. It rested on the principle that the causes of our sensa must resemble the sensa which they cause. But now it turns out that the causes are to resemble a mere abstraction from our sensa—an abstraction which is so far from being what we experience, that we cannot even imagine what experience of it would be like.

Now there seems no ground for the distinction between primary and secondary qualities. If what we perceive of the secondary qualities of anything varies from time to time, and from observer to observer, so also does what we perceive of the primary qualities. If what we perceive of the primary qualities exhibits a certain uniformity from time to time, and from observer to observer, so also does what we perceive of the secondary qualities.

Shall we, then, drop the distinction, and say that matter has not only size, shape, position, mobility, and impenetrability, but also colour, hardness, smell, taste, and sound? This change certainly avoids some of the objections to the more ordinary theory. It does not make an arbitrary and gratuitous difference in the treatment of two sets of qualities. And it gives matter a nature not utterly unlike our sensa, and not utterly unimaginable by us.

But the inconsistency has not been removed. For the sensa which, if matter exists, I receive from the material object change from moment to moment. If I look at a thing under one set of conditions of light and shade, I perceive one colour; if I change the conditions next minute I perceive quite a different colour. And if two men look at it simultaneously under the different conditions of light and shade, they will perceive simultaneously the two colours which I perceived successively. Now it is impossible to suppose that the object has two different colours at once. And if it has only one, then that colour must differ from one of the two perceived by the two observers, since these two colours differ from one another.

The same difficulty arises with all the other qualities which, on this theory, are attributed to matter, whether they are those which the other theory classes as primary, or those which it classes as secondary. Two men who look at a cube simultaneously from different positions perceive sensa of quite different shape. Yet a body cannot have two shapes at once, and each of these men would, under normal circumstances, agree about the shape of the body, though they started from dissimilar sensa. It is clear therefore that the shape of the body cannot resemble the sensa of both the observers, since they do not resemble each other.

367. This line of argument for the existence of matter, then, must be rejected. For not only does it rest on a principle—the similarity of cause and effect—which we are not justified in adopting, but it can only be reached from that principle by means of great inconsistency. There is, however, another ground on which the existence of matter has been maintained. This has never, I think, been put more clearly or forcibly than by Dr Broad. "It is, of course," he says, "perfectly true that a set of conditions —and, moreover, a set which is only one *part* of the total conditions—of a sensum, must not be *assumed* to resemble in its properties the sensum which it partially determines." On the other hand, it were equally unreasonable to assume that the two *cannot* resemble each other. There can be no inner contradiction in the qualities of shape and size, since *sensa*, at least, certainly have shape and size and certainly exist. If such qualities involved any kind of internal contradiction, no existent whatever could possess them. Hence it is perfectly legitimate to postulate hypothetically any amount of resemblance that we choose between sensa and the permanent part of their total conditions. If now we find that by postulating certain qualities in those permanent conditions, we can account for the most striking facts about our sensa, and that without making this hypothesis we cannot do so, the hypothesis in question may reach a very high degree of probability.

"Now we find that the visual sensa of a group which we ascribe to a single physical object are related projectively to each other and to the tactual sensum which we ascribe to the same object. If we regard their common permanent condition as having something analogous to shape, we can explain the shapes of the various sensa in the group as projections of the shape of their common permanent condition. If we refuse to attribute anything like shape to the permanent conditions, we cannot explain the variations in shape of the visual sensa as the observer moves into different positions. This does not, of course, *prove* that the common and relatively permanent conditions of a group of sensa do have shape, but it does render the hypothesis highly plausible."[1]

368. In the first place let us consider Dr Broad's contention that there can be no inner contradiction in the qualities of shape and size. His argument is perfectly valid against anyone who does not admit the possibility of erroneous perception. For there is no doubt that we do perceive certain things as sensa having shape and size. But if we admit the possibility of erroneous

---

[1] *Scientific Thought*, p. 278.

perception—as I have done—then it is possible that there should
be an inner contradiction in the qualities of shape and size. For
then it is possible that nothing has those qualities, although some
things appear as sensa having them[1].

And it does not seem correct to say that "if we refuse to attribute anything like shape to the permanent conditions, we
cannot explain the variations in shape of the visual sensa as the
observer moves into different positions." For it seems clear that
we can explain them, in the only sense in which anything can ever
be explained—by bringing these variations under a general law.

Let us suppose that a self A should perceive two spiritual
substances, B and C[2]. B and C would be in relation to that other
spiritual substance, A', which appears as A's body, and these
relations could not, if we are right, be spatial. Neither could they
change, if we were right in our previous conclusion that nothing
is really temporal. But if A misperceives B, C, and A' as bodies,
and misperceives the relations between them as being spatial
relations, and as changing, then it is just what is to be expected
that he should perceive B and C as having shapes, which vary
with the apparent changes of spatial relations between them and
A', and which vary by the same laws of spatial nature to which
Dr Broad appeals to afford an explanation on his hypothesis. If
the nature of the characteristic of space is such that it would
account for the changes in the apparent shapes of real bodies,
with real shapes, which were really changing their real spatial
relations, then it is such that it will account for the changes in
the apparent shapes of substances which appear to be bodies,
and which appear to change their apparent spatial relations. For
the substances in question, since they appear as being in space,
will have their appearances connected together by the laws of
space, as much as real substances in space would have their real
natures connected by the laws of space.

369. And thus we should have explained the variations in
shape, since we should have shown that they were connected

[1] Cp. p. 43.

[2] It does not follow directly that what is not material should be spiritual, but
it is sufficient for our purpose that what was spiritual would be non-material.
"Spiritual substances" would include, not only selves, but parts and groups of
selves.

with each other according to a law. The explanation, indeed, is neither so simple nor so complete as that which could be given if we had been able to accept the view that any substance could be really spatial. It will not be so simple, for on Dr Broad's view the same laws—those of projective geometry—will connect the shapes of the sensa both with one another, and with their permanent conditions. But, if reality is not really spatial, then, while the laws of projective geometry will connect the various spatial appearances with one another, it must be a very different law which determines why particular non-spatial and non-material realities determine things to appear to us as having particular spatial and material qualities.

Moreover, although it follows from our results that there must be such a law as this last, we do not know what it is. We do not know what non-spatial qualities in the reality cause one thing to appear as a square sensum, and another as a circular sensum. And it does not seem very probable that we shall ever find out. Dr Broad's hypothesis, on the other hand, only requires the laws of projective geometry, which are known. And thus his explanation is certainly more complete.

But neither simplicity nor completeness are decisive in favour of an explanation. There is so much in the world that we do not yet know that an explanation is not necessarily to be rejected as inferior to another because it leaves more unexplained. And, if simplicity were decisive, then, I imagine, Dr Einstein's theories would have very little chance against the older views which they are now so generally recognized as superseding. At any rate, no simplicity or completeness in an explanation could be of any force against a demonstration that it involved a contradiction. And, if our results in the earlier part of this chapter are correct, a contradiction is involved in any theory which holds that anything real is spatial.

And thus, even apart from the conclusions which we have based on the impossibility of simple substances, there is no reason to believe that matter does exist. For we have seen that the argument from the qualities of sensa to the qualities of their causes is untenable, and we have seen that the facts of experience can be explained on the hypothesis that there is no matter.

This would, by itself, compel us to refrain from believing that matter did exist, but could not, of course, justify us in believing that matter did not exist. To reach this result we require the arguments put forward in the earlier part of this chapter.

370. The belief in the non-existence of matter does not compel us to adopt a sceptical attitude towards the vast mass of knowledge, given us by science and in everyday life, which, *primâ facie*, relates to matter. For that knowledge holds true of various perceptions which occur to various men, and of the laws according to which these occurrences are connected, so that from the presence of certain perceptions in me I can infer that, under certain conditions, I shall or shall not have certain other perceptions, and can also infer that, under certain other conditions, other men will or will not have certain perceptions.

It will be objected that this is not what common experience and science profess to do. When we say that this bottle contains champagne, and that bottle vinegar, we are not talking about our perceptions, but about bottles or liquids. And physical science deals with such things as planets, acids, and nerves, none of which are either perceptions, or the sensa which are perceived.

It is quite true that it is usual to express the conclusions of common experience and of science in terms which assume the existence of matter. Most people in the past have believed that matter did exist, and our language has been moulded by this belief. The result is that such statements as "this bottle holds champagne," "all lead sinks in water," can be expressed simply and shortly, because they are statements of a type which has always been in frequent use; while the corresponding statements about perceptions would be elaborate and long, because they have been made so rarely that it has not been worth while to form language in such a way that they can be expressed simply and shortly. Thus even people who do not believe in the existence of matter find it convenient to speak of bottles and of lead rather than of actual and possible perceptions. And the fact that most people do believe in the existence of matter renders it, of course, still more natural for them to speak in this way.

But what is meant, in common experience or in science, remains just as true if we take the view that matter does not exist. Something has been changed, no doubt, but what has been changed is no part either of common experience or of science, but of a metaphysical theory which belongs to neither. And so we sacrifice neither the experience of everyday life nor the results of science by denying the existence of matter.

I say, in ordinary language, that this is champagne and that is vinegar. Suppose that there is neither champagne nor vinegar, but that it remains true that the perception of a certain group of sensa of sight and smell is a trustworthy indication that I can secure a certain taste by making certain volitions, and that the perception of another such group is a trustworthy indication that I can secure a different taste by making similar volitions. Does not this have a perfectly definite and coherent meaning in the experience of everyday life, which fits every detail of that experience as well as the more common view does, and only differs from it on a question of metaphysics?

It is the same with science. Every observation made by science, every uniformity which it has established, every statement which it has asserted, whether about the past or the future, would still have its meaning. The observations would inform us of what had been experienced, the uniformities would inform us of the connection of various experiences, the statements as to the past and the future would tell us what has been or will be experienced, or would be so if the necessary conditions were present. What more does science tell us, or what more could it desire to tell us? If the language in which scientific results are generally expressed does seem to tell us more, and to imply the existence of matter, that is not science, but metaphysics—the unconscious metaphysics of ordinary language—and its rejection does not involve rejecting or distrusting a single result of science.

Science requires, no doubt, that experience should exhibit certain uniformities, so that a certain experience can safely be taken as an indication of what experience will follow it under certain conditions. But this proves nothing as to the existence of matter. If I myself have a constant nature, and the external causes of my experience have also a constant nature, the experience which is their joint effect will exhibit uniformities. And a non-material cause can have a constant nature just as easily as a material cause could have.

Science also requires that experience should have a community of nature between different persons, so that it shall be possible for us to infer from my experience what the experience of another person would be under conditions more or less similar. This, again, can be obtained without matter as easily as it could be obtained with it. If my nature and that of other persons were not more or less similar, our experience would present no similarities, whatever the nature of its external cause. But if our natures are more or less similar, then it is obvious that the action of the same external cause on each of us would produce results in each of us which would present similarities.

371. The denial of the existence of matter, it must also be noted, does not lead us towards solipsism—the denial by each individual of the existence of anything but himself. The arguments which prove that my experience must have causes which are not myself, nor part of myself, but some other reality, lose none of their force if we decide that these causes are not of a material nature. And the other arguments against solipsism, which will be discussed in Chapter XXXIX, are just as strong on the hypothesis that matter does not exist[1].

We have rejected the view that the causes of our sensa resemble those sensa, but, in doing this, we have not deprived ourselves of all chance of proving more about them than the fact that they are causes of our sensa. Such causes must be substances, and we have already arrived at various conclusions as to the nature of all substances, which will apply, amongst others, to those which are causes of our sensa. And in the course of our argument we shall come to the conclusion that there is good reason to believe that all substances must be of a spiritual nature.

372. Before concluding this chapter it may be well to point out that, although most philosophers, and, I believe, everyone who is not a philosopher, have used "matter" in much the same sense in which I have used it, yet some philosophers have used

---

[1] Pages 44–48 and 52–54 of this chapter are reprinted, with certain alterations and revisions, from my earlier book *Some Dogmas of Religion*, Sections 69–76.

it in very different senses. It is sometimes used to denote any cause of my sensa which is not myself or part of myself, whatever the nature of that cause may be. Thus, if the word is to be used in this way, it would be possible to assert the existence of matter and to assert, also, that nothing existed which was not spiritual. For my sensa may be caused by some spiritual substance which is not myself or part of myself. Again, it has sometimes been said that groups of sensa are to be called matter, provided that they are connected by certain relations, and by certain laws[1].

It is obvious that none of the arguments in this chapter would be valid against the existence of matter, if the word is to be taken in the first of these senses. And, if it is to be taken in the second sense, only some of them will be applicable, and these would require restating[2]. But then it is not in either of these senses that I have been using the word.

---

[1] Both usages may be found in Mr Russell's works—the first in *Problems of Philosophy*, the second in *Our Knowledge of the External World*.

[2] This will be done in the next chapter, in which I shall endeavour to prove that no sensa exist, from which, of course, it follows that no groups of sensa exist.

CHAPTER XXXV

SENSA

373. The objects which we perceive are called Perception Data or Percepta. They are divided into two classes. The first is the class of those data which the percipient perceives by introspection. It is generally admitted that we can perceive our own mental states in this manner, and I shall give reasons in the next chapter for thinking that in this manner each of us can perceive himself. The second class consists of those data which appear *primâ facie* to be given us by means of the sense organs of our bodies—data of sight, touch, hearing, smell, and taste, together with those given in motor and organic sensations. The members of this class are called Sense Data or Sensa. Whenever we perceive sensa we have a spontaneous and natural tendency to believe in the existence of some piece of matter, corresponding to and causing each sensum, though, as I have endeavoured to show in the last chapter, such beliefs are erroneous.

The percepta which are not sensa are *primâ facie* spiritual. If spirit did not really exist, they would not, of course, be really spiritual. But, if anything is spiritual, then the percepta of this class are spiritual.

But how about the sensa? It was, till recently, a common view that the sensa perceived by any percipient were part of that percipient, and were therefore spiritual. But the view that the sensum was part of the percipient seems to have arisen from a confusion between the sensum which was perceived, on the one hand, and the perception of it, on the other. The latter was judged correctly to be a part of the percipient, and the distinction between the perception and the perceptum was not clearly realized. And, when it is realized, there seems no reason to regard the sensum as part of the percipient[1].

---

[1] That the sensum was not part of the percipient would not involve that it was independent of the percipient. It is possible that a sensum should only exist when it is perceived, and yet not be part of the percipient.

In this case we have, so far, no reason to suppose that the
sensa are in any sense spiritual. And, at the same time, it is
clear that they must *primâ facie* be distinguished from matter.
For example, two men who were, in ordinary language, looking
at the same plate from different points of view, would, *primâ
facie*, be perceiving sensa which were dissimilar, and which must
therefore be numerically different. But the ordinary view would
be that both sensa were caused by, and justified us in inferring
the existence of, the same piece of matter.

Thus the world, in which we tend *primâ facie* to believe, is
divided, not, as is often said, into spirit and matter, but into
spirit, sensa, and matter. But we found reason in the last chapter
to conclude that this *primâ facie* appearance was illusory in the
case of matter, and that matter does not really exist. In this
chapter I propose to argue that the appearance is also illusory
in the case of sensa, and that they do not really exist.

374. Sensa are, *primâ facie*, perceived by us. The view which
I shall put forward is that, when we appear to perceive a sensum,
we do really perceive something, but that we misperceive it.
The object which we perceive has not the nature which it appears
to have. And as "sensum" is generally taken to mean something
which has this nature, it seems better to say, not that we misperceive sensa, but that sensa do not exist, though some percepta
are misperceived as having the nature of sensa.

I shall endeavour to show later on that the objects which we
do perceive when we appear to perceive sensa are all spiritual.
In this chapter I shall confine myself to arguing that they cannot
have the nature of sensa, because sensa cannot exist.

If sensa did exist, they must have parts within parts to infinity.
If so, sufficient descriptions of such parts must be determined by
determining correspondence. And it seems to me that this is impossible, for much the same reasons as led to a similar conclusion
in the case of matter.

375. What are the qualities which sensa appear as having,
and which, if they really exist, they really have? The first question
which arises is as follows. Among these qualities are some which
cause us to attribute certain qualities to the material objects
whose existence we tend to infer from the existence of the sensa.

If I say that this table is red, and that chair blue, the ground of
my judgment is the differing qualities perceived in data appearing
as sensa. Now when I say the table is red, is this because I perceive the datum as having the quality of being red, and do I
attribute to the table the same quality which I perceive the datum
as having? Or is it the case that the datum is perceived as having
a different quality—the quality "being a sensum of red"—and
from this I am led to attribute to the table another quality, the
quality of being red?

Both these views have been maintained. To me it seems clear
that it is the same quality of redness which I perceive the datum
as having, and which I attribute to the table. We are certainly
aware of the quality of redness which we attribute to the table,
for otherwise a statement that the table is red would mean
nothing to us, and, whether it is right or wrong, it certainly has
a meaning. But we cannot be aware à priori of such a quality as
redness, and therefore the only way in which we can be aware of
it is by perception. Now if we perceive any datum as being red,
we are aware of redness by perception, and there is no difficulty.
But if no datum is perceived as being red, how could we be aware
of redness by perception? There would be only one possible way.
We might perceive the datum as having the complex quality
"being a sensum of red," and so might be aware of that complex
quality. And then, being aware of the complex quality, we might
by analysis be aware of redness, as one of its elements.

In that case the quality which we perceive the datum as having
is a complex quality, and a complex quality of whose complexity
we are aware, since otherwise we could not be aware of its
elements. Now I think that introspection makes it clear that the
quality which we perceive the datum as having is not complex
but simple. And, more directly, I think that introspection also
makes it clear that the quality which we perceive the datum as
having is the same quality which we attribute to the table.

Indeed, I do not think the other view would ever have been
maintained, if it had not been for the belief, mentioned above,
that the sensum must be part of the percipient. Thinkers who
believed this, but who saw that redness could not be a quality
of anything spiritual, were driven to deny that the datum had

the quality of redness, and to substitute the quality "being a sensum of red." But when we see that the datum need not be part of the percipient, the only ground for denying it to be red has gone.

376. Thus the qualities which we perceive the data as having will include qualities attributed to matter. They will not, however, be attributed to the data in the same way in which they are attributed to matter. In the first place those qualities which are called secondary are thought by most people not to belong to matter, while there is no doubt that they belong to the data as much as the primary qualities do. In the second place material objects are held to be, for example, both coloured and hard, while it is admitted that one sensum cannot have both these qualities. In the third place it is held that the same matter can have, at different times, qualities which it could not have simultaneously. The same piece of lead may be a cube at Easter and a sphere at Michaelmas. But no one would suggest that a sensum of square-ness at Easter, and a sensum of triangularity at Michaelmas, could be the same sensum.

Besides the primary and secondary qualities, the data will also be perceived as having the quality of duration in time. For every perceptum is perceived as existing simultaneously with the perception of it, and perceptions are in time.

So far we have found no qualities, which the datum is perceived as having, which are not also attributed to matter, and we have seen in the last chapter that these qualities are not such as to determine by determining correspondence sufficient descriptions of a series of parts of parts to infinity. Are they perceived as having any other qualities? They are perceived as being percepta, and as being the particular sort of percepta which are called sensa, but these qualities clearly cannot give the determining correspondence required, because each of them is a single quality with no sub-divisions or series of terms, and so could not give a plurality of sufficient descriptions.

377. It may be said that the data are perceived as having two qualities which are not attributed to matter. The one is in-tensity, in the sense in which we say that a bright light is more intense than a dull light. The other is extensity, in the sense in which we say that the data perceived when we are hungry, or are in a hot bath, have a certain massiveness.

It might be objected that these are qualities of the perception and not of the perceptum, but this would, I think, be erroneous. But even as qualities of the perceptum they will not give us what is wanted. For both intensity and extensity, in the sense in which we are using them here, are examples of intensive quantity, and not of extensive quantity[1]. In extensive quantity the difference between a greater and a less quantity of the same sort is a third quantity of the same sort. But this is not the case either with this intensity or with this extensity. The difference between a brighter light and a less bright light is not a light of another brightness. And the difference between a more massive pain and a less massive pain is not a pain of another massiveness.

It is impossible, therefore, that a datum should be divided into parts in respect of either of those qualities. It is as impossible as it would be for a temperature to be divided into two other temperatures; and therefore the infinite series of parts within parts cannot be reached in this way.

378. None, therefore, of the qualities which these data are perceived as possessing can give sufficient descriptions for the infinite series of parts within parts. And the sufficient descriptions must be given somehow, since the perception data exist. The data must therefore have other qualities which they are not perceived as possessing, and which cannot be deduced by us from any of the qualities which they are perceived as possessing.

Their nature, therefore, would be very different from the nature hitherto assigned to sensa. And, if they possessed those other qualities, and did not possess the qualities they are perceived as possessing, it would be misleading to call them sensa, and we should have to say that, although substances certainly existed which were percepta, yet none of them were sensa.

379. But, it might be objected, there would remain two alternatives analogous to those which were mentioned in the last chapter (p. 43) as to matter. In the first place, the data might possess the qualities which they are perceived as possessing,

---

[1] I failed to see this when I wrote Section 163, in my first volume. The error, however, does not affect the argument of the section.

but might also possess other qualities, which were such as to determine sufficient descriptions of the infinite series of parts, and to determine them as possessing spatial qualities, or those qualities which appear as temporal, or both. In that case the data would be infinitely divisible in space, or in apparent time, or in both.

Or, secondly, the data might possess the qualities which they are perceived as possessing, but might not be infinitely divisible in respect of any of those qualities. That is, they might consist of units which were not divisible in space and apparent time— the only dimensions which they are perceived as possessing. But these units might, in addition to their perceived qualities, have other qualities such as to determine by determining correspondence sufficient descriptions of an infinite series of parts within parts.

A substance which, in addition to its perceived qualities, had such non-perceived qualities as these, would certainly not be the sort of sensum whose existence has hitherto been asserted by anyone. It will be more convenient to postpone the consideration of the possibility of such a substance to Chapter XXXVIII (pp. 116–119). We shall then find reason to reject it. But at any rate it cannot affect our conclusion that sensa, with the nature ordinarily ascribed to sensa, cannot exist.

380. Nor would our conclusion have been different if we had held that a sensum cannot have, for example, the quality of redness, but only the quality "being a sensum of redness." Qualities which are of the type "being a sensum of x," where x is a quality attributed to matter, can obviously only vary as x varies, since the element "being a sensum of" is constant. Thus, in order to form an infinite series of qualities of this type, there would have to be an infinite series of variations of x, determined by determining correspondence. And we saw in the last chapter that there cannot be an infinite series, so determined, of variations of the qualities attributed to matter.

CHAPTER XXXVI

SPIRIT

381. It is impossible, then, that matter or sensa should exist. Is it possible that Spirit should exist? In the first place, how are we to define spirit?

I propose to define the quality of spirituality by saying that it is the quality of having content, all of which is the content of one or more selves. Nothing can have this quality except substances, and so nothing but substances are spiritual. Selves, of course, will answer to this definition, and so will parts of selves, and groups of selves, however trivial or arbitrary, and groups whose members are selves and parts of selves. The content of any such substance will be called spirit. But, in accordance with usage, I shall not use the phrase "*a* spirit" of any spiritual substance except a self.

382. We have defined spirituality by means of the conception of a self. What then do we mean by a self? I should say that the quality of being a self is a simple quality which is known to me because I perceive—in the strict sense of the word—one substance as possessing this quality. This substance is myself. And I believe that every self-conscious being—that is, every self who knows that he is a self—directly perceives himself in this manner.

The greater part of this chapter will be devoted to the support of this view. Its establishment would not by itself prove that selves did exist. It would only prove that something was perceived as being a self. And we saw in Chapter XXXII that we must admit that perception could be illusory—that a thing could be perceived as being something, which, in reality, it was not. But we shall see in the next chapter that spirit, unlike matter and sensa, can have parts within parts to infinity, and that therefore there is not the same ground for rejecting the existence of spirit that there is for rejecting the existence of matter and sensa. And in Chapter XXXVIII we shall see reasons for holding that spirit does exist, and that no substances exist which are not spiritual.

The reasons which have led me to accept the view that the self is known to itself by direct perception were suggested to me by a passage of Mr Russell's article "Knowledge by Acquaintance and Knowledge by Description."[1] Mr Russell did not work out his position in detail—which was not essential for the main design of his paper. And he has now ceased to hold the position at all. I remain, however, convinced of the truth of the view, the first suggestion of which I owe to him.

The argument is as follows. I can judge that I am aware of certain things—for example, of the relation of equality. I assert, then, the proposition "I am aware of equality." This proposition, whether true or false, has certainly a meaning. And, since I know what the proposition means, I must know each constituent of it. I must therefore know "I." Whatever is known must be known by acquaintance or by description. If, therefore, "I" cannot be known by description, it must be known by acquaintance, and I must be aware of it.

Now how could "I" be described in this case? The description must be an exclusive description, in the sense which we have given to that phrase, since I do not know "I" by description unless I know enough about it to distinguish it from everything else. Can I describe "I" as that which is aware of equality? But it is obvious that this is not an exclusive description of "I." It could not be an exclusive description of "I" unless I was the only person who was ever aware of equality. And it is obvious that this is not certain, and that it is possible that some one besides me was, is, or will be aware of equality. (In point of fact, I have, of course, overwhelming empirical evidence for the conclusion that some other persons *are* aware of equality.) Thus we cannot get an exclusive description of "I" in this way.

383. It may be thought that an exclusive description could be reached by going a step further. I am not only aware of equality, but I am also aware, by introspection, of *this* awareness of equality—the particular mental event which is my consciousness of equality here and now. Now if "I" were described as that which is aware of this awareness of equality, should we not have reached an exclusive description? For no one else, it may

[1] *Mysticism and Logic*, p. 211.

be argued, could be aware of *this* awareness of equality, except I myself who have it. Of course, in order that this should be an exclusive description of "I," I must know what I mean by *this* act of awareness. But this would not require a description, because the act of awareness would itself be known by awareness. It would be a perceptum given in introspection. Thus, it is said, we can dispense with the necessity for awareness of self.

This argument, as has been said, depends on the assertion that no one can be aware of an awareness of equality except the person who has the awareness of equality. To this point we shall return later. But first it must be pointed out that, even if this assertion were correct, the argument would not be valid.

The judgment that we are now considering is the judgment "I am aware of this awareness."[1] This is not merely a judgment that a particular person is aware of this awareness. It also asserts that the person who is aware of the awareness is the person who is making a judgment. Now how am I entitled to assert this identity, if "I." can only be known by description? In that case I am aware of this awareness, and of this judgment, but not of myself. I may be entitled to infer that there is someone who is aware of this awareness, and that there is someone who is making this judgment about it, since awarenesses and judgments require selves to be aware and to judge. And it may be the case that "the person who is aware of this awareness" is an exclusive description of the person to whom it is applied. But how do I know that the person thus described is the person who makes the judgment? If I am not aware of myself, the only thing I know about the person who makes the judgment is just the description "the person who makes this judgment." This is doubtless an exclusive description, but I am still not entitled to say "I am aware of this awareness" unless I know that the two descriptions apply to the same person. And if the person is only known by these descriptions, it does not seem possible to know anything of the sort. Thus if "I" can only be known by description, it seems impossible that I can know that I am aware, either

---

[1] It must be remembered that this phrase does not mean that I *have* the awareness in question. It means that the awareness (in this case the awareness of equality) is the object of a fresh act of awareness.

of this awareness of equality, or of anything else, since the judgment "I am aware of X" always means that the person who is
aware of X is also the person who is making the judgment.

If, on the other hand, I do perceive myself, there is no
difficulty in justifying either the judgment, "I am aware of this
awareness," or the judgment, "I am aware of equality." There is
no need now to find an exclusive description of "I," because I
am aware of myself, that is, know myself by acquaintance, and
do not require to know myself by description. And I can now
justify the assertion, implied in the use of "I," that the person
who is aware (whether of this awareness or of equality) is the
person who makes the judgment. For in perceiving myself, I
perceive myself as having some of the characteristics which
I possess. And if "I," which is a term in the judgment, and which
is known by perception, is perceived as having the awareness,
then I am justified in holding that it is the same person who is
aware and who makes the judgment.

384. And thus the attempt to describe the self as that which
is aware of a particular awareness has broken down, even if we
grant the premise which is assumed—namely, that "that which
is aware of this awareness" is an exclusive description of the
substance to which it refers. But we must now examine into the
truth of this premise, for, although the argument would not
hold, even if the premise were valid, the question of its validity
is important in itself, and will be of special importance with
reference to some of the results at which we shall arrive in the
next chapter.

It is very commonly held that it is impossible for any person
to be aware of any mental state, except the person who has the
state, and that, therefore, only one person can be aware of any
mental state.

It is only of mental states that it is held that the awareness
of them is thus restricted to a single person. With regard to the
awareness which is not perception—the awareness of characteristics as such—it is universally admitted that it is not confined
to a single person, but that more than one person may be aware
of yellowness, sweetness, or goodness. It is only by awareness
that we can know what any simple characteristic means—for,

being simple, it cannot be defined—and the meaning of compound characteristics depends on the meaning of their simple components. If, therefore, two people could not be aware of the same simple characteristic, it would be impossible for one person ever to communicate his thoughts to another.

With regard to sensa, it is, as we have seen, held by many thinkers that the sensa perceived by any mind are not parts of the mind which perceives them. And it would be held by many of these that the same sensum could be perceived by more persons than one—that is, that more persons than one could be aware of it.

But with mental states it is generally held to be different. As we have said, the ordinary view is that a mental state can only be perceived by the self who has that state. Thus those thinkers who hold, as some do, that sensa are states of the mind, also hold that each sensum can only be perceived by one person—who is, of course, that person of whom they are states. And in the case of those percepta which are not sensa—the percepta which are given in introspection, and are admittedly states of the mind—it is generally held, and indeed generally tacitly assumed, that they can have no other percipient than the mind in which they fall. Among these latter percepta are of course all states of awareness. And thus it is held that no one can be aware of a state of awareness except the one person of whom it is a state, and that "that which is aware of this awareness" is an exclusive description.

Now it does not seem to me that we are justified in asserting this as an absolute necessity. It is true, no doubt, that in present experience I do not perceive the state of mind of any person but myself. And I have good reason to believe that no one of the persons whom I know, or who have recorded their experience in any way which is accessible to me, has perceived the states of mind of any other person than himself. Nor have I any reason to believe that any person in the universe has done so[1].

[1] The statements in this paragraph refer only to our present experience. I shall endeavour to show later that metaphysical considerations lead us to the conclusion that, in absolute reality, selves perceive each other, and the parts of each other.

It may be said that the statement in the text is not true, even as to present experience, since we know that various persons have mystical experiences in

But the fact that there is no reason to suppose that it *does* happen is very far from being a proof that it could not happen. And I can see no reason for supposing that it could not happen. Even if it should be held that in present experience no self perceives anything but its own states (a position which would involve the improbable view that sensa are parts of the percipient self) I can see no impossibility in its doing so. That relative isolation of the self (of course it is not complete isolation), which would prevent it from entering into a relation of perception with anything outside itself, need not be essential to the self because it is found in it throughout our present experience. And if sensa are not parts of the percipient self, then the isolation, even in our present experience, would be less than if they were such parts.

385. It must be remembered that, if A should perceive a state of B, that fact would not make it a state of A, or any less exclusively a state of B. To have a state and to perceive that state are two quite different things. In our present experience, as we have just said, no one does the second who does not do the first. But the first often occurs without the second. In my present experience I often have a state, even a conscious state, without being aware of that state[1]. And this does not make it any the less my state. I believe the confusion here has had a good deal to do with the prevailing belief that one self cannot perceive a state of another self. The real impossibility of a state of one self being also a state of another self has been confused with a supposed impossibility of a state of one self being perceived by another self.

which they claim to have direct experience of other selves. It would take us too far to endeavour to interpret the significance of mystical experiences in this respect. But I do not think that any of the accounts known to me lead to the conclusion that one self does really perceive another and still less that he perceives parts of another. If, however, I should be wrong in this, such a result would strengthen my argument in the text, that there is nothing intrinsically impossible in the perception by one self of the states of another self.

[1] If this were not so, every conscious state would start an infinite series of perceptions, since a perception is itself a state, and I should have to perceive that also and so on infinitely. And we know that in present experience this is not the case. We do not generally perceive a perception, and I suppose that we scarcely ever perceive the perception of a perception, except possibly when we are engaged on epistemological or psychological investigations.

It is therefore not intrinsically impossible that a state of a self may be perceived by two or more selves (one of whom may be the self of which it is the state). We cannot, therefore, be certain that "the self who is aware of this awareness" is an exclusive description of any self of which it is true. And, therefore, if "I" can only be known by means of this description, I cannot be certain who "I" is, and cannot be certain that I know the meaning of the proposition "I am aware of equality" (since the "I" in the latter proposition has to be described by means of the former). Thus, for a second reason, the attempt to show that "I" can be known by description in this manner, has broken down.

386. An attempt might be made to know "I by description, which would not be liable to this second objection. For it might be said, and I think truly, that, while it is not impossible for more than one self to be aware of a particular awareness, it is impossible for more than one self to *have* the same awareness. If I am aware of X, it is not impossible that you, as well as I, should be aware of my awareness of X, but it is impossible that my particular awareness of X should also be your awareness of X, since what is a state—that is, a part—of one self can in no case be a state of another self.

The view that two selves cannot have the same awareness has been denied, but, as I have said, I believe it to be true. But it will not give knowledge of "I" by description.

The attempt to know it by description on this basis would, I conceive, be as follows. If we start from "I am aware of equality," and wish to describe the "I," we must proceed to the further proposition "I have *this* awareness of equality," which will always be true if the other is. Then the "I" in the latter proposition can be described as the self which has *this* acquaintance with equality. This description cannot apply to more than one thing, and is therefore an exclusive description of it. And the thing so described is the "I" in both propositions.

And in this way we do avoid the second objection. But the new attempt is still open to the first objection. It involves that the two descriptions apply to the same self, and this is an assumption which we have no right to make. For when I assert the proposition "I have this awareness," it means that the self who has this awareness is the same as the self who asserts the proposition. Now I can only describe the one—if it is to be described at all— as the self which has this awareness—and the second as the self which makes this judgment. Both these descriptions are exclusive descriptions. But I have no reason to suppose that they refer to the same self, and therefore I am not entitled to say "I have this awareness."

If, on the other hand, I am aware of myself, I am entitled to say "I have this awareness" because the "I" which is a term in the judgment, and which is known by perception, is perceivable as having the awareness. Once more, then, we are brought back to the conclusion that, if I am entitled to make any assertion about my awareness of anything, I must be aware of myself.

The same line of argument will show that, unless "I" is known by acquaintance, I am not justified in making *any* statement about myself, whether it deals with awareness or not. If I start with the judgment "I am angry," and then, on the same principle as before, describe "I" as that which has *this* state of anger, my assertion will involve that it is the same self which has this same state of anger, and which is making this judgment. And, if "I" can only be known by description, there is no reason to hold that it is the same self which both has the state and makes the assertion.

387. It is not, of course, impossible for us to have good reasons for believing that two descriptions both apply to some substance which we only know by description. I only know other people by description, but I may have good reason to believe that one of my friends is both a Socialist and a Cubist. But the case before us is not analogous to this. In the latter we arrive at our conclusion because we have reason to infer certain facts about the self who is a Socialist, and certain facts about the self who is a Cubist, which are incompatible with their being different selves[1]. But in the case before us I am certain that it is I who am angry, even though I am aware of no characteristic of the anger from which

---

[1] For example, we might observe that sounds which were a confession of Socialism, and sounds which were a confession of Cubism, proceeded from the mouth of the same body. And from this we might infer that they were due to the volitions of the same self, and that they expressed the opinions of that self.

it could be inferred what particular self had it, and even though
I am aware of no characteristic of the judgment from which it
could be inferred what particular self made it. My knowledge
that it is the same self, which is angry and which makes the
judgment, is as immediate and direct as my knowledge that
some self is angry, and that some self makes the judgment. Unless, therefore, I perceive myself, and perceive myself as having
the anger and as making the judgment, what will be the data
before me? Only the awareness of a state of anger, the awareness
of a judgment, and the general principle that every state of anger
and every judgment must belong to *some* self. This will not justify
the conclusion that the anger and the judgment belong to the
same self, and therefore I shall not be entitled to assert, "I am
angry."

388. An attempt has been made to describe "I" in another
manner. It is no longer described as that which is aware of something, or which has a mental state. It is described as a whole of
which certain mental states are parts. The classical statement of
this view is Hume's. "I may venture to affirm of...mankind, that
they are nothing but a bundle or collection of different perceptions
which succeed each other with an inconceivable rapidity and are
in a perpetual flux or movement."[1]

This gives, of course, a very different view of the self from that
which is generally held. In the first place, the knowledge of the
self is logically subsequent to the knowledge of the mental states.
We can know the states without knowing the self, but we can
only know the self by means of our knowledge of the states. In
the second place, it would seem that the theory holds that this
relation of knowledge corresponds to a relation in the things
themselves. The ultimate realities are the mental states, and the
selves are only secondary, since they are nothing but aggregates
of the states. In the third place we must no longer say that the
self perceives, thinks, or loves, or that it has a perception of thought

---

[1] *Treatise* I. iv. 6: The strict grammatical meaning of Hume's expression
seems to be that all mankind are one bundle. But it is evident from the context
that he holds that there is a separate bundle for every separate self. Hume, it
will be noticed, uses "perception" in a wider sense than that which we have
adopted. In the following argument I have used "state" as the equivalent of
Hume's "perception."

or an emotion. We can only say that the bundle includes a perception, a thought, or an emotion, as one of its parts[1].

On this theory, then, when I use the word "I," I know what "I" means by description, and it is described as meaning that bundle of mental states ,of which my use of the word is one member. Is this satisfactory?

389. In the first place we must note that it is by no means every group of mental states which is a bundle in Hume's sense of the word, that is to say, an aggregate of mental states which form a self[2]. For any two mental states form a group by themselves. And there are an infinite number of groups, of each of which both G and H are members. All these groups are not bundles. The emotions of James II on the acquittal of the seven Bishops, and the volitions of William III at the Boyne, are to be found together in an infinite number of groups. But no one supposes— neither Hume nor anyone else—that they belong to the same self. They are therefore not in the same bundle.

But, since every group is not a bundle, we say nothing definite when we say that two mental states are in the same bundle, unless we are able to distinguish bundles from other groups. How is this to be done? Can we distinguish them by saying that the members of bundles have relations to one another which the members of groups which are not bundles do not have? But what would such relations be?

They could not be spatial relations, nor relations of apparent spatiality. For in many cases—as with emotions and abstract thoughts—the states have no special relation to anything which is or appears as spatial. And in cases in which they do have those relations, I can judge, for example, that I have seen Benares and Piccadilly and that Jones has seen Regent Street. Or again I can judge that I have seen Piccadilly and Regent

---

[1] We shall see in the next chapter (pp. 92–97) that it is really the case that the mental states of the self are parts of it. But if, unlike Hume, we hold that the self can be known otherwise than as the aggregate of its parts, it can be seen to have other relations to its parts, beside the relation of inclusion, and it can be seen to be true both that the perception of H is a part of the self, and also that the self perceives H.

[2] The definition of "group" was given in Section 120. For the rest of this discussion I shall, for brevity, use the word "bundle" to indicate exclusively those "bundles or collections" to which Hume reduces selves.

Street, and that Smith has seen Benares. Thus perceptions of sensa which appear as related to objects close together may be in the same bundle or in different bundles, and the same is true of sensa which appear as related to objects distant from one another.

Neither can they be temporal relations, or relations of apparent temporality. For in some cases we say that experiences separated by years belong to the same bundle, and in some cases to different bundles. And in some cases we say that simultaneous experiences belong to the same bundle, and in some cases to different bundles.

They cannot be relations of similarity or dissimilarity. For in every bundle there are states which are similar and dissimilar to other states in that bundle, and which are similar and dissimilar to states in other bundles. Nor can it be causation. For my happiness to-day may have no causal connection with my misery yesterday, whereas, if I am malignant, it may be caused by the misery of Jones to-day.

Again the relation cannot be the relation of knowledge. For I can know both my own misery and that of Jones. Nor can it be the relation of apparent perception. For, of my state of misery yesterday and my state of happiness to-day, neither apparently perceives the other. Nor can they be apparently perceived by the same state, for one has ceased some time before the other began.

The relation we are looking for, then, cannot be any of these. Nor do I see any other *direct* relation between the states which could determine the bundle to which they belong. There seems only one alternative left. The relation must be an indirect relation, and it must be through the self. We must say that those states, and those only, which are states of the same self, form the bundle of parts of that self.

There is no difficulty about this, if, as I have maintained, a self is aware of himself by perception. But it is fatal to the attempt to know "I" by description. It would obviously be a vicious circle if I described "I" as being that bundle of states of which my use of the word is a member, and then distinguished that bundle from other groups by describing it as that group of mental states which are states of "I."

390. One more attempt to know "I" by description must be considered. It might be admitted that, if we adhered to a purely presentationist position like Hume's, the bundles could not be described except by their relations to selves. But, it might be said, if we admit the existence of matter (or of some substance which appears as matter), they could be described in another way. For then, it might be considered, we could say that states belong to the same self when, and only when, the same living body (or what appears as such) stands in a certain relation of causality to both of them. In that sense the meaning of "I am angry" would be that the same living body stood in that relation of causality both to the state of anger and to the judgment about it.

I have said "a certain relation of causality" because it is clear that not all relations of causality would do. The movements of an actor's body may cause aesthetic emotions in each of a thousand spectators, but these emotions admittedly belong to different selves. It might perhaps suffice if we say that the relation between the living body and the mental state must not be mediated by the intervention of any other living body.

The view that *every* mental state has a cerebral state which stands in such a relation to it, is by no means established, and is rejected by many eminent psychologists. But, even if it were accepted, the theory which we are here considering would break down.

It is to be noticed that all that makes states part of the same self is the indirect relation through the body. It is not any direct relation between the states, which is *caused* by the indirect relation, but which would perhaps be perceived even if the indirect relation was not known. It could not be this, for we have seen that no direct relation can be found such that each state in a self has it to all other states in the self and to no other states.

But if there is no relation but the indirect relation, then no man has any reason to say that any two states belong to the same self unless he has a reason to believe them to be caused by the same body. And this means that the vast majority of such statements as "I was envious yesterday" are absolutely untrustworthy. In the first place, by far the greater number of them have been made by people who have never heard of the doctrine that emotions and judgments are caused by bodily states. They could not, therefore, have any reason to believe that the envy and the judgment were caused by the same body. And therefore they could have no reason to believe that they belong to the same self. But, as we have seen, in asserting "I was envious yesterday" I am asserting that the envy and the judgment belong to the same self.

In the second place, even those people who have heard of the doctrine that all mental states are caused by bodily states and who accept it, do not, in far the greater number of cases, base their judgments that two states belong to the same self on a previous conviction that they are caused by the same body. And, indeed, in the case of an emotion and a judgment it is impossible that they should do so. For it would be impossible for any man to observe his brain, and to observe it in two states which he could identify as the causes of the emotion and the judgment respectively. And his only ground for believing that they were caused by the same living body would depend on his recognizing them as belonging to the same self. It is impossible therefore that he can legitimately base his belief that they belong to the same self on the ground that they were caused by the same body.

Thus this theory would involve that every judgment of the type "I am x," or "I was x," or "I did x," where x is anything that a substance can be or do, is totally untrustworthy. Such scepticism, even if not absolutely self-contradictory, which I think it is, is so extreme that it may be regarded as a *reductio ad absurdum*.

391. But we may go further. "I was envious yesterday" has no meaning for anyone who does not know the meaning of "I." Now if "I" can only be known by description, and the only description which is true of it is "that group of mental states, caused by the same living body, of which the envy and my judgment are members," it follows that anyone who does not describe "I" in that way, will not know what "I" means, and so will mean nothing when he says "I was envious yesterday." But the assertion that the meaning of "I was envious yesterday" depends on the acceptance, by the man who makes it, of the doctrine of the cerebral causation of all mental states, is clearly preposterous.

We may now, I think, conclude that the meaning of "I" cannot be known by description, and that, since the meaning of "I" is certainly known—or all propositions containing it would be meaningless—it must be known by acquaintance. Each self, then, who knows the meaning of "I" (it is quite possible that many selves have not reached this knowledge), must do it by perceiving himself.

392. It has however been maintained—and notably by Mr Bradley[1]—that whatever becomes an object becomes *ipso facto* part of the not-self, and what is part of the not-self cannot be the self or part of it. If this were correct, it is obvious that a self could not perceive himself and so could not know himself by acquaintance. As it has just been shown that he cannot know himself by description, the result would be that no self could know himself at all, and that all statements containing "I" as a term would be unmeaning.

But Mr Bradley gives, as far as I can see, no reason why a self cannot be his own object, remaining all the time the self which has the object. And I am unable to see any reason why this should not be so. The presumption certainly is that a self can be his own object. For, as we have just seen, if he could not, no statement with "I" as a term could have any meaning. And there is a strong presumption—to put it mildly—that some statements of this sort have some meaning.

And, again, it is certain that a thing can stand in some relations to itself. A thing—in the widest sense of "thing"—can be its own square root, its own trustee, its own cousin. And, if it has a duration, it can be equal to itself in duration. What is there in the case of knowledge which should lead us to reject the *primâ facie* view that knowledge is one of these relations? I can see nothing. On the contrary, I think that the more closely we contemplate our experience, the more reason we find for holding that it is impossible to reject knowledge of self.

393. The direct perception of the self has not been accepted by the majority of recent philosophers. The explanation may be

---

[1] *Appearance and Reality*, Chapter IX, Div. VI.

partly that they have seen that "the self which has this state" is an exclusive description of a self, when the state is known by awareness, and that they have not seen the further point that the description gave us no ground to identify the self which has the state with the self making the assertion, and that this identity is implied in the use of "I."

But the chief reason is, I think, that they looked for the awareness of the self in the wrong way. They tried to find a consciousness of self which had the same positive evidence for being an awareness as is found in an awareness of equality, or in an awareness of some particular sensum. And this attempt failed. For the "I" is much more illusive than those other existent realities of which we are aware by perception. It is divided into parts which are not themselves selves. And these parts we can perceive, and we generally do perceive some of them whenever we perceive the "I." It is easy, therefore, to suppose that it is only the parts—the mental states—which we perceive, and that the "I" is only known by description, and the belief in it can only be justified by inferences from our knowledge of the states.

Thus, if we merely inspect our experience, the fact that we are aware of the "I" by perception is far from obvious. The only way of making it obvious is, I think, that suggested by Mr Russell and employed in this chapter. We must take propositions containing "I," and, to test the view that "I" is known by description, we must endeavour in these propositions to replace "I" by its description. Not till then does it become clear that it is impossible to know "I" except by acquaintance.

394. We have thus, as it seems to me, justified the statement at the beginning of this chapter, "the quality of being a self is a quality which is known to me because I perceive—in the strict sense of the word—one substance as possessing this quality. This substance is myself." And this quality is simple. We can perceive no parts or elements of which it is composed, any more than we can with the quality of redness. Like redness it is simple and undefinable.

We must now proceed to examine certain questions which arise as to the nature of the self. The first of these is the relation of the self to time, or to that real series which appears as a time-series. Does the self persist through time[1]?

My knowledge of the nature of selves depends, as we have said, on my awareness of myself by perception. I can therefore only be certain of qualities of the self which I perceive myself as possessing, or which are involved in others which I perceive myself as possessing.

395. Now I perceive myself as persisting through time. For a perception lasts through a specious present. At any moment of time, then, I may perceive myself both at that moment of time, and also at any other moment within the limits of a specious present. And if between those two points I begin or cease to perceive something else, I can perceive myself as existing both while the perception exists, and also before or after the perception exists. And in that case I shall perceive myself as persisting in time.

This period of time is, of course, very short even relatively to the life of a human body. Have we any reason to suppose that the self which is perceived through a specious present persists through any longer time? It has been maintained that, for certain periods which are earlier than any part of the specious present, but yet comparatively near, our memory gives us absolute certainty that the things which we remember did occur. If this is so—it is not necessary to discuss here whether it is or is not the case—I can have absolute certainty that I existed at a certain time, provided that it falls within the limits within which memory is absolutely trustworthy. If at the present moment I remember that I was aware of myself in the past, then the "I" who now remembers, and the "I" who was then aware, must be the same "I," unless the memory is erroneous, which, by the hypothesis, it cannot be within these limits. And therefore the same "I" must have persisted from the moment of the remembered awareness to the moment of the remembrance.

Beyond these limits there is no *certainty* of the persistence of the self. If I remember that I did, or that I was, certain things in the past, that professed memory may be deceptive in two ways.

---

[1] In the following discussion I shall, for the sake of brevity, use "time" as equivalent to "time, or the real series which appears as a time-series."

It may be false in the ordinary sense, as when, in a dream, I remember that I rode upon a dragon last year. Or, in the second place, even if the events I remember did happen to someone in my body, I may be mistaken in thinking that *I* experienced them. There may then have been another self related to the body which is now mine. And I may know the experience of that self, and mistakenly judge it to be my own. This last alternative is not I think at all probable, but I cannot see that it is impossible.

396. But although there is no absolute certainty that my present self has lasted longer than the specious present, and the short preceding period of certain memory—if there is such a period—yet there may be very good reasons for holding that it is extremely probable that it has done so. There is, I think, very little reason to doubt that the feelings with which, as I now remember, I saw Benares, really did occur more than thirty years ago, and that the self who experienced them was the same self who is now remembering them. And there is very little reason to doubt that the same self which I am now perceiving did have various experiences ever since the birth of my present body, of which I have now no remembrance. And on similar grounds, there is very little reason to doubt that, unless my body dies within the next week, the self which I now perceive will still exist at the end of that week[1].

The grounds on which such conclusions are reached will, of course, be empirical. But the results to which our arguments have led us, as to the nature of the self, and as to my own certainty, by perception, of my own existence, will have an important bearing on the validity of such conclusions. For when objections have been offered to the ordinary view that each self—at any rate under normal circumstances—persists through the whole life of a living body, they have generally rested, either on the ground that we do not know what the self is, which is said to persist, or else on the ground that its persistence is incompatible with the changes in the "bundle" of mental states.

---

[1] The question whether there is any reason to believe that myself existed before the birth of my present body, and will exist after the death of that body, will be discussed later.

But we are now able to say that by "self" we mean something of which the "I" which each of us perceives is an example. And so the question, as to myself, who exist to-day, whether I existed twenty years ago, is a perfectly definite question, whatever the true answer to it may be. This answers the first objection. And to the second we may now reply that each of us is aware, within the specious present, of a self which remains the same while changes occur among the mental states which are its parts.

397. The second question which we have to discuss is the relation of the quality of selfness to the qualities of consciousness and self-consciousness. When we say that the self is conscious, we mean, I suppose, that it is conscious of something, that is, that it knows something. It would be a difficult question to decide at this stage whether the possession of selfness necessarily involved the possession of consciousness, and, if so, whether a self had to be conscious at all times when it was a self, or whether its selfness could continue during intervals when it had not consciousness. When we have proceeded further, however[1], we shall see reason to believe that all selves are conscious at all times when they exist[2].

A self-conscious self is one which knows itself, and which therefore, by our previous results, perceives itself. Must a self be self-conscious? It has been maintained that it must be. Sometimes it is said that, as consciousness is essential to a self, and as no being can be conscious without being self-conscious, all selves must be self-conscious. Sometimes it is admitted that a being might be conscious without being self-conscious, but then, it is said, it ought not to be called a self.

398. I disagree with both these views. It seems to me perfectly possible for a being to be conscious without being self-

---

[1] Chapter XLIX, p. 248.

[2] Besides selves, the states of selves are sometimes said to be conscious. If the phrase were used of the states in a sense as near as possible to that in which it is used of selves, a conscious state would be a state of knowledge. Even then the word would be used differently of the selves and of the states, for the characteristic of being a self that knows is not the same as a characteristic of being a state of knowledge. But, by a rather inconvenient usage, "conscious" is generally used of the states of selves in another sense—as meaning a state of the self of which the self is aware, or, sometimes, a state of the self of which the self *might* be aware. (Cp. Chapter LXIV, p. 406.)

conscious. It is true that the only conscious being of whom, in present experience, I am aware, is necessarily self-conscious when I am aware of him, since he is myself. But I do not think that I am always self-conscious when I am conscious. It seems to me that memory gives me positive reason to believe that there are times when I am conscious without being aware of myself at all. I am not speaking of states which are mystical, or in any way abnormal, nor of states in which in any sense I *am* not a self, or am less a self than at other times. I am speaking of perfectly normal and usual states, in which I am conscious of other objects, but in which I am not conscious of myself, because my attention does not happen to be turned that way. It seems to me that I remember such states. And, even if I did not remember them, it would still be perfectly possible that there should be such states, though I might then have no reason for supposing that there were. Nor can I see any reason why there should not be beings who are always in this condition, in which I am sometimes, of being conscious without being self-conscious.

In answer to such considerations as these it is sometimes said that self-consciousness always exists where consciousness exists, but that the self-consciousness is sometimes so faint that it escapes observation when we try to describe the experience which we remember. If there were any impossibility in the existence of consciousness without self-consciousness, we should, no doubt, be driven to this hypothesis. But I can see no reason whatever why I should not be conscious of something else without being conscious of myself. I can therefore see no reason why we should accept the existence of this faint self-consciousness, of which, by the hypothesis, we have no direct evidence.

Or again, it is said that there is always implicit or potential self-consciousness. This means, I suppose, that a conscious self could always be self-conscious if circumstances turned his attention to himself, instead of away from himself. In other words, it is maintained that no conscious being is intrinsically incapable of self-consciousness. It is doubtless true of me, and of other selves like me, that we are not intrinsically incapable of self-consciousness, even at the times when we are not self-conscious. But this does not alter the fact that, at those times, we are just as really not self-conscious as at other times we are self-conscious. Nor can I see any reason why there should not be beings who were conscious, but whose nature was such that they could not under any circumstances be self-conscious.

399. It has also been maintained, as was said above, that, even if there could be beings who were conscious without being self-conscious, the name of self should be reserved for those who are self-conscious. This usage would not, I think, be as convenient as that which I have adopted. To call a conscious being a self only when it was self-conscious would involve that each of us would gain and lose the right to the name many times a day. It would be less inconvenient, no doubt, if the name were applied to all beings who were ever self-conscious, even at the times when they were not so. But even this limitation would be undesirable. There is a quality—the one which we have called selfness—which can only be perceived by me, in present experience, when I am self-conscious, since, in present experience, I can only perceive it in myself, but which is a quality which can exist without self-consciousness. This quality wants a name, and it seems best to appropriate the name of selfness to it.

400. We defined spirituality as the quality of having content, all of which is the content of one or more selves. But some thinkers who might agree with our treatment of the self might think this definition of spirituality too narrow. It would be admitted that whatever falls within the content of a self is certainly spiritual, but spirit, it is said, can include content which is not content of a self. For, it is said, there is, or there may be, experience—knowledge, volition, or emotion—which does not fall within any self, and so is not part of the experiences of any self. And such experience, it might be said, would be spiritual. This brings us to the third question about the nature of the self which we have to discuss. Is there any experience which is not part of a self?

It seems to me that it is impossible that there should be any experience which is not part of a self. This is not a question about words. I believe that I mean the same thing by the words self, experience, knowledge, volition, and emotion, that is meant by the advocates of this view. (Or, at any rate, any slight differences there may be in the meaning of the words would not account for the difference of opinion about impersonal experience.) I believe that there cannot be experience which is not experienced by a self, because that proposition seems to me evident, not as a part of the meaning of the term experience, but as a synthetic truth about experience. This truth is, I think, ultimate. I do not know how to defend it against attacks. But it seems to me to be beyond doubt. The more clearly I realize—or seem to myself to realize—the nature of experience in general, or of knowledge, volition, or emotion, in particular, the more clearly does it appear to me that any of them are impossible except as the experience of a self.

Nor are we led to doubt this conclusion by finding that it leads us into any difficulties. For nothing that we know, so far as I can see, suggests to us the existence of impersonal experience. We never perceive it, since none of us perceives at present any experience, except, by introspection, his own experience. And none of the facts which we do perceive can be better explained on the hypothesis that there is impersonal experience than on the hypothesis that there is only personal experience.

The view that there is impersonal experience, although, as we have said, it is compatible with such a view of the self as we have adopted, is generally held by thinkers who deny the reality of the self, and, consequently, the reality of personal experience. Since they are not prepared to deny the reality of all experience, they are driven to the acceptance of experience which is impersonal. But this ground for accepting it fails, of course, for those who admit the reality of the self.

401. All content of spirit, then, must fall within some self. But another question arises, and this is the fourth question we have to discuss. Can any of this content fall within more than one self? In that case either one self would form part of another, or two selves would overlap, having a part which was common to both. Is this possible?

Both alternatives seem to me to be impossible. When I contemplate what is meant by a cognition, an emotion, or any other part of my experience, it seems as impossible to me that such a state should belong to more than one self, as it is that it should not belong to a self at all. And this impossibility, like the other, seems to me to be an ultimate synthetic proposition.

It might possibly be objected that it does not follow that, because none of those parts of a self which are known to us could be parts of another self, therefore no parts of a self could be parts of another self. It is possible, it might be said, that there should be parts of a self of which that self is not aware[1], which had a nature so different from the parts of which we are aware, that one of them could be a part of two selves.

I do not think that this particular view—that unperceived parts can be common to two selves, though parts which are perceived cannot—has ever been maintained. But in any case it seems to me to be false. Of such parts of the self we should only know that, if they did exist, they would be unperceived parts of the self. But this, I think, is sufficient to show that they could not exist. For we should know that they would be parts of a self. And when I consider what is meant by a self, it seems clear to me that a self is something which cannot have a part in common with another self. The peculiar unity which a self has, puts it into a relation with its parts which is such that two selves cannot have it to the same part. Or, to put the same thing the other way round, any relation which a substance can have to each of two wholes, of each of which it is a part, cannot be the relation of the state of a self to a self.

In addition to this, we shall see in the next chapter that we must accept the view that there is a set of parts of each self, the members of which are perceptions, so that all the content of the self falls within one or other of these perceptions. In that case, every part of the self, whether it is itself perceived or not, will either be a perception, or a part of a perception, or a group of perceptions or parts of perceptions. And if we consider the nature of perception, it seems evident that no perception, and no part of a perception, could be a part of more selves than one.

The impossibility that any part of any self should also be part of any other self cannot be proved, since, as was said above, it is ultimate. But it can be supported indirectly by discussing

---

[1] Such parts would often be called unconscious parts of the self. (Cp. p. 79, footnote.)

various ways in which it has been supposed that it is possible that two parts should belong to more than one self.

402. In the first place, it is often said that one self (and so the parts of it) can be parts of another, if the included self is a manifestation of the inclusive self. This view has always been popular, because one of the chief grounds for wishing to show that one self can be part of another has been to make it possible for men to be parts of God. Many people have been anxious to combine theism and pantheism, and to hold that a personal God—a God who is a self—is the whole of what exists, or is the whole in which all spiritual life falls. In that case each man must be part of God. And if a man is part of God, it is a natural and attractive view to regard him as a manifestation of God. If a self could be part of another on condition of being its manifestation, it would cover those cases in which people are generally most desirous to show that one self is part of another.

Now it is no doubt true that a self can manifest a whole of which it is the part. Thus we may say that Dante was a manifestation of the society of the Middle Ages, and that Chatham was a manifestation of England. But then England and the society of the Middle Ages are not selves. And, again, one self can perhaps be said to manifest another. Thus a theist, who was not a pantheist, might say of a good man that he was a manifestation of God. But then the self who manifests is not part of the self which is manifested.

And it seems to me that in many cases in which it is said that one self can be part of another, the assertion is based on a confusion about manifestation. It is said that the inclusion can take place if the included whole is a manifestation of the other. And because it may be possible that a self should be a manifestation of a whole of which it is a part, and also possible that a self should be a manifestation of another self, it is held that it must be possible that a self should manifest something which is *both* a whole of which it is a part, and also another self. But this is, of course, an illogical inference.

403. In the second place, it is suggested that, if a self A should perceive a self B, and all its parts, and should have other contents besides these perceptions, then B would be a part of A, and the parts of B would also be parts of A. This suggestion, like the last, applies chiefly to the inclusion of man in God. For we know of no case in present experience where a man can perceive another man, or his parts, and it is generally held that this would be impossible. But in the case of God it is often thought that there would be no such impossibility, and that he could perceive other selves and their parts.

It seems to me, as I have explained above[1], quite possible that B and his parts could be perceived by A, whether A was God or not. But this will not make B and his parts into parts of A. B perceives his own parts, or some of them, but the relation of having them as parts, and the relation of perceiving them, are quite different relations; and if A should have the second to the parts of B, it does not follow that he will have the first. The confusion of the two relations is probably due to the fact that, in our present experience, no part of any self is perceived except by that self of which it is a part. And it is therefore mistakenly supposed that, in any circumstances, a self which perceived a part of a self must be a self of which that part was a part.

These considerations indirectly support our view that it is an ultimate certainty that the inclusion in any self of another self, or of part of another self, is impossible. It might be asked why, if this is an ultimate truth, so many thinkers have believed that it was not true at all. But any force there might be in this objection is diminished, if it turns out that many of the people who had supposed that the inclusion was not impossible, had confused it with one of various other things, which are quite possible, but which are not the inclusion in question.

404. It is sometimes assumed, not only that such an inclusion is possible, but that we have empirical evidence that it does occur in those comparatively rare cases which are usually spoken of as exhibiting "multiple personality."[2] But it does not seem to me that the facts in any of the cases which I have read are incompatible with another view. That view is that in each case only one self is concerned with all the events happening in connection with any one body, but that the character of that self, and the

---

[1] p. 67.

[2] Cp. for example Dr Morton Prince's *The Dissociation of a Personality*.

field of its memory, suffer rapid oscillations due to causes not completely ascertained. That such oscillations do take place has been certain since the time of the first man who got quarrelsome or maudlin when drunk, and reverted to his ordinary character when sober. The oscillations in such cases as we are now considering differ in degree, no doubt, from those seen in everyday life, but they introduce no qualitative difference.

Whether all the facts of this class which have been recorded can be explained in this way is a question which we cannot discuss here. But if any of them were of such a nature as to be incompatible with the theory which I have put forward, they would necessarily, I think, be of such a nature as to be compatible with the theory that they were caused by two selves, neither of which included the other, or any part of the other, but both of which happened to be connected with the same body—a concurrence which we do not come across in any other part of our experience, but which has no intrinsic impossibility.

Thus any of the facts which have been explained by "multiple personality" could be accounted for without requiring the hypothesis of inclusion, and these facts, therefore, can cause no doubt as to the correctness of our view that the inclusion of a self, or of a part of a self, within another self is an impossibility.

Since such an inclusion is an impossibility, it follows that, unless I am the whole universe, the universe cannot be a self. For I am a self, and, if I am not the whole universe, I am part of it. And a whole of which a self is part cannot be a self. This result follows whether, of that part of the universe which is not me, all, some, or none, consists of other selves.

As all the content of spirit falls within some self, and none of it falls within more than one self, it follows that all existent selves form a set of parts of that whole which consists of all existent spirit[1].

[1] The contents of this chapter, in a rather different form, appeared as an article on "Personality" in the *Encyclopaedia of Religion and Ethics*.

CHAPTER XXXVII

COGITATION

405. We have now determined what is to be meant by spirit. And the question arises, firstly, whether spirit does exist, and secondly, whether there is any reason to suppose that all substance is spirit.

Primâ facie, spirit, like matter, exists. And the claim of spirit to existence is stronger than that of matter, since as was shown in the last chapter, we certainly perceive something as being spirit—each of us admittedly perceiving various things as parts of himself, and also, as I have endeavoured to show, perceiving something as himself. And, if this perception is correct, then existent things are selves and parts of selves, and therefore spirit, according to our definition of spirit, exists.

But we have seen that any existent substance must have parts within parts to infinity, determined by determining correspondence. If we came to the conclusion that nothing which had the nature of spirit could fulfil this requirement, we should be forced to hold that nothing with the nature of spirit could exist, and that when anything was perceived as spirit, it was misperceived. If it should be impossible to adopt this alternative, we should be left with a hopeless antinomy.

When, however, we consider the nature of spirit, we find that there is one class of spiritual realities which, on certain hypotheses, may furnish a series of parts of parts to infinity, determined by determining correspondence.

406. The spiritual realities which *primâ facie* occur within selves may be classed as perceptions, awarenesses of characteristics, judgments, assumptions, imagings, volitions, and emotions. Of these, the first two, as we have said, may be classed together as awarenesses, and awarenesses and judgments may be classed together as cognitions. Cognitions, assumptions, and imagings may again be classed together as cogitations. In this chapter I shall endeavour to prove that perceptions can form an infinite series of the type required, and that no other cogitation can do so. The position of volition and emotion will be considered in Chapters XL and XLI.

407. The definition of perception was given in Section 44. In Chapter XXXII, Sections 299—302, it was pointed out that every perception is awareness of a substance, but that perception gave us knowledge of the qualities of the substance perceived, including the relational qualities arising from its relationships with other substances. We must not, indeed, say that it is perceived *that* the substance has such and such qualities, since this would be a judgment, not a perception. What we must say is that the substance is perceived, and that it is perceived *as having* such and such qualities.

Is it necessary that, when we perceive a substance, we should perceive it as having all the qualities which it actually does have? In that case every perception would present us with every fact in the universe, since, as we saw in Section 221, a complete description of any substance would include descriptions of all other substances.

There is, so far as I can see, no reason why we should doubt the possibility of perceiving a substance without perceiving it as having all the qualities which it does have. Nor can I see any reason to make a distinction here between original and relational qualities, and to hold that, when a substance is perceived, it must be perceived as having all the original qualities which, in fact, it has.

But, it may be asked further, is it necessary that, when a thing is perceived, it should be perceived as having qualities which are sufficient to constitute a sufficient description of that thing? For this, also, I can see no necessity. It would be possible, I think, to perceive a thing as having the qualities XYZ, and not to perceive it as having any other qualities, although there should be something else in the universe which has the qualities XYZ.

The further question then arises, whether our perception must be accurate as far as it goes. When we perceive a thing as having the quality X, is it certain that it really has the quality X? We have seen that we must allow somewhere for erroneous perception, since our theory of the nature of absolute reality involves that some perceptions must be erroneous. But, when we come to consider this subject in the next Book, we shall find reason to believe that there is no place for error among those perceptions which give us, in the first instance, an infinite series of parts within parts, determined by determining correspondence. Error will have to be found elsewhere.

408. We must now consider how it is possible for perceptions to have this infinite series of parts within parts, determined by determining correspondence.

Let us begin by making three assumptions. The first is that a self can perceive another self, and a part of another self. The second is that a perception is part of the percipient self. The third is that a perception of a part of a whole can be part of a perception of that whole. Later in the chapter I shall endeavour to show that all three propositions are true. But it will be convenient to expound our whole view before we discuss the truth of these propositions, and therefore we will treat them for the present as assumptions

Let us then suppose a primary whole, all the primary parts of which are selves. And let us suppose that each of these selves has a separate perception, and only one such perception, of each self, and of each part of each self. And let us suppose, as we have just assumed to be possible, that when any one of these percepta is part of another perceptum, then any perception of the first will be a part of a perception of the second. We shall then have a series of parts within parts to infinity, determined by determining correspondence. Let us take, for simplicity, a primary whole consisting of two primary parts, B and C, which are selves. Then B will perceive himself and C, and will perceive the perceptions which he and C have of themselves and of one another, and the perceptions which they have of these perceptions, and so on to infinity. And B's perceptions of this infinite series of percepta will form an infinite series of perceptions, since he has a separate perception of each perceptum. And since the perceptions of the parts will be parts of the perceptions of the wholes, the infinite series will be series of parts within parts. A similar series, of course, will occur in C.

409. Such an infinite series could still be obtained if the conditions were rather different from those which we have just taken. If we consider what was said in Section 201, as to determining correspondence in general, it will be clear that the conditions could be modified in three respects. In the first place, it is not necessary that each self should perceive all the selves in the primary whole. It is sufficient if each self has a differentiating group, consisting of two or more selves, and if it perceives each self in this group, and each part of each of them. In the second place, it is also possible that *some* selves should be only determinate and not determinant—that is, that they should be percipient without being perceived. In the third place, it is possible that *some* selves in any primary whole may have, instead of a differentiating group of selves, a single determinant self—that is, that there may be some selves each of whom perceives nothing but one other self and his parts. But it will always be necessary that in each primary whole there should be at least one group of selves in which determination is reciprocal in the way defined in Section 201.

410. The relation of determining correspondence, then, will be that, for example, the determinate $B\,!\,C$ is a perception of the determinant C. And we shall find that this relation complies with all the five conditions with which we saw, in Section 229, that a relation of determining correspondence must comply. The first of these is that there must be a certain sufficient description of C (including the fact that C is in that relation to *some* part of B) which shall intrinsically determine a sufficient description of the part of B in question, $B\,!\,C$. Now a sufficient description, XYZ, of a self C, which includes the fact that it is perceived by B (and therefore includes a sufficient description, UVW, of B, without which C's perception by B would not be sufficiently described), will intrinsically determine a sufficient description of $B\,!\,C$. This sufficient description of $B\,!\,C$ will be "the perception by the only self which is UVW of the only self which is XYZ." For, by our hypothesis, B has only one perception of C, and therefore this description will apply to one substance only. Similar sufficient descriptions can be found for secondary parts of lower grades, for example, for B's perception of C's perception of B ($B\,!\,C\,!\,B$).

The second condition is that the relation must be such that each determinant term could determine more than one determinate term. And with this we have complied, since it is possible for many different selves to perceive the same self, or part of a self, in which case that which is perceived will determine a part in each of the selves which perceives it.

The third condition is that it should be a relation such that $B \mathbin{!} C$ corresponds to only one determinant, C, while C, though it may be the determinant of many parts of the primary whole, A, is the determinant of only one of the parts of A which fall within B. This condition is satisfied. For our hypothesis was that B had a separate perception of C, and only one such perception. Therefore B's perception of C cannot also be a perception of anything but C, and there cannot be more than one perception of C in B. On the other hand, there can be other perceptions of C in A, for there could be such a perception in C itself, which is a part of A, and there could also be such perceptions in the other primary parts of A, in those cases in which a primary whole is divided into more than two primary parts.

The fourth condition is that the determining correspondence must be, in some cases at least, reciprocal. And, obviously, if one self can perceive another at all, there is no reason why each of a group of selves should not perceive all the others.

The fifth condition is that the correspondence should be such that it is possible to have a whole divided into parts of parts infinitely, sufficient descriptions of all which parts are implied, by means of determining correspondence, in a sufficient description of the whole. And we have seen that B and C are divided in this way. So also is A. For we could sufficiently describe A as a primary whole which has a set of parts, B and C, each of which perceives itself, and the other, and their parts. And this would imply the division of A in the manner required.

If then there is such a mutual perception of selves as we have described, it would fulfil all the five conditions which have been laid down as necessary for determining correspondence. Is such a mutual perception of selves possible? In considering this we shall have to discuss four questions. The first three relate to the three assumptions which we made provisionally on page 89. Firstly, is it possible that a self should perceive another self, or part of another self? Secondly, are perceptions parts of the percipient self? Thirdly, can a perception of a part of a whole be part of the perception of that whole? The fourth question is whether there is any difficulty in the existence of a perception whose parts of parts to infinity are again perceptions.

411. In the last chapter we enquired whether "the person who is aware of this awareness" is an exclusive description, and came to the conclusion that it was not necessarily exclusive (Chapter XXXVI, pp. 65–68). And the reasons which led us to this conclusion will also lead to the conclusion that there would be no impossibility in one self perceiving another self and its parts. The first question, therefore, may be considered as settled.

412. We now pass to the second question—are perceptions parts of the percipient self? The natural view, that which would be adopted by most people, if not by all, on first being asked the question, is, I think, that they *are* parts of it. But this has been denied. It has been said that perception—and also all other awareness—is not a state, but a relation. My perception of a datum would then be a relation between myself and the datum, and would not involve the existence of any part of myself which is a state of perception.

Of course, on any theory, there would be a special relation between a self and any object which the self perceived. If a self, *B*, perceives *M*, that fact involves a relation between *B* and *M*, of such a nature that it only holds between a percipient and its perceptum. But the question is whether there is, besides such a relation, a state of perception which is part of the percipient self.

It seems to me, as far as I can trust my introspection, that there is a state of perception within the percipient self. But I cannot place much reliance on this, in view of the fact that other people interpret their experience differently. There are, however, several reasons which seem to lead to the conclusion that a perception is a state of the percipient.

In the first place, when my perceptions, whether simultaneous or in rapid succession, are many in number, the condition of myself, when compared with its condition when the perceptions are few in number, differs in a way which seems to be appropriately expressed by the metaphor of being fuller. Now if this is an appropriate metaphor, it can only be so because the perceptions are parts of the self. For one thing is only fuller than another if it has more content, which means more substance. If the only difference in the condition of the self was that it had more relations of a certain kind, it would never have occurred to us to speak of it as fuller.

Another way of putting the same consideration is that, as it seems to me, we feel, when we contemplate our cogitations, volitions, and emotions, that, taken together, they do in some sense exhaust the self, so that it is completely comprised in them. Now the self could not be said to be exhausted or comprised in its relations, even if we took all its relations into account. Still less could it be exhausted in its relations to the objects of its cogitation, volition, and emotion, since these are not the only relations in which it stands. And thus, if cogitation, volition, and emotion are only relations, we must reject—and it seems to be very difficult to reject—the view that they do in any way exhaust the self. On the other hand, if they are states, and therefore parts, of the self, it is easy to see that they do exhaust it, since there is a very real sense in which a substance is exhausted in a set of its parts.

Again it is almost universally admitted that B's knowledge of C makes more *direct* difference to B than it does to C—that the direct difference between B who knows C, and B if he did not know C, is greater than the direct difference between C which is known by B, and C if it had not been known to B. This view may be rejected by some of those who do not hold that the truth of knowledge consists in correspondence to the object known—though it is not rejected by all of them. But I think it is always accepted by those thinkers who do hold that the truth consists in such a correspondence—a view which we found reason to accept in Chapter II.

I have spoken above of direct difference. If indirect difference is included, the matter may be quite different. If a detective knows that a man has committed a murder, the knowledge will have a greater effect on the future of the murderer than on the future of the detective. If a man knows that the potato on his plate is rotten, this will affect the destiny of the potato more than the destiny of the man. But such indirect differences to the object of knowledge depend on the action which the knowledge determines in the subject who knows it. When we take the immediate difference caused to each of the terms—the knower and the known—by the simple fact of the knowledge, it is true, as was said above, that this is greater for the knower than for the known.

This applies to perception as well as to other knowledge. If, indeed, there were any percepta whose existence was dependent on being perceived, it is obvious that perception would make more difference to them than to the percipient, since his existence does not depend on his perceiving a particular perceptum. We shall, however, find reason to hold that the only things which are perceived are selves and parts of selves. And those do not depend for their existence on being perceived.

Now if knowledge were only a relation between the self and its object, it seems very difficult to see why there should be this greater difference. In this case, all that knowledge would involve about the knowing self is that it would be the term of the relation which knows, while the object would be the term of the relation which is known. And there is nothing, as far as I can see, in these two characteristics which could possibly account for one of them making a greater difference to the term which has it than the other does.

But if my knowledge is a part of myself, the difficulty is removed. For then, while the knowledge involves nothing in the object known except a relation to the knowing self, it involves in the knowing self, not only a relation to the object known, but also the presence of a part with certain characteristics. And since the knowledge makes a difference of parts, as well as of relations, to the knower, and only a difference of relations to the object known, it makes a greater difference in the one case than it does in the other.

We may add one more consideration. Many of those thinkers who would deny that cognitions, including perceptions, are states of the self would admit that pleasures and pains are parts of the self. This is incompatible, at any rate, with any argument which should reject the view that perceptions are parts of the self on the ground that selves can have no parts (or, at least, no simultaneous parts). But we may go further. Pleasure and pain are often very closely connected with cognitions. Now if these cognitions are not parts of the self, but its relations to other things, then the pleasure or pain connected with the cognition can only be connected with it in this way—that it is a state which is excited in the self when the self is in a certain relation of cognition. Closer than this the relation cannot be. So far we can only have this result. But we shall find in Chapter XLI that such a view of pleasure and pain cannot be maintained, and we shall thus have additional reason for rejecting a view of cognition which implies it.

All these reasons tend to confirm the theory that cognitions, and among them perceptions, are parts of the self. Nor do I see any valid argument against this. No doubt, as we have seen, a cognition does imply a relation between cognizer and cognized. But it does not follow that the cognition and the relation are identical, and our view, as has been pointed out, does admit that the relation is there.

I believe that the view that cognition is only a relation has partly sprung from confusion on this point. And I believe that it has partly sprung from an unwillingness to admit that the self has parts, due to a supposed incompatibility between having parts and being a real unity. There is, however, no such incompatibility. And we saw at the end of Book III that every substance must have parts, so that, if the self exists at all, it must have parts of some sort, whether its cognitions are parts of it or not. And indeed, as we have said, it would generally be admitted that pleasures and pains are parts of the self, and it would also be generally admitted that selves existed in time, and had parts in the dimension of time.

It seems to me, therefore, that we are entitled to conclude that our perceptions are states, that is, parts, of ourselves. Of course this does not imply that, when I perceive M, it is only the perception which I know directly, and that my knowledge of M is mediated by my perception of my perception of M. This would be quite incorrect. My knowledge of M is immediate, and consists in my perception of M. It is not in any way dependent on my perception of my perception of M—which may or may not accompany my perception of M.

413. The third question which we had to discuss was whether the perception of the part of a whole could be part of the perception of the whole. It seems to me that this is possible, and, indeed, that our experience assures us that it is sometimes true.

We often make judgments that various wholes with parts do exist, for which we have no warrant but our perception of corresponding data. For example, I judge that there is at this moment a carpet in this room with a pattern on it, when I have no reason to do so except that, in ordinary language, I see the carpet. It would, I think, be generally admitted that I am perceiving a sensum which is a whole with parts, and that my judgment that the carpet is a whole with parts depends upon my perception of the whole sensum, and of the parts of the sensum. But the question remains whether, when both the whole and the parts of the datum are perceived, the perceptions of the parts are parts of the perception of the whole.

In what other way could we perceive the whole as having parts, or the parts as making up a whole? It may be argued that we might perceive the whole in one perception, that we might perceive each of the parts in other perceptions, and that we might perceive the relation between the things perceived. But, even if this does sometimes happen, it seems clear that there are cases where it does not happen. If I can trust my introspection, there are cases in which I perceive a whole and its parts, in which my perceptions of the parts are parts of my perception of the whole. It is possible, of course, that my introspection may be faulty, but in this case I am inclined to trust it. In particular, I would direct attention to what happens when we gradually perceive the parts of a datum of which we only perceived the whole before—as when, with a gradual increase of light, more details appear in the pattern of the carpet. If we had separate perceptions of the whole and of the parts of the datum perceived in such an experience, the change ought to appear as a change to a state with more perceptions, whereas it seems quite clear in my case that it appears as a change from a relatively simple perception to one which is relatively complex.

We must therefore come to the conclusion that it is possible that perceptions of parts should be parts of perceptions of wholes. In our present experience this does not always happen. For it is possible to perceive a whole without perceiving any of its parts, and it may be possible to perceive both a whole and its part without the perception of the part forming part of the perception of the whole. But it is sufficient for our purpose that it is possible that perceptions of parts should be parts of perceptions of wholes.

But this is only possible when—as was the case with the example which we have just taken—the whole, W, is perceived as being a whole of which the part, P, is a part, and P is perceived as being a part of W[1]. If, for example, I perceive a circular sensum, and a square sensum (perhaps marked out by a surrounding line) which is, in fact, a part of the circular sensum, then my perception of the square surface can only be a part of my perception of the circular surface, if the square surface is perceived as being part of the circular surface. For no one would suggest that my perception of Q, which is not a part of W, could be part of my perception of W. And the fact that P is, in fact, part of W, could not make the case any different with regard to the perceptions of P and W, if the perceptions do not perceive P and W as being part and whole.

414. Our theory requires, not only that perceptions shall have parts which are perceptions, but that this shall be the case with every perception, so that the series of parts within parts will be infinite. And the fourth question we had to consider was whether the existence of such an infinite series presents any difficulty.

I can see no reason why any difficulty should be introduced by the series being prolonged to infinity. We have seen that there is no contradiction involved in a substance having parts within parts to infinity, if the series is determined by determining correspondence, as it is in this case. It is true that we have here the additional fact that each perception will not only have such an infinite series of parts within parts, but an infinite series

---

[1] It is not, of course, necessary that there should be judgments *that* W is a whole of which P is a part, and P a part of the whole W. It is sufficient that they should be perceived as standing in those relations.

in which each part is also a perception. But I cannot see that this could affect the possibility of the series.

We cannot of course imagine such an infinite series—our imaginations, in present experience, are never able to reach an infinity, though our thoughts are. But, in considering what the nature of such an infinitely compound perception would be, we must remember that it is the perception about which we are speaking, and not a series of judgments about the different parts of the perception. With such judgments we should have an infinite series of terms, each outside the other, since a judgment about a part is not a part of a judgment about the whole. But with perception's the infinite series is one of parts within parts. And thus, though infinite, it all falls within the limit of a single perception. This, I think, makes it nearer to our present experience than an infinite series of judgments would be, and makes it easier to imagine a state of consciousness approximately like it.

415. We come to the conclusion, then, that the infinite series of parts within parts can be determined if the primary parts are selves which perceive selves and their perceptions. But is it also necessary that the selves should be perceived as being selves, and the perceptions as being perceptions? This also can be shown to be necessary.

In the first place, let us enquire whether the general nature of perception, without reference to the infinite series of perceptions within perceptions, makes it possible that the same self, at the same time[1], can have separate perceptions of two percepta without perceiving them as having different qualities[2].

This, I think, is not possible. We saw above[3] that a self need not perceive a perceptum as having all the qualities which it actually has, nor as having qualities which form a sufficient description. But we have now a different question before us. Must the self, if he is to have a separate perception of the perceptum, perceive it as having qualities which differentiate it from all other percepta perceived by that self at that time? And this does seem to be necessary.

---

[1] The relation which appears as "being at the same time" is really, as we have seen, "being at the same stage in the C series."

[2] A difference in relational qualities is, of course, a difference in qualities.

[3] p. 88.

It is not necessary in order to give sufficient descriptions to the perceptions. Supposing that it were otherwise possible for the same self at the same time to perceive two percepta as each having the qualities VW, and no others, the perceptions would have, not only sufficient descriptions, but sufficient descriptions determined by their percepta. If, for example, one of the percepta were sufficiently described as VWX, and the other as VWY, then the perceptions would be sufficiently described as being B's perceptions, at a given time, of a substance which was, in fact, VWX, and of another substance which was, in fact, VWY. For this would be true, although B only perceived each substance as being VW.

But, although the perception of the two percepta as having different qualities is not necessary for this reason, it is necessary for another. For, if two perceptions were not separated from each other by being perceptions by different selves, or by the same selves at different times, and if, further, they were not separated by difference of content, then there would be nothing to separate them, and they would not be separated at all. That is, there would not be separate perceptions in B of C and D.

This conclusion has nothing in it which is inconsistent with present experience. It must be noted that we are not asserting that it is necessary that the percipient self should judge the two perceptions to be different, but only that he should perceive them as having qualities which do make them different. He may not make any judgment about either of them. Or, if he does make judgments about both of them, still he may make no judgments as to their difference from one another. Of course the perceptions will afford a basis for a judgment of difference, if the percipient's attention should be directed towards them simultaneously, and if he should have sufficient capacity to analyse the nature of his perceptions. But if these conditions are not complied with, he will not make a judgment of difference.

And it must also be remembered that two percepta would be perceived as having different qualities if they were perceived as having different relations, since they would then be perceived as having different relational qualities. It would be sufficient, therefore, if they were perceived as standing in any unreflexive relation to each other, or in different relations to a third thing.

If we pay attention to these two points, it is clear that there is nothing in present experience to suggest that any self ever has two perceptions at the same time which do not perceive the percepta as having different qualities.

416. But, it may be objected, although there is no evidence that there are such separate perceptions, yet our contention that there *cannot* be any such is fallacious. For a perception is not only a perception of the qualities of its perceptum. Indeed, it is not strictly of the qualities at all. It is of the perceptum— the substance. The perceptions, it may be said, of C and D are, after all, perceptions of C and D. And since C and D are separate things, this will be sufficient to discriminate the perceptions of them, even if the perceptions do not perceive them as having different qualities.

But this would be erroneous. For, as we have seen in Section 95, substances are not things in themselves, in the Hegelian sense of the phrase, with an individuality apart from their qualities. They are individual, but only through and by means of their qualities, and therefore, when we perceive them, we can only be aware of them individually in so far as we are aware of them as having qualities which are different in the case of each substance.

417. It might, however, be further objected that B's perceptions of C and D as each being VW might be differentiated by means of qualities of the perceptions which did not depend on perceiving their percepta with different qualities. For example, the perception of C might be more intense than the perception of D. This might be caused by some difference in the qualities of C and D—that one was X and the other was Y—although they were not perceived as being X and Y.

On the whole, I do not think this objection valid. It seems to me that the absence of difference in the qualities with which the percepta are perceived positively involves that there cannot be a plurality of perceptions, and that therefore there is no opportunity for the difference in intensity to take place. But the question is no doubt difficult, and perhaps the possibility ought not to be excluded.

418. So far, however, as was said on p. 98, we have been discussing the subject with regard only to the general nature of
perception, and without taking into account the further fact that
we are dealing here with infinite series of perceptions within
perceptions. And it is certain that, even if some perceptions could
be differentiated otherwise than by perceiving their percepta as
having different qualities, yet this could not happen to all the
perceptions in the infinite series of determining correspondence.
The reason of this is that, as we saw in Section 225, if any substance is divided into parts of parts to infinity, it is impossible
that every one of those parts should have any qualities other
than those determined by determining correspondence, though
it will be possible for all or any of the parts for any finite number
of grades to have such other qualities. No parts in any grade
below the last grade, M, of that finite number can have any
qualities except those determined by determining correspondence.

It follows from this that no perceptions in the grades below
M could be differentiated from each other by a difference in their
own intensities, or in some other quality, in the way suggested
on p. 100. Nor can they be differentiated by being at different
points in the C series, for, as we shall see in Chapter LI, p. 275, the
whole system of perceptions determined by determining correspondence is at the same point in the C series. And, as these two
alternatives are eliminated, the conclusion is that they can only
be separate perceptions by perceiving their percepta as having
different qualities.

What qualities, then, must the percepta of such perceptions
be perceived as having? These percepta are, of course, themselves
perceptions. And these perceptions in grade N, and in every other
grade which is lower than grade M, can have no qualities except
those determined by determining correspondence. Every perception is one grade lower than that other perception which is its
perceptum. It follows that all perceptions in grade O and in every
lower grade, must perceive their percepta as having qualities
which differentiate them from all other percepta perceived in
the determining correspondence system by that self. Thus if
$C!D!E$ is such a part, it must be perceived by B as the perception which a self with the qualities VW has of the perception

which a self with the qualities ST has of a self with the qualities QR. Here VW, ST, and QR are not necessarily *sufficient* descriptions of C, D, and E, respectively—that is, descriptions which distinguish them from all other substances—but they are descriptions which are adequate to discriminate C, D, and E from all other substances perceived, directly or indirectly, by B.

But, it may be asked, is it necessary to perceive $C!D!E$ as specifically a perception? Could it not be perceived only as that part of a substance with the qualities VW, which was determined by determining correspondence with that part of a substance with the qualities ST, which was determined by determining correspondence with a substance with the qualities QR? This would give a description of it by determining correspondence without bringing in the fact that perception was the particular relation of determining correspondence in question.

This, however, would not be possible. For determining correspondence is what Mr Johnson has called a determinable—a generic characteristic which, whenever it occurs, must occur in some specific form. Now it is quite possible to *judge* that a thing possesses a determinable, without judging what determinate of that determinable it possesses. I may judge that the eyes of the first Bishop of Rome had some colour, without having any opinion as to what colour they had. But with perception it is different. I cannot perceive a thing as having colour without perceiving it as having some particular colour[1]. And, in the same way, if I perceive one thing as being determined by determining correspondence with another, I must perceive it as having that particular sort of determining correspondence which it actually has.

419. It is impossible, then, to perceive all the members of the infinite series of parts, unless we perceive some of them (and the infinitely greater number of them) as being, what they are, perceptions by selves of selves or of parts of selves. But, as was said above, it is possible that, for any finite number of grades, all or any of the secondary parts might have qualities not determined

---

[1] It is possible that I may learn from perception that a perceptum is coloured, and be unable to learn whether it is blue or green. But this would not mean that I perceived it as coloured, but not as having any particular colour. It would mean that the particular shade, which I did perceive it as having, was intermediate between a typical shade of blue and a typical shade of green.

by determining correspondence. And it might be possible that, by means of these qualities, the secondary part in question could be discriminated from all others known to B. In that case, it might be asked, could not B perceive the part in question as having these qualities, and not perceive it as having any others, and so not perceive it as being a perception of its perceptum, and, further, not perceive it as being a perception at all?

In the same way, might it not be possible that C could be discriminated from all other selves perceived by B by means of qualities which did not include the quality of being a self? And, in that case, could not B perceive C without perceiving him as a self at all?

It is not easy to conceive what qualities selves could have which should discriminate them from each other, and which were such that the selves could be perceived as having them without being perceived as selves. And it is equally difficult to conceive what qualities perceptions could have which should discriminate them from each other, and which were such that the perceptions could be perceived as having them without being perceived as perceptions. But there are more positive reasons for rejecting the hypothesis.

Let us suppose that the grades of secondary parts which have no qualities except those determined by determining correspondence are all those below the first grade[1]. Then $C!D!E$, which is a secondary part of the second grade, would have no qualities except those determined by determining correspondence. It must therefore be perceived by B as being the perception which a self with the qualities VW has of the perception which a self with the qualities ST has of a self with the qualities QR.

$C!D!E$, then, is known to B as being *inter alia* a perception whose percipient is a self with the qualities VW. If B knows this, he knows the self which has the qualities VW, and knows it as having these qualities. (He can only know of one such self,

---

[1] Of course the earliest grade in which the parts have no qualities except those determined by determining correspondence may not be, as in our example, the second grade of secondary parts, but the millionth or any lower grade which has a finite number of grades above it. But the argument would be just the same whatever the grade was, and it can be stated more simply if we take it as the second grade.

since VW, as we have seen, must include qualities which discriminate C from anything else known to B.) But he cannot know this self except by perception. For, as we shall see later in this chapter, there cannot be any judgments in absolute reality, since they could not be differentiated into parts of parts to infinity. Therefore B must perceive C as a self with the qualities VW.

Thus $B!C$ will be B's perception of C as a self with the qualities VW. But how about $B!C!D$? Must this be B's perception of $C!D$ as a perception which the self with the qualities VW has of a self with the qualities ST? This also must be so. For, as we saw on p. 97, in any case B must perceive $C!D$ as a part of C, and as a whole of which $C!D!E$ is a part. And B, as we have just seen, must perceive C as a self, and $C!D!E$ as a perception. Now is it possible to perceive anything as being a part of a self, and as having a perception as its own part, without perceiving it as having itself the nature of perception?

I think that this is clearly impossible, and that $C!D$ must be perceived as having the nature of perception. A group of perceptions or a part of a perception has also the nature of perception. But it will be necessary that $C!D$ should be perceived as being, as it is, a single perception. For it is not possible to perceive anything as a perception without perceiving its perceptum. To be a perception is, no doubt, a quality. But it is a relational quality—it is generated in the manner described in Section 85, by the relationship in which that which has the quality stands to its perception. And we cannot perceive it as standing in this relation unless we perceive the other term of the perception. We shall therefore perceive D, to which $C!D$ stands in this relation, and, since D is a single object, we shall see that $C!D$ is a single perception. And we have thus justified the statement made on p. 98, that in determining correspondence the selves must be perceived as selves, and the perceptions must be perceived as perceptions.

420. Perceptions, then, can give an infinite series of parts within parts determined by determining correspondence. Is this possible with any other of the sorts of spiritual reality which *primâ facie* do occur? And, in the first place, is it possible with any form of cogitation except perception?

We saw on p. 87 what those forms are. The first is the awareness of characteristics. Such awareness could not give us the
required series. For in order that the infinite series of parts
should not be vicious, it is necessary that some determinant
terms in such a determining correspondence should not only
determine awarenesses, but should have awarenesses as its parts.
As we saw in Section 201, it is only by such reciprocal determination that a valid infinite series can arise. Now, if the
determining correspondence were awareness, this requirement
takes the form that objects of which there is an awareness should
have awarenesses as their parts. All these objects must therefore
be existent, since awarenesses are existent. And therefore the
awareness of them will be perception, and not awareness of
characteristics.

And, again, if the awareness of characteristics could yield a
series of parts within parts to infinity, this would involve that
each of the characteristics should have parts within parts to
infinity. And we saw in Section 64 that this was impossible.

421. Judgment, also, is incapable of giving us parts of parts
to infinity by means of determining correspondence. For in that
case it would be necessary that a judgment about a whole, W,
could be made up of a set of parts which are judgments about
each of a set of parts of W. Now, as we have seen, a *perception*
of a whole can be made up of a set of perceptions of the parts
of that whole. But nothing corresponding to this can happen
with judgments. The only case in which one judgment can be
part of another, is the case where something is judged about a
judgment, as when we say "the judgment that all swans are
white is false." And in that case there are parts of the inclusive
judgment, "is," "false," and "judgment," which are not judgments, nor made up of judgments. It is impossible then that
all parts of judgments should be judgments, and therefore the
required series of parts of parts to infinity cannot consist of
judgments.

Moreover, the determination of judgments as parts of judgments to infinity, even if it were otherwise possible, would be
impossible because the infinite series would in this case be
vicious. A judgment has meaning. And when anything which

has meaning is a complex whose elements have meaning, the meaning of that complex is dependent on the meaning of the elements. The meaning of a judgment, therefore, ultimately depends on the meaning of those of its terms which have meaning themselves, and which have no parts which have meaning. And if a judgment should consist of judgments within judgments to infinity, its meaning would depend on the meanings of the final terms of series which have no final terms. Thus it would have no meaning, and could not be a judgment.

Nor will assumption give us a relation of determining correspondence. The internal structure of an assumption is the same as that of a judgment. The only difference is that a judgment is an assertion, while an assumption is not. It is true of an assumption that it cannot have a set of parts which are all assumptions, just as it is true of a judgment that it cannot have a set of parts which are all judgments. And it is true of an assumption, as it is of a judgment, that it has meaning. And so, for analogous reasons to those which applied to judgments, it is impossible that an assumption should be divided to infinity into parts of parts, all of which were assumptions.

422. There remains imaging. It is possible for me to picture—the phrase, though loose, is helpful—something which I do not perceive now, or which I never have perceived, or which does not exist. I can picture a red disc on a white ground. I can picture toothache felt by me in the past, which did exist, or toothache felt by me in the present, which does not exist. I can picture Cromwell's distrust of Charles I, which presumably existed, or Cromwell's contempt for the Young Pretender, which certainly did not exist.

I propose to call the process of doing this by the name of imaging, and to speak of the mental states involved as imagings, or states of imaging. That which is imaged I shall call the imaginatum.

I use "imaging" instead of "imagining" because the latter term is ambiguous. It would be said, for example, of a conceited man that he imagined himself to be the equal of Shakespeare, when what is meant is that he believed it, but believed it falsely. In other cases what is called an imagination is really an assumption. When it is said that a man imagines what he would do if he should become a millionaire, it is often meant that he is considering what propositions would be made true if the assumption that he became a millionaire were true. And again, while in these cases we speak of imagination where there is no imaging, there are cases of imaging which we should not call imagination. I may remember a cat I saw yesterday, and this memory may have as an element an imaging of the cat. But it would not be said that I imagined the cat.

I have avoided the use of "image" because that also is ambiguous. It is sometimes used for the state of imaging, and sometimes for the imaginatum. Indeed, it is often used for both of them at once, because the distinction between them is ignored, in the same way that perception and perceptum have so often not been distinguished from one another.

Imagings, in their internal structure, resemble perceptions, and are dissimilar to judgments and to assumptions. A judgment or an assumption is a proposition, in the sense given in Section 45. We judge or assume *that* something is true of something. But there is no "that" about imaging, any more than there is about perception. The imaginatum, like the perceptum, is a substance, and what we image is the substance. But an imaginatum is imaged as having characteristics, just as a perceptum is perceived as having characteristics. This is proved by the fact that we can make judgments asserting the presence of those characteristics. I can judge, for example, that what I am now imaging has the characteristic of being a red disc on a white ground.

And only those sorts of things can, in our present experience, be imaged, which can, in our present experience, be perceived; that is to say, sensa and mental realities of the sort which can be perceived by introspection. It is true that we commonly talk about imaging Westminster Abbey, which cannot be perceived. But then we talk as commonly of seeing Westminster Abbey, though it cannot be seen. What we image in the one case, and perceive in the other, are sensa. If I say that I image the execution of George III on the Tower Bridge, what is meant is that I image sensa such that a person who perceived them (instead of only imaging them, as I do) would normally judge that it was, at any rate, phenomenally true that such an execution was taking place[1].

But in another respect imaging differs fundamentally from perception and judgment, and resembles assumptions. Perceptions and judgments are cognitions. They profess to give knowledge, and in so far as they are not erroneous they do give knowledge. But imaging does not profess to give knowledge. If I image something as answering to the description "Cromwell's contempt for the Young Pretender," that act of imaging may perhaps be called false, on the same principle that the assumption "that Cromwell despised the Young Pretender" may be called false. But I am not in error in imaging the one, any more than I am in error in assuming the other. Where there is no claim to give knowledge, there can be no error.

423. Can imaging give us a series of parts within parts to infinity, determined by determining correspondence? The two obstacles which prevent judgments and assumptions from giving us such a series do not apply here. For the same considerations which led us to believe that the parts of a perception can be perceptions would lead us to believe that the parts of an imaging could be imagings. And an imaging, like a perception, has no meaning (though, of course, the *description* of either of them will have meaning). An image, therefore, could have an infinite series of parts within parts, all of which were images, since the series would not be vicious.

Suppose, for example, that B and C imaged themselves, and each other, and all their parts. Then B would image C, and C's imagings of B and of C, and C's imagings of B's and C's imagings of B and C, and so on infinitely. And B will also have a similar series of imagings of its own imagings.

But a difficulty arises which will compel us to modify our view of the nature of imaging. And this difficulty is connected with the nature of the imaginatum. It seems clear that there *is* an imaginatum—something which is imaged, as distinct from the mental state of imaging it. This is, I think, evident from intro-

---

[1] Another connection between imaging and perception is that in our present experience no one can image anything as having any simple characteristic which he has not previously perceived.

spection. My state of imaging images something other than itself.
And this is confirmed by the analogy of perception, where there
is certainly a perceptum distinct from the perception. It need not
surprise us if the reality of the imaginatum, as distinct from the
imaging, has not always been realized, since, till comparatively
recently, it was exceptional to realize the distinction of the perceptum from the perception, although in this case the distinction
is easier to recognize than in the case of imaging.

There is, then, an imaginatum wherever there is an imaging.
But of what nature is the imaginatum, and of what is it a
part?

It is certain that in some cases nothing exists which has the
qualities which I image something as having. There is nothing
existent in past, present, or future, which has the quality of being
contempt entertained by Cromwell for the Young Pretender. Nor
is there anything existent which has the quality of being a group
of sensa which would normally suggest to the percipient of them
that George III was being executed on the Tower Bridge.

Nor could the difficulty be removed by the suggestion that
such things could be real without being existent. All imaginata,
as we have seen, are substances. And it is clear that a substance
can only be real by existing. If Henry VIII, and my table, are real,
they exist. If King Arthur, and the Round Table, do not exist,
they are not real[1].

An attempt has sometimes been made to remove this difficulty
by placing the imaginatum inside the mind of the imaging self.
But the difficulty cannot be removed like this. For, in the first
place, if a thing does not exist at all, it cannot exist within a self.
And, in the second place, it is clear that in many cases where a
thing can exist outside the imaging self, it could not exist in that
self. I can image Cromwell, and his distrust of Charles I. And it is
very possible that Cromwell, and his distrust, did exist outside
me. But it is quite certain that neither Cromwell, nor his distrust
of Charles I, nor his contempt of the Young Pretender, can exist

---

[1] I am speaking, of course, of the actual tables, and not of the descriptions of
them as a table belonging to me, or a table round which King Arthur's knights
sat. These descriptions are not substances but complex characteristics. The reality
and existence of characteristics were discussed in Chapter II.

as parts of me. And I can image my pride if I had destroyed a hostile airship. But, as I never did destroy one, the pride in question cannot be a part of me[1].

Nor would it be more tenable to say, as I believe some people would say, that the imaginatum was not Cromwell's contempt, but a representation of Cromwell's contempt, and that this could be in the imaging self. For, if this theory were true, I should never be able to image Cromwell's contempt, but only a representation of Cromwell's contempt. And this is not the case. I can image the contempt itself.

424. What solution remains? I think that only one solution is possible. This is that the imaginatum always exists, but not always with the qualities which it is imaged as having, and that imaging is really perception which, in the first place, is itself sometimes erroneous, and, in the second place, is in its turn misperceived in introspection, so that it appears to be imaging, while it is really perception. The first error—the error *in* the state— would allow for something appearing as Cromwell's contempt, though it was not really that contempt. The second error—the error *about* the state—would allow for that which was really a perception appearing as an imaging.

This view may seem at first sight paradoxical. But what other view is possible? There must be an imaginatum (that is, there must be an object of the state which appears as a state of imaging). The nature of an imaginatum includes the quality of being a substance. If it is the imaginatum of an existent state of imaging, it must be real; and a substance cannot be real unless it is existent. The state of imaging is an awareness of its imaginatum, and what is an awareness of an existent substance if it is not a perception? If it is a perception it must be in some cases an erroneous perception, since in some cases it perceives its perceptum as possessing qualities which it certainly does not possess.

---

[1] Thus it is more obvious that all imaginata cannot be within the imaging self than that all percepta cannot be within the percipient self. For, in present experience, I perceive nothing but myself and parts of myself on the one hand, and sensa on the other. Parts of myself are, of course, really within myself. And although there is no reason to hold that the sensa are within myself, yet it is not so obvious that they are not within myself as it is that neither Cromwell nor his contempt can be within myself.

And if the state is perceived as an imaging, when it is really a perception, it must itself be misperceived.

Our position will be strengthened when we find, as we shall find in Chapter LVI, that it is possible to explain in detail how a misperceived perception can appear as being an imaging. And this will again be incidentally strengthened when we see, in Chapter XLIV, that even perceptions which appear as being perceptions are in many cases erroneous perceptions, and when we see, in Chapter XLV, that there is no reason to regard the existence of erroneous perceptions with suspicion.

And it will again be incidentally strengthened by seeing that those states which are *primâ facie* judgments, assumptions, or awarenesses of characteristics are in reality perceptions, and that their appearance as being something different is due to their being misperceived. But it must be noted that the course of the argument about these three is different from the argument about imaging. The conclusion is the same in each case—that the states are really misperceived perceptions—but it is reached in a different way. In the case of judgments, assumptions, or awarenesses of characteristics, there is no reason to suppose that they are not what they appear to be, except a reason which depends on results reached previously in this work. No substance can exist, we have decided, unless it has parts within parts to infinity, and we have seen in this chapter that nothing which was really a judgment, an assumption, or an awareness of a characteristic, could have parts within parts to infinity. This does not apply to imagings, for we have seen that a state which was really an imaging could have parts within parts to infinity. The reasons why that which appears as an imaging must really be a perception are those which have just been given.

It may be objected that, in spite of this, we have not got rid of the reality of Cromwell's contempt for the Young Pretender. For, granting that I misperceive something which has not the quality of being that contempt, and misperceive it as being that contempt, I am after all thinking of that contempt. And can I think of anything which is not real? This is a question about erroneous perceptions in general, and not specially about such of them as appear as imagings. It will be discussed in Chapter LII, and we shall find that such a perception would not involve the reality of such a contempt.

425. What are we to say about memory? I do not think that memory includes any element which we have not already considered. It seems to me that memory is a cognition which appears as a judgment about something else which appears as an imaginatum. It is clear that it is a cognition. For it professes to give, and, when faithful, actually does give, knowledge about the past. And it clearly does not appear as being a perception. For everything which, in our present experience, appears as being a perception, only gives information about what is simultaneous with itself, while memory gives information about what is earlier than itself.

Since it appears as a cognition of the existent without appearing as a perception, it can only appear as a judgment. And on introspection it seems clear that this is the case. It appears as a judgment that a present imaginatum has been perceived in the past.

All such judgments, however, are not memory. I may image myself as doing something when I was a child, and I may believe that I did do it, and this belief may be true and well-grounded. But if I believe it exclusively on the ground that I have been told by someone else that I did do it, then there is no memory. The question as to what distinguishes memory judgments from such judgments as these is interesting and important, but does not concern our present purpose. For if memory is a judgment, it is clear that it cannot give us the required series of parts within parts to infinity.

426. We have now spoken of all the different forms of cogitation which *primâ facie* exist within selves. But how about volition and emotion? As to these, I shall endeavour to show, in Chapters XL and XLI, that states of volition and emotion are really states of cogitation, which are distinguished from other states of cogitation by the possession of certain additional qualities.

The result of this chapter is that spirit, unlike matter and sensa, can really exist. But it can do so only if it contains no parts except perceptions and groups of perceptions. For, as has just been said, volitions and emotions will be found to be cogitations. And of the four sorts of cogitation, other than perception, which *primâ facie* are found in spirit, we have seen that judgments, assumptions, and awarenesses of characteristics could not have parts of parts to infinity. What appears as being any of these sorts then, must really be one of the others. This leaves perceptions and imagings as the only possible states of selves. And as what appear as imagings have been shown to be really perceptions, perceptions are left as the only possible states.

CHAPTER XXXVIII

IDEALISM

427. We have seen that nothing which exists can have such qualities as would justify us in calling it matter. And we have also seen that nothing can exist which has such qualities as would justify us in calling it a sensum. Anything, therefore, which is perceived as having the qualities of matter or of sensa, must be misperceived—perceived as having qualities which it has not.

What are we to say about spirit? The fact that each of us perceives various substances—himself and the parts of himself—as being spiritual, does not settle the question. For each of us also perceives certain substances as being sensa, although no sensa exist. Shall we be forced to say that what is perceived as spiritual is not really spiritual?

There would be special difficulties about saying anything of this sort. For would it not imply that the reality in question is really perceived? And in that case must not the perception, at any rate, be spiritual? But we need not consider this question, because there is no reason why we should endeavour to take up the position that spirit is unreal. We were forced to this conclusion in the case of matter and sensa, because we found that their natures were incompatible with the determination by determining correspondence of an infinite series of parts within parts, and because no substance could exist without such a series. But we found that it was possible for a spiritual substance to have parts within parts to infinity, provided that these parts were perceptions. There is, therefore, no reason to suppose that, when anything is perceived as spirit, it is misperceived.

We have, then, good reason to suppose that something exists with the nature of spirit. For each of us perceives himself, and some of his parts, and each of us has good, though empirical, reasons for believing that other substances exist which resemble himself in being spiritual. But can we go further, and say that every existent substance must be spiritual?

428. There appear *primâ facie* to be three sorts of substance—spirit, sensa, and matter. We have seen that neither of the two last exists. And we cannot even imagine any substance which is neither spiritual, material, nor of the nature of sensa. If, therefore, there is any other sort of substance, it must be one of which we have no experience, and which we cannot even imagine. This does not amount to a positive proof that all substance is spiritual. For there remains the possibility that there is some other form of substance, whose nature is such as to allow of the determination in it of an infinite series of parts of parts. If there is such a form of substance, we know nothing of it in our present experience—either because we have had no opportunity of observing its existence, or because mankind have not yet been sharp-sighted enough to avail themselves of the opportunity. In this case there might be, by the side of spirit, one or more other sorts of substance existing in the universe.

But, although we have not a positive proof that nothing exists but spirit, we have, I think, good reason to believe that nothing but spirit does exist. There are certain conditions to which every existent substance must conform. Of all forms of substance which have ever appeared to be experienced, only one conforms to these conditions, and not only our experience but our imagination fails to suggest any further form. Under these circumstances it seems to me that we are entitled to hold all substance to be spiritual, not as a proposition which has been rigorously demonstrated, but as one which it is reasonable to believe and unreasonable to disbelieve.

We may notice, too, that if any form of substance, besides spirit, should exist, it will have an important and fundamental point of agreement with spirit, and of disagreement with matter and sensa. For if it exists, its nature must be such that it is possible for determining correspondence to determine within such substances a series of parts within parts to infinity; and we have seen that this is possible with spirit, and impossible with matter or sensa. But we must not lay too much stress on this, since those qualities of spirit which have value and practical interest are the qualities which we learn by perception, not those which can be shown to be *à priori* necessary. And we can say

nothing as to any resemblance between spirit and this hypo-
thetical other substance, except in the qualities shown to be
à priori necessary.

429. We are entitled, then, to believe that all substance is
spiritual. But does this exclude the possibility that some, or all,
substance should also have a nature which is material or sensal?
We rejected the existence of matter and of sensa, because
material and sensal qualities, as ordinarily defined, would not
permit the determination, within the substances possessing them,
of an infinite series of parts within parts. But suppose, in the
first place, that something which had the qualities of matter, as
ordinarily defined, had also other qualities, which might determine an infinite series of parts within parts. Or suppose, in the
second place, that this was true of something which had the
qualities of sensa, as ordinarily defined. Or suppose. in the third
place, that matter should consist of a number of units which
were *materially* simple and indivisible—that is, were not divisible in the dimensions of space or time— but which, in addition
to their material qualities, had others which would determine
the required infinite series. Or suppose, in the fourth place, that
this was true of sensa which were *sensally* simple and indivisible—that is, were not divisible in the dimensions of space and
time. If any of these suppositions were true, would it not be
possible that substances could have material or sensal qualities?

We spoke of the first and third of these alternatives in
Chap. XXXIV, p. 43, and of the second and fourth in Chap. XXXV,
p. 61, but did not then consider their possibility, as such substances would certainly not be matter or sensa, as the words are
commonly used. Now, however, we must consider whether any
of them are possible.

It seems clear that none of them are possible. The other
qualities in question would have to be spiritual qualities, since
we have found no others which can determine infinite series of
parts within parts.

Now there is nothing in our empirical experience which
suggests that anything which has spiritual qualities could also
have material or sensal qualities. Also, it is impossible, I think,
to see any way in which the spiritual qualities of a determinant

could determine the material or sensal qualities of a determinate. And this would be required for the first and second hypotheses, since they take the substances as materially or sensally divided to infinity. But, apart from these negative considerations, we must hold, I think, that all four hypotheses must be rejected on the ground of positive incompatibility between spiritual qualities, on the one hand, and material or sensal qualities on the other.

430. Let us take the first hypothesis. According to this, something can be both spiritual, and a piece of matter divisible to infinity. Then a self or a perception can be a divisible piece of matter. If it is a divisible piece of matter, it must have a size and a shape. Now I submit that it is clearly impossible that a self or a perception could be, for example, six inches across and globular. The more I try to accept as possible a self which is globular, the more I find that I slip away to one of two other ideas—the idea of two closely connected substances, of which one is a self and one is globular, and the idea of a substance which really is a self, and is misperceived as being globular. And neither of these, of course, is the idea of a globular self. When I do keep to this idea, the impossibility of there being anything corresponding to it seems manifest[1].

According to the second hypothesis something can be both spiritual, and a sensum divisible to infinity. Then a self or a perception can be a divisible sensum, and have a size and shape. And, once more, it is clearly impossible that a self or a percep-

---

[1] The name of hylozoism, if taken in a wide sense, might be used of all the four hypotheses which we are now discussing. The most common form of hylozoism, I suppose, is that which asserts that the same substance which, in respect of one set of its qualities, is my body (or perhaps my brain only) is, in respect of another set of its qualities, my mind. And this comes under the first hypothesis. It follows from this theory, if I have understood it correctly, that if a bear eats the brain of an Arctic explorer, and if the brain of the bear is subsequently eaten by an Esquimaux, then the same substance is at first a part of the mind of the explorer, then part of the mind of the bear, and finally part of the mind of the Esquimaux. (Or if it should be held that a bear has no mind—which seems improbable—then it would be at first a part of the mind of the explorer, then would not be a part of any mind, and would then be a part of the mind of the Esquimaux.) It seems, however, sufficiently obvious that anything which is ever part of one mind can never become a part of another mind, or exist without being part of a mind at all.

tion could be circular, or could have half the area of another self
or perception.

The third hypothesis does not require that selves and perceptions should have size or shape. For the selves might, on this
hypothesis, be units which were spatially indivisible, and the
perceptions might be parts of them in some dimension which
was not spatial. But groups of these selves would be spatially
divisible, and would therefore have size and shape. And it is as
impossible that a group of selves should be six inches across, or
globular, as that a single self should be so.

On the fourth hypothesis, the sensa must either be spatial or
non-spatial. If they are spatial, then it must be possible for a
group of selves to be circular, and to have half the area of another
group of selves. And this is impossible. If, on the other hand, the
sensa are not spatial, then it must be possible that at any rate
groups of selves, if not also selves and perceptions, could be sounds,
tastes, or smells. And this is equally impossible[1].

431. All the four hypotheses, then, break down. It is true that
there is, as Lotze has pointed out, a certain sense in which a self
may be said to have a spatial position—in which it may be said
to be in the body, and, more specifically, in the brain. But to say
this implies that there is a world of matter, occupying space.
Then, if a self has direct causal relations with some matter, and
not with all, we may say that it has its seat in the part of matter
with which it has direct relations, and that it occupies the same
position. The assertion is not that a self occupies a spatial position
directly, but that it does so by its relation to matter, which
occupies that space in its own right. And, even then, the self is
not asserted to occupy the space in the way in which the matter
occupies it. If the seat of a self was a tract in the brain which
was three inches across, and globular in shape, no one would say
that this made the self three inches across, or globular.

Our conclusion that nothing which is spiritual is also material

---

[1] The third and fourth hypotheses require, of course, that there should be
indivisible points of space. This would not be incompatible with our view that
there can be no simple substances, since, as was said above, the substances
occupying these points would be divisible non-spatially. But the reality of
indivisible points of space has recently been challenged by distinguished
mathematicians.

or sensal leaves it possible that what is really spiritual may *appear* as being material or sensal. Indeed, this must be the case. For it is beyond doubt that something does appear to us as material and as sensal, and what appears thus must be spirit, if nothing but spirit exists.

432. No substance, then, has material or sensal qualities, and all reality is spirit. This conclusion I propose, following general usage, to call by the name of Idealism. This usage has not been unchallenged. It is sometimes said that the name of Idealism should be reserved as a name for a position in epistemology, rather than in ontology. In that case, while Kant would be called an idealist, Berkeley would not, although Kant did not assert that all reality was spiritual, and Berkeley did. But it seems more convenient not to restrict the term to epistemology, because there is no other term which could conveniently be substituted for it in its ontological use. Spiritualism would be intrinsically better, since the position we are considering deals with spirit, and not specially with ideas. But the name is already appropriated to a very different belief. Psychism might also be intrinsically better than idealism, but the word would be new, and difficult to introduce, and it might have misleading associations with psychology. Let us say, therefore, that our position is idealist, in that sense in which Leibniz, Berkeley, and Hegel were idealists.

We have, it will be remembered, made no attempt to give a rigid demonstration of Idealism. It would be impossible to base any rigid demonstration of Idealism on the results obtained in the previous Books, or, indeed, as far as I can see, on anything else. It is possible by rigid demonstration to lay down conditions to which the existent must conform. And it is possible to show by rigid demonstration that some asserted forms of substance do not conform to those conditions, and therefore do not exist. It is also possible to show rigidly of some asserted form of substance that it does conform to those conditions, and that it is the only form, hitherto suggested, which does so. But it can never be shown rigidly that it is the only form which does so. For the various asserted forms of substance are all suggested to us by our perception. And it would be impossible to be certain that our perception has suggested to us all possible forms of substance.

CHAPTER XXXIX

FURTHER CONSIDERATIONS ON SELVES

433. We have come to the conclusion that all that exists is spiritual, that the primary parts in the system of determining correspondence are selves, and that the secondary parts of all grades are perceptions. The selves, then, occupy an unique position in the universe. They, and they alone, are primary parts. And they, and they alone, are percipients. This distinguishes them from their own parts, which are all secondary parts in the system of determining correspondence, and which are perceptions and not percipients.

We have now to consider what further conclusions as to selves, and as to determining correspondence, can be deduced from these results. And, in the first place, is it true, not only, as we have just said, that all primary parts are selves, but also that all selves are primary parts?

This also must be the case. For the primary parts form a set of parts of the universe, and they contain between them all the content of the universe. If there were any other selves, then the content of each of these would have to fall within one or more primary parts which are selves. In that case one self would include another, or two selves would have a part in common. But we came to the conclusion in Chapter XXXVI that this is impossible.

434. Since all selves are primary parts, it follows that I myself, and any selves whom, in present experience, I know empirically, are primary parts. But this raises considerable difficulties. For such selves are far from appearing to correspond to the description which we found must be true of selves which are primary parts. They appear as selves, no doubt, and as having perceptions, and, in some cases, their perceptions appear as having parts which are again perceptions. But there are four important respects in which they do not appear as having the nature which our theory requires. It requires that the whole content of each self should consist of perceptions only. Our selves, however, appear to contain parts which are not perceptions, and which are not made up of perceptions. The theory also requires that the selves, since they exist, should be timeless. But our selves appear to be in time Again, the theory requires that the selves should perceive other selves and their parts, and should perceive nothing except selves and their parts. But our selves do not appear to perceive other selves and their parts, and they do appear to perceive many things which are neither selves nor parts of selves. Finally the theory requires that every perception should consist of other perceptions, and that, therefore, every perception should be infinitely divided. But in many cases our perceptions appear not to have perceptions as their parts, and in no case do they appear to be infinitely divided. The question whether these appearances can be explained in a way compatible with the truth of our theory, will, like other questions of this nature, be considered in Book VI.

A second consequence which follows from the fact that all selves are primary parts is that the universe cannot be a self. The universe cannot be a primary part, for it is either a primary whole, or a group of primary wholes. Moreover the universe contains primary parts, and therefore contains selves; and no self can be part of another self.

A third consequence which follows from this fact is that Solipsism must be false. Solipsism is the belief that no substance exists except the person who is holding that belief, and the parts of that person. But the universe must contain more than one primary part, and since the solipsist is a self, he is a primary part, and there are one or more primary parts outside him[1].

435. We have seen that a self can perceive himself. It is interesting to note the effect of this on the most important argument which has been offered against solipsism in the past— that put forward by Mr Bradley. That argument rests on the contention that the self "involves and only exists through an intellectual construction. The self is thus a construction based on, and itself transcending, immediate experience."[2] Thus, if

---

[1] Of course this leaves it possible that the universe should consist only of the solipsist and one other self, but then the solipsist is mistaken in his solipsism.

[2] *Appearance and Reality*, Chapter XXVII, p. 524 (2nd ed.). Cp. also Chapter XXI.

I understand Mr Bradley rightly, it is unjustifiable to base the possibility of solipsism on the ground that I am more certain of the existence of myself than of the existence of anything outside myself. My self and the external world have each to be reached by inference, and by inference substantially of the same kind. If that inference does not give me justification for believing in something existent outside myself, it does not give me justification for believing in myself.

If, however, the self can be perceived, the matter is somewhat different. For I certainly do not perceive anything at present, except myself, the parts of myself, and sensa. As we have seen, it is sometimes held that all sensa are parts of the perceiving self. In that case I should perceive nothing but myself and my parts, and the solipsist would be justified in saying that the existence of his self and its parts was more immediately certain to him than the existence of anything else.

But the spirit of Mr Bradley's argument, though not its exact form, would remain valid. A solipsism which did not admit the obligation to treat experience as a coherent whole would be self-condemned. For, unless that obligation is accepted, there can be no reason to believe or disbelieve any theory of the nature of reality. But it is quite impossible to make a coherent whole of what I perceive at any one moment without bringing into the explanation things which I do not perceive at that moment. The solipsist, who does not accept the existence of anything outside himself, has to introduce into his explanation unconscious states of mind, and states of mind which occurred in the past. But what is called an unconscious state of mind—that is, a state of which the mind is not conscious—is not, of course, perceived. Nor are past states of mind perceived in the present. Thus the solipsist has to assert the existence of substances which he does not perceive, just as much as the believer in external existence does. And so it is still the case that he has to trust to inference for his belief in substances. And, if inference is not to be trusted when it leads me to believe in other selves besides myself, it is just as little to be trusted when it leads me to believe in states of myself which I am not perceiving.

We may add a further consideration on the subject of solipsism.

If the sensa, or whatever appear as sensa, are not parts of the percipient self—and this seems much the most probable view[1]—then solipsism is clearly false, since those things which I perceive as sensa do unquestionably exist.

436. When we considered in Book IV the nature of determining correspondence in general, we found that there were many questions as to the precise nature of that correspondence which we were unable to answer. Now, however, we have come to the conclusion that the only relation of determining correspondence is perception, and that the primary parts in the system of determining correspondence are all selves, and that there are no other selves. Shall we now be enabled to solve any of the questions which were left unsolved before? In the rest of this chapter we shall consider eight such questions.

The first of these is the question whether the primary parts in the universe are finite or infinite in number. (I omitted to discuss this question in Book IV, though I assumed in Section 197 that the number of primary parts in each primary whole might be either finite or infinite.) I cannot see that there is any reason to reject either alternative. It is perfectly possible that the number of primary parts may be finite. But there seems no impossibility in its being infinite. In that case, no doubt, each of the infinite number of primary parts would have to have a sufficient description which was an ultimate fact. But this does not, as far as I can see, involve the difficulties mentioned in Section 190, because it would not involve ultimate coincidences with other descriptions. Both possibilities, then, are open, so far as the general nature of determining correspondence goes; nor does the fact that the primary parts are now determined to be selves make any difference.

437. We saw in Section 201 that it was not necessary for determining correspondence that each primary part should have as its differentiating group all the other primary parts in the same primary whole. *A fortiori* it was not necessary that it should have as its differentiating group all the other primary parts in the universe. At the same time this was perfectly possible—though only, of course, on the hypothesis that the universe formed

---

[1] Cp. Chapter xxxv, p. 56.

a single primary whole. And our second question will be whether, now that the primary parts are determined to be selves, it remains possible, or becomes necessary, that each of them should have all the others in its differentiating group. In this case, of course, each self would directly perceive all other selves, as well as itself.

I do not think that there is any reason to deny that this is possible. In my present experience, indeed, taking it as it appears to be, I certainly do not perceive all the content of the universe— which would of course be the case if I perceived all the selves. But then, if my present experience is taken as it appears to be, I do not perceive any self except my own self. And if our theory is true at all, I must perceive at least one self which is not my own. The difference between the appearance and the reality will have to be overcome in this respect. And, if it can be overcome in this respect, I do not see why it should be impossible to over-come it so as to enable us to accept the view that every self perceives all selves.

If each of us perceived all the selves in the universe, and if, as we have just seen to be possible, the number of selves in the universe were infinite, the perception of each self would be in-finitely extended. But there seems no reason why this should not be so. It must be remembered that, however limited the differentiating group may be, the amount of perception in each self is infinite, since each self perceives the parts of all its per-cepta to infinity. It is true that, as was said in Chapter XXXVII, p. 98, the perception of the parts of any one primary part is to be called infinitely compound, rather than infinitely extended, since it is not unbounded. And this, as was there pointed out, seems to bring it nearer to our present experience, and makes it easier to imagine a state of consciousness approximately like it. But although an infinitely extended field of perception may be in some ways less like our present experience, it seems to present no more logical difficulties than the other, which, as we saw in Chapter XXXVII, p. 97, did not present any.

Our present emotional relations to other selves, again, suggest that our real relation to all of them is not of the same nature. And this could be accounted for if each self did not directly perceive all the others. But, here again, the true explanation may lie in the difference which must exist in any case between present experience and absolute reality.

It is therefore not impossible that each self should directly perceive every other self. But there is no reason to hold that this is actually the case. It would be possible for each self to have a differentiating group which did not comprise the whole universe. Thus the question must remain undecided.

438. The third question to be considered is whether it is necessary that all selves should be perceived. We saw in Section 201 that, while it was necessary that every primary part should be determined by determining correspondence, it was not necessary that every primary part should be a determinant in a determining correspondence. It was necessary for determination to be reciprocal in some cases, as without this we should not have any infinite series of determinations. But when once such an infinite series had been established (for example, by B, C and D determining one another, or by B and C determining D, C and D determining B, and D and B determining C), there could be other primary parts which were merely determined. E could get its infinite series of determinations by being determined by B, C, and D, or by any one of them, since B, C, and D had each already got its infinite series of determinations. And so it would be possible for E not to be a determinant at all.

Is this possibility removed, now that we know the primary parts to be selves, and the determination to be perception? I do not see that it is. There seems no difficulty in a self existing unperceived by anyone, including himself. If he does not perceive himself, he will not, of course, be self-conscious. But we saw in Chapter XXXVI, p. 81, that a self can be conscious without being self-conscious—that is, can perceive other things without perceiving himself.

It may also be noticed that, even if every self was perceived by some other self, it would not follow that the selves which he perceived were the same as those by which he was perceived. E might be perceived by F and G, which he did not perceive, and not be perceived by B and C, which he did perceive.

439. We now pass to our fourth question. We have already seen (Section 437) that it is not necessary that every self should

perceive every self directly. But the question still remains whether
it is necessary that each self should perceive each self either
directly or indirectly. In Section 203 we came to the conclusion
that there was nothing in the general nature of determining
correspondence which required that every primary part should
determine every other primary part in the same primary whole,
whether directly or indirectly. *A fortiori* there could be nothing
in the general nature of determining correspondence which
required that every primary part in the universe should determine every other, either directly or indirectly. Are we able to
say any more, now that we know that the primary parts are
selves?

When one self, B, is determined by another self, C, the specific
relation, as we have seen, is perception. If C is determined by,
that is, perceives, D, what is the specific form of the indirect
determination of B by D? I think we must answer that it is
indirect perception of D. For B will have to perceive C's perception of D as having the quality of being C's perception of D.
(Chap. XXXVII, p. 104.) And then it seems impossible to deny that
his relation to D is of the nature of perception. For, as we have
just said, he will know $C! D$ as being $C! D$—as being C's perception of D. And he cannot know this unless he knows D. His
perception of $C! D$, then, must give him knowledge of D, in
order to give him knowledge of $C! D$. Since the state is thus a
state of knowledge of D, and a state of perception of a perception
of D, we cannot deny it the name of perception of D. But there
is certainly a difference between the relation of such a perception to B, and the relation in which the perceptions of C and of
$C! D$ stand to B. And this can be best expressed by calling it
an indirect perception, since it is certainly an indirect determination.

We shall speak in future then both of direct and indirect
perception, and shall use the word perception, when unqualified,
to include both. The name of direct perception has sometimes
been used in the past by writers who did not recognize the
indirect perception of which we have been speaking, but employed direct perception as equivalent to perception. This usage
probably came from the belief that perception may be regarded

as a more direct form of knowledge than judgment—which is true. But the fact that perception is the most direct form of knowledge would not be a valid reason for speaking of direct perception, if there were no indirect perception from which it was to be distinguished.

We can recognize indirect perception in our present experience. If I perceive my perception of a sensum, then this second and introspective perception, which is a direct perception of my perception of the sensum, can be an indirect perception of the sensum. But in our present experience the importance of such indirect perceptions is very small. For in present experience I perceive no self but myself, and so the only objects which I perceive indirectly are those which my own self perceives directly. And thus indirect perception of anything does not give me knowledge of any object which I do not know otherwise. But it would be quite different when my perception extended to other selves, for then I could perceive indirectly things which those other selves perceived directly, but which I did not perceive directly.

It would be quite possible for the proportion of direct to indirect perception to be very small. Even supposing that no self perceived more than two selves directly, it would still be possible that the differentiating groups, though each consisting only of two selves, should interlace in such a manner that each self should indirectly perceive all the rest.

It is therefore possible that each self perceives all other selves indirectly, even if he does not do so directly. But, on the other hand, it is equally possible that he should not. We have seen that there is nothing in the general nature of determining correspondence which requires that every primary part should determine every other, either directly or indirectly. And I do not see that any difference is made by the fact that the primary parts are selves, and the determination is perception.

440. It might be objected to this that my knowledge of the universe proves that I must perceive all other selves. "The existence of the universe" it might be said "is certain (cp. Section 135) and, whether anyone else knows it or not, I know it. But all knowledge is perception, and therefore I must perceive the universe. The universe, again, is not a primary part, but a group of many primary parts, and I can therefore perceive the universe only by perceiving all the primary parts which make up the universe—that is, all the primary parts which exist. But every self is a primary part, and therefore I, and everyone else who knows anything about the universe, must perceive all other selves."

I do not think that this argument is valid. Let us consider the proposition "the universe exists." This follows from any proposition which asserts that any substance exists. A universe was defined as a substance which contains all existent content. If any substance exists, there is existent content, and therefore there is a substance which contains all existent content. In that case a universe exists, and since it follows from the definition of a universe that there can only be one, we may express our result as "*the* universe exists."

It follows then from the existence of any substance which I perceive as existing that the universe exists. And it is a quality of that substance that its existence implies the existence of the universe. Of course it is not necessary that I should perceive it as having this particular quality which it does have, but it would be possible to perceive it as having it. For substances, as we have seen, are perceived as having qualities, and there is no ground on which we can say of any quality which a thing has that it would be impossible to perceive the thing as having it. And if, in perceiving a particular self, or a particular perception, I perceive it as having this quality, then this perception would give me knowledge of the existence of the universe.

It must be noticed that this would not be an indirect perception of the universe. For the self or perception, C, which I am perceiving, is not necessarily itself a percipient of the universe, or a perception of the universe. The existence of C implies the existence of the universe, but this, of course, can happen without C knowing anything about the universe.

Nevertheless, my perception of C as implying the existence of the universe does give me knowledge of the universe, though no perception of it. And this accounts for my knowledge that the universe exists.

"The universe exists" is, no doubt, a judgment about the universe, and not a perception of some other substance as implying the existence of the universe. We are thus forced to the conclusion that, while the reality is such a perception of *C*, that perception appears as such a judgment. The possibility of this will be considered in Chapter LIV, where I shall endeavour to show that it is possible. All that we can say here is that unless it is possible that what is really a perception can appear as a judgment, our whole theory breaks down, since some things certainly appear as judgments, which, if our theory is true, are really perceptions. And, if this is possible at all, there is no special difficulty about it in the case before us[1].

My knowledge that the universe exists, then, need not disturb our previous conclusion that it is not necessary that I should perceive all other selves.

441. Our fifth question is whether all selves belong to the same primary whole. There is nothing in the general nature of determining correspondence which either requires or forbids that all primary parts should belong to the same primary whole. Does the fact that the primary parts are selves make any difference?

It is possible that all selves should belong to the same primary whole, even if every self did not perceive, directly or indirectly, every self. For each primary part in a primary whole need not be determined by each part in that whole, though it must either be determined by it, or determine it, or both. (Cp. Section 203.)

On the other hand two selves, one of which perceived the other, must be in the same primary whole. And, therefore, if any one self should perceive all other selves, it would follow that all selves were in the same primary whole. But we have just seen that there is no reason to hold that any self does perceive all other selves. Nor does there seem any other consideration which would enable us to determine either that all selves were in one primary whole, or that they were not.

As to these five questions, then, we are no more able to determine them now, than we were before. They could not be decided

---

[1] It may be shown by a similar argument that the fact that I know something about the British nation does not imply that I perceive every one of the selves who make up that nation. (Cp. Chapter LIV, pp. 308–309.)

from the general nature of determining correspondence, and they remain undeterminable now that we know that determining correspondence is perception, and that selves are primary parts. But there are three other questions as to which it is possible to be more definite now than was possible when we had only the general nature of determining correspondence to start from.

442. The sixth and seventh relate to systems of determining correspondence. "In the first place," I said in Section 227, "I cannot see that it can be shown to be necessary that the same sort of determining correspondence should occur everywhere. It seems to me quite possible that, if there is more than one primary whole, the relation might be a different one in each of them. And, even within one primary whole, there seems no reason to deny that B might have a different sort of determining correspondence with the parts of C from that which D has with the parts of E, or even from that which B itself has with the parts of F, or, again, from that which G has with the parts of B." This possibility is now eliminated. For we have decided that we have good reason to believe that perception is the only sort of determining correspondence which exists at all.

"In the second place," I continued in Section 228, "we cannot at present exclude the possibility that there might be more than one species of determining correspondence extending over the whole universe, or over a part of it. It might, for anything we can see yet, be possible that the universe should have two sets of parts, which were such that none of the members of either were directly determined by determining correspondence, and also such that from sufficient descriptions of all the members of either set there followed, by determining correspondence, sufficient descriptions of the members of all sequent sets. In that case the universe would have two sets of parts, each of which was by our definition a set of primary parts, and each of which would start a system of determining correspondence extending over the whole universe. And the number of such sets of parts need not be confined to two." But now we have seen that there is only one sort of determining correspondence which can exist. And since, in that determining correspondence, all primary parts are selves, there cannot be more than one set of them. For each set of primary parts would contain all the content of the universe, and therefore, if there were two sets of selves which were primary parts, one self would have to overlap, or be part of, another. And this is impossible.

443. There remains the eighth point. We saw in Section 225, that, though it is impossible for all the parts of B to infinity to have any qualities not determined by determining correspondence, yet it is possible for all or any of the parts for any finite number of grades to have qualities not determined by determining correspondence. This result remains unchanged, but we are now able to limit the qualities which it is possible for the parts to have. For now we know that the primary parts are selves, and all other parts are perceptions. We know that they all perceive other selves and perceptions, and that they have no content except what falls in such perceptions. We know that they perceive their percepta as being selves and perceptions, and perceive them as perceiving their own percepta. And we know that they are not in time. Any other qualities which they can have must be compatible with these.

CHAPTER XL

VOLITION

444. We have seen that a self can have parts within parts to infinity, if those parts are determined by determining correspondence and are themselves perceptions. We must now consider the question whether such a self could possess volitions.

The name volition is sometimes applied exclusively to states of will. I propose to use it in a wider sense—as synonymous with desire. Thus there is no will which is not volition, but there is much volition which is not will. Any man might desire that London should not be destroyed by an earthquake to-morrow, or that he himself had behaved better yesterday. But no one would will either of them, unless he believed that he had the power of controlling earthquakes, or of altering the past.

A volition is a part of that self who desires. The considerations which support this conclusion are analogous to those which are given above (Chap. XXXVII, pp. 92–96) for the conclusion that a perception is a part of that self which perceives. They need not, therefore, be given here.

Can we define the quality of being a desire? It seems to me that it is simple and ultimate, and therefore cannot be defined. Attempts have been made to define it in terms of cogitation. It has been said to be an idea which tends to realize itself, and it has sometimes been confused with a judgment that something is good, or is pleasant. Some volitions tend, no doubt, to realize themselves, though I do not see how this could be the case with volitions that London should be preserved from future earthquakes, or that I had behaved better yesterday. But even if it were the case that all volitions tended to realize themselves, it seems clear to me that the quality of being a volition is a different quality from that of being an idea tending to realization of itself. And although judgments about what is good are often closely connected with volition, the judgment and the volition are different things.

445. Although desire cannot be explained in terms of cogitation, it is obviously very closely connected with cogitation. All desire is for something. It is true that there are states of desire which are commonly described as states of wanting something, while not knowing what we want. But in these cases what really happens is that the object of the desire is wide and vague— it may be only a desire that *some* change should take place in an environment which we find oppressive—and that we do not know what particular change would realize this wide and vague want. But it is always a desire for something, however vague, and however negative.

I cannot, then, desire without desiring something. And can I desire anything unless I have some cogitation of that which I desire—unless I perceive, judge, assume, or image it? It seems plain that I cannot do so, and that my desire of X involves my cogitation of it[1].

It follows from this that cogitation is in a more independent position than volition, since it is intrinsically necessary that no volition can exist except in a certain intimate relation to a cogitation, while it is not intrinsically necessary that every cogitation should be in the corresponding relation to a volition. For there is, at any rate, no obvious impossibility in the supposition that we can have cogitations of things as to which we entertain no desire.

Whatever is desired, then, must be given in a state of cogitation. But would it be sufficient that there should be two mental states, a state of cogitation which is not a state of desire, followed or accompanied by a state of desire which is not a state of cogitation? This would not be sufficient. For then there would be no cogitation, in the state of desire, for that which is desired, and how could it be a state of desire for that rather than for anything else?

It might be replied that it is a desire for that object because of some special relation which exists between the state of desire

---

[1] I am not, of course, maintaining that I cannot desire X unless I know that I desire X. The latter knowledge is in any case logically subsequent to the desire, and is often not present at all, since we often desire without reflecting on the desire.

and the state of cogitation. But this seems inconsistent with the facts. If this view were correct, then when, besides desiring X, I was aware that I was desiring X, I should be aware of two states, A and B, with a relation between them. A would be a state of cogitation of X, and the fact that it was so would be independent of its relation to B: if we abstracted from the relation, A would still be a cogitation of X, B, on the other hand, would be only a desire, without being a desire for anything. Now it seems clear that this is not the case. A state of desire of X is as directly and immediately a desire of X, as a judgment or assumption of X is a judgment or assumption of X. It does not require anything outside itself to make it a desire of X.

446. How then can we reconcile the two results at which we have arrived? On the one hand, we have decided that there can be no desire without cogitation of what is desired. On the other hand, we have decided that this cogitation cannot be a state of cogitation separate from the state of desire. There seems only one alternative, which I will state in the words of Dr Moore, by which it was first suggested to me. He says that the view he inclines to adopt is "that the 'founding cognitive Act' is always not merely simultaneous with but a constituent of the Act which is founded on it; and further that the other constituent of the founded Act was not another complete Act, directed in a different specific way on the object, but merely a quality of the cognitive Act."[1]

This view seems to me to be correct. The cogitation of what is desired and the desire of it are one and the same mental state, which has both the quality of being a cogitation, and the quality of being a desire.

So far as I know, this view originated with Dr Moore. It is, I think, opposed to the general opinion on the subject. But any appearance of paradox which it may present is due, I believe, to a confusion. Most states of desire are connected with states of cognition which are not states of desire, and it is erroneously supposed that these non-volitional states of cogitation are all the cogitation which the desire requires.

---

[1] *Mind*, 1910, p. 400, in a review of Dr A. Messer's *Empfindung und Denken*. The expression "founding cognitive Act" is taken from Dr Messer.

It generally happens that an object of desire is cogitated before it is desired. We first realize what it is, and then proceed to desire it. When we are realizing it and have not yet desired it, there is a state of cogitation which is not yet a state of volition.

This is always the case when a thing is desired as a case of something else which is desired. If I start by desiring everything which is an X, and then desire Y because it is an X, it is clear that, before I desire Y, I must know that Y is an X. Again, if I desire X for its own sake, and Y as a means to X, any desire of Y must be preceded by the knowledge that it is a means to X. In these cases a cogitation which is not a desire *must* precede the desire. But even when something is desired directly for itself, it is possible and common, though not necessary, that a cogitation of the thing desired should precede the desire.

Since, then, there are so often states of cogitation which are not states of desire, and which are connected with states of desire, it is easy to fall into the mistake of supposing that they are the only states of cogitation concerned, and that the states of desire are not themselves states of cogitation. But the supposition is mistaken, for, as we have seen, a state of desire, which is always a desire for something, must itself be a cogitation of that thing.

We conclude, then, that among cogitations there are some which have the additional quality of being desires, just as, among desires, there are some which have the additional quality of being states of will.

Our previous conclusion (p. 133) was that cogitation was in a more independent position than volition, because it was intrinsically necessary that no volition could exist except in a certain intimate relation to a cogitation, while it was not intrinsically necessary that every cogitation should be in a corresponding relation to a volition. But now the independence of cogitation relatively to volition is still more marked. For, as we have seen, every volition must be a cogitation, while there is no corresponding necessity that every cogitation should be a volition.

447. We must now consider various other questions about states of volition. The first of these is whether we can say anything general about the things which are desired. Has the group of such things any exclusive common quality belonging to its members, other than the qualities of being desired, and of being cogitated by the selves that desire them? I do not think that it has. It has sometimes been held that I cannot desire anything unless I believe that it will afford me pleasure. To this it is generally added that I cannot will anything unless I believe that it will afford me greater pleasure than any known alternative to it would have done. This is the doctrine of Psychological Hedonism. It is not, I think, necessary to consider here the various arguments against it, since it is seldom maintained by writers of the present day.

Another rather similar view is that a man can only desire what he believes to be good. This seems to me to be as inconsistent with facts as is the case with psychological hedonism. I find continually in my own experience, and in the accounts which other men give me of their experience, that objects are desired which are known to be bad at the time when they are desired.

It is sometimes said that in such cases the badness of what is desired is not realized with sufficient clearness. But the judgment that they are bad is made with perfect clearness and perfect certainty. To say that it is not vivid enough seems to come to nothing more than that it does not prevent our desiring the things which are judged to be bad. And this surrenders the position that the things can be desired only if judged to be good.

Again, it is sometimes said that, if a man desires what he knows to be bad, it is only when he does not regard it as bad *for him*. In the strict sense of the words, it is impossible that a thing should be good for one man and bad for another. Goodness is a quality of the thing judged good, and a thing can no more be good for one man and bad for another, than twice four can be eight for one man and nine for another. (It is quite possible that one man may believe it to be eight and another believe it to be nine, but that does not make it nine for the second man. All that happens is that he believes it to be nine, and that he is wrong.) In a somewhat loose sense of the words, no doubt, it is possible to say that something is good for one man, though not good as a whole. In this case what is meant is that those of its results which affect him are good, although all its results, when taken together, are more bad than good. If it is maintained, however, that no one can desire anything unless its effects on himself are believed by him to be good, the theory is again in conflict with the facts, which show clearly that a man can sacrifice, either to duty or to a blind passion, what he believes to be his true welfare.

We must conclude, then, that no common relation can be established between desire on the one hand, and either pleasure or goodness on the other[1]. So far as I know, it has never been suggested that such a relation could be established between desire and any other quality. We must therefore answer our first question in the negative.

448. Our second question is whether desire has any necessary relation to change. It is sometimes held that any desire must either be a desire that some change should take place, or a desire to resist some change which is or may be attempted. I do not, however, think that this is true of all desires. It is true of many of them, but there seem to be cases where desire has no relation to change. Take the case of a man who believes that God exists. If he accepts the usual theistic view of God's existence, he will believe that it is impossible that God should ever cease to exist. A desire for God's existence cannot, for such a man, be a desire for a change, since he believes that God exists. Nor can it be a desire to resist change, since any change in the fact of God's existence would be impossible. But cannot such a man have a desire for God's existence? It seems to me that he certainly can. If he holds God's existence to be good, or to be advantageous to him, and if he desires whatever is good, or whatever is to his own advantage, then he will desire God's existence. Or, again, he may desire it ultimately, and without a reason.

We must hold therefore that desire is not necessarily directed towards change[2]. It is primarily acquiescence. The word is not altogether suitable, as it seems inappropriate for cases of

[1] Nor can any such relation be established between will and either pleasure or goodness. The view that a man can will only what he believes to be good, or good for himself, is incompatible with the many cases in which men not only desire, but finally choose, things which they know to be intrinsically bad, and bad for themselves who desire them.

[2] Cp. Lotze, *Microcosmus*, IX. v. 3.

passionate desire. And I do not suggest that we should habitually
substitute it for the word desire. But it is useful to employ it
occasionally as a synonym, because it is universally admitted
that acquiescence does not involve any relation to change, and
we shall thus make it clear that no such relation is involved by
desire.

449. We have spoken of desire as acquiescence, without the
introduction of any negative correlative of acquiescence. This
brings us to our third question. Have some desires the quality
of being positive, and some the quality of being negative? There
is no doubt that there is a sense in which some desires are
negative—that is, they are desires that something should not
be. And this negation is not merely a consequence of the desire,
but a part of it. It is not only that I desire Y, which is, in point
of fact, incompatible with X, so that the gratification of my
desire involves that X should not occur. Of course this does
happen in some cases, but in others the desire is directly for something negative. I desire that A should not be X, and that is all
I do desire. It may be that, if A is not X, it will have to be Y.
But I may not know this, and, even if I do know it, my desire
may not be directed to it. It is directed exclusively to A's not
being X.

In this sense, then, some desires are positive and some negative.
But, more strictly, I think we must say that there are no such
things as negative desires. The quality of being a desire is not
a genus with two species, one of which has the quality of being
positive, and the other the quality of being negative. In the
cases which we distinguished above as positive and negative,
there is no difference in the desire itself. The difference is only
in the object desired. One is a desire for A to be X, the other
is a desire for A not to be X. The nature of that which is desired
is different, but the nature of the desire is the same.

The only evidence which I have for this view is introspection.
But I do not know that it has ever been explicitly denied, and
I think that it will not be regarded as strange or paradoxical,
if it is remembered that we have admitted that what is desired
can be really negative[1].

---

[1] This suggests that the difference between affirmation and denial is more

We should therefore say, if we speak strictly, that desires are neither positive nor negative. But it should be noticed that our result involves that all desires are of that class which would be called positive by anyone who accepted the distinction of positive and negative desires. For they would hold that a positive desire was one which accepted something, and that a negative desire was one which rejected something. And our view is that all desires accept something, though that which they accept is often itself of a negative nature.

450. We now come to our fourth question. We saw above (p. 134) that every desire must also be an act of cogitation. That is to say, it must either be a perception, a judgment, an assumption, or an imaging. But this still leaves it possible that all desires should fall within one, two, or three of those classes, and that there might be one or more of those classes the members of which could not be desires. This is the question which we have now to consider.

In our present experience most of our desires are for something which is expressed in a proposition. That is, the desires themselves must be cogitatively either judgments or assumptions. And most of them are assumptions. This must be the case when I desire something which I do not believe to exist. If I desire that it should be fine to-morrow, or that I had remembered my umbrella to-day, these desires are cogitatively assumptions. I do not know that it will be fine to-morrow, and with regard to to-day's remembrance of my umbrella, I know that it did not happen, and cannot believe that it did happen. These desires must be assumptions. And again, when I desire in a general way the good, or the pleasurable, or what is pleasurable for myself it is clear that I am desiring much of which I do not know whether it exists or not. I do not know how far the universe is good or pleasurable, or how far my own future life may be pleasurable. And in these cases, also, the desire must be an assumption.

The desire cannot be cogitatively a judgment except in those

properly described as a difference between the affirmation of a positive content and of a negative content. And this, also, seems to me to be supported by introspection.

cases in which a man believes that what he desires is already real. In present experience most of our desires are nôt of this sort; they are either desires for something which we believe not to be real, or for something of which we do not know whether, or how far, it is real. And even when a man does believe that what he desires is already real, it is not necessary that his desire should be cogitatively a judgment. It is quite possible that he should have both a judgment that the thing is so, and also an assumption that it is so; and it may be the latter which is the desire. Nevertheless, it would seem that there are cases in present experience in which the judgment "A is B," or the judgment "A exists," has also the quality of being an acquiescence in the fact that A is B, or that A exists, and is therefore a desire.

451. Can a desire be a perception? It seems to me that it can be so, even in our present experience. In present experience, indeed, the examples of such desires will be few in number. For in present experience I have no perceptions except of sensa, of myself, and of parts of myself. It follows that no desire for anything which is, or includes, anything material can be a perception. Nor can any desire for anything which is, or includes, anything in any other self. Nor can any desire for any abstract result— such as that π should or should not be 3. The only desires which can be perceptions are desires for myself, or some part of myself, or some sensum, as having certain characteristics. And they can, of course, only be desires for the substance as having some characteristic which it is perceived as having. If it were not perceived as having the characteristic, the desire could not be a perception, but must have some other cogitative character[1].

And even when a desire is directed to a characteristic which a perceptum is perceived as having, it need not be a perception, though it can be one. When I perceive a thing as being X, it is possible that at the same time I should judge that it exists, or assume that it exists, and in such cases it might be the judgment or the assumption, and not the perception, which was the desire.

Thus most of our desires in present experience cannot be per-

[1] It could be an assumption or an imaging. Or it could be a judgment, if the substance, though not perceived as having the characteristic, was believed to have it.

ceptions, and none of them need be so. It is therefore to be
expected that the existence of perceptions which are desires
could easily be overlooked. But when we look carefully, I think
that we can see that they do exist—that there are perceptions
which have the same quality of acquiescence in their content
which we find in those assumptions and judgments which are
desires. Of course we should not know *that* we were desiring this
or that, or desiring at all, unless, besides the perceptions, we had
judgments about the perceptions. We cannot know *that* any fact
is true about perceptions, except by judgments about them, but
that does not prevent perceptions from being desires, any more
than it prevents them from being perceptions.

In the same way, it seems to me that an imaging can, and
sometimes does, have the characteristic of being a desire. And
so, in present experience, a desire can belong to any of the four
species of cogitation; it can be an assumption, a judgment, a perception, or an imaging.

452. A desire is either fulfilled or not fulfilled, according as
what is desired is or is not the fact. A desire which is cogitatively
an assumption may be either fulfilled or unfulfilled; and the same
is the case with a desire which is an imaging.

If a true judgment is a desire, it must be a desire which is
fulfilled. If the judgment "A is X" is also a desire, what is desired
is that A should be X. And, if the judgment is true, A is X, and
the desire is fulfilled. If the judgment is false, the desire is not
fulfilled, but it is believed to be fulfilled. If it is afterwards seen
that A is *not* X, then, of course, the judgment that it is X is no
longer made. Instead of the judgment, there is an assumption,
"that A is X," which is a desire. And that desire is known not
to be fulfilled.

When a desire is cogitatively a perception, it must be fulfilled,
if the perception is correct. For a perception of A as being X will
be a desire for A as being X, and, if the perception is correct,
A is X. But if it is a misperception, then perhaps A is not X,
and the desire is unfulfilled. We have seen that we must admit
that some perceptions are misperceptions, and therefore there is
a possibility that a desire which is a perception, like a desire
which is a judgment, can be unfulfilled.

453. We have thus reached certain results as to the nature of volition, and we have now to apply these results to our immediate problem—the place of volition in absolute reality. In absolute reality, as we have seen, every self has parts within parts determined to infinity by determining correspondence, and all these parts must be perceptions. If there are any desires in absolute reality, then, they must be perceptions. And we have seen that perceptions can be desires.

But it will not be possible that they should be unfulfilled desires. For, as we have seen, an unfulfilled desire must be cogitatively either an assumption, an imaging, a false judgment, or a perception which is a misperception. We shall see (Chap. XLVII, pp. 228–232) that the perceptions which are parts in the system of determining correspondence are not misperceptions. And therefore they cannot be unfulfilled desires.

Although a correct perception cannot be an unfulfilled desire, yet it may, in our present experience, be the object of an unfulfilled desire. I may have a desire that under certain circumstances I might have a perception with the quality X. If under these circumstances I had a perception without that quality, the perception would involve that the desire was unfulfilled. And if I made a judgment that my perception had not the quality X, I should know that my desire was unfulfilled. But in such cases the desire itself would be an assumption and not a correct perception, though the perception which was its object might be correct. And since the system of determining correspondences contains no cogitations except correct perceptions, it follows that nothing in absolute reality is either an unfulfilled desire or the object of an unfulfilled desire.

454. But are there any desires at all in absolute reality? And, if there are, are all perceptions in absolute reality desires, or is it possible that only some of them should be?

I do not see that at this point we can answer either of these questions. There is nothing to prevent some or all of the perceptions being desires. For a perception can be a desire, and there is no difficulty in all the perceptions in the system of determining correspondence being desires. On the other hand, I can see, so far, nothing to exclude the possibility that none of the perceptions in that system should be desires. It is true that the perceptions in that system exhaust the whole content of myself, and that I certainly appear to have desires. But we have already seen that much of what appears as being true of our nature is in reality not true. Can we be certain that this is not the case with the appearance of desires?

There is, however, a further fact to consider. We know more about what is perceived by each self than the fact that it consists of primary parts and their secondary parts. For every primary part is a self, and every secondary part is a perception. And thus we know that what is perceived by a self are selves and their perceptions. The results of this on the emotions will be considered in the next chapter, and we shall see that those results will afford us grounds for deciding that all perceptions in the system of determining correspondence are also desires.

CHAPTER XLI

EMOTION

455. We have now to consider what relation the series of perceptions determined by determining correspondence bears to emotion. What is emotion? The first point to be noticed is that emotion has many species. This is a marked difference from volition, which has no such species. Cogitation, indeed, is divided into species—awareness of characteristics, perceptions, judgments, assumptions, and imagings. But there are only five of these, and no one, as far as I know, would suggest that they do not cover the whole extent of cogitation, or that any of them are not fundamental. With emotion we have a much larger number of species, nor is it always clear which of them should be taken as fundamental, and which can be treated as varieties of others. We may, however, form a list which, with no pretence to systematic completeness, contains no important omissions. I suggest, as such a list: liking and repugnance, love and hatred, sympathy and malignancy, approval and disapproval, pride and humility, gladness and sadness, hope and fear, courage and cowardice, anger, wonder, curiosity[1]. It will be noticed that most, though not all, of these are grouped in pairs of polar opposites.

Emotion itself, I think, cannot be defined. Like cogitation and volition, it is an ultimate conception. But with so many examples of it there is no difficulty in identifying what is meant.

456. Every emotion is directed towards something[2]. I am

---

[1] Love and hatred are varieties of liking and repugnance, but, for reasons which will appear in the course of this chapter, it is convenient to mention them separately. Approval and disapproval are distinguished from liking and repugnance by the fact that they are for qualities, or for substances in respect of their possession of those qualities, while liking and repugnance are for particular substances as wholes, though they may be *determined* by the qualities of the substances. Regret is, I think, to be taken as a variety of sadness, while remorse is a species of humility—one which is determined by a particular sort of cause. The question of loyalty will be considered later.

[2] It may be objected that in states of general elation or depression we have emotions of gladness or sadness which are not directed to anything. I think, however, that in such states the emotions are directed towards everything, or almost

proud of myself, love someone, hope something, am anxious about something. And I cannot have an emotion, unless I have some cogitation of that to which the emotion is directed.

Thus the relation of cogitation to emotion is analogous to its relation with volition. Cogitation is in a more independent position than emotion, since it is intrinsically necessary that no emotion can exist except in a certain intimate relation to a cogitation, while it is not intrinsically necessary that every cogitation should be in the corresponding relation to an emotion. For there is, at any rate, no obvious impossibility in the supposition that we can cogitate something which excites no emotion.

But here, as with desire, we must go further. It would not be sufficient that there should be two mental states, a state of cogitation which is not a state of emotion, followed or accompanied by a state of emotion which is not a state of cogitation. For, if so, there would be, in the state of emotion, no cogitation of that towards which the emotion was felt. And how then would it be an emotion towards that, rather than towards anything else?

It might be replied that it is an emotion towards it because of some special relation which exists between the state of emotion and the state of cogitation. But an objection arises here, similar to that which compelled us to reject the analogous theory about volition. For, on this theory, when I loved X, and was aware that I did so, I should be aware of two states, A and B, with a relation between them. A would be a state of cogitation of X, and the fact that it was such a state would be independent of its relation to B. B, on the other hand, would be a state of love, which was only a state of love for X because of its relation to A. If we abstracted from the relation, A would still be a cogitation of X, while B would be a state of love, but not a state of love for anybody. Now it seems clear that this is not the case. A state of love for X is as directly and immediately love for X, as a judgment or assumption about X is a judgment or assumption about X. It does not require anything outside itself to make it love for X.

everything, which falls within our cogitation at that moment. Its presence is not due to any special characteristic in the things, but to a special characteristic possessed, for the time, by the self.

We must therefore adopt for emotion a theory analogous to Dr Moore's theory of volition. We must hold that the cogitation of that to which the emotion is directed, and the emotion towards it, are the same mental state, which has both the quality of being a cogitation of it, and the quality of being an emotion directed towards it.

457. A cogitation which is also an emotion can, in its cogitative aspect, be either a perception, a judgment, an assumption, or an imaging. In our present experience, an emotion is more often a perception or a judgment than a volition is. For our present volitions are more often for what is cogitated as not existent, than for what is cogitated as existent. And volitions of the former class must be assumptions or imagings. But our emotions are excited at least as much by what is cogitated as existent as by what is cogitated as not existent.

A cogitation can have both the quality of being a volition and the quality of being an emotion. I can simultaneously hope for and desire some future event, or love X and acquiesce in his existence. And there seems no reason to suppose that in such a case there must be two separate cogitations of the event, or of X.

In absolute reality, as we have seen, there are no cogitations except perceptions. All our emotions will therefore be cogitatively perceptions, and we shall have no emotions except for what exists. And as, in absolute reality, nothing exists except selves, parts of selves, and groups of selves, and they are perceived as being such, it is only towards selves, parts of selves, or groups of selves, perceived as such, that we can feel emotions. Can we determine what emotions we feel?

458. Since everyone must perceive more than one self, it follows that he must perceive at least one other self. What emotions do we feel in absolute reality towards other selves? In our present experience the emotions which can be felt towards other selves are of many different kinds. But then our cognitions of other selves in our present experience differ from those in absolute reality by being indirect. In our present experience no one perceives any other self. He only knows him by description—as having such or such qualities and relations—and even these qualities and relations of the other self are not known directly, but only by means of the knower's perceptions of sensa. Thus the knowledge of another self is doubly indirect. And our knowledge of the parts of other selves is doubly indirect in the same way.

In absolute reality, on the other hand, our knowledge, both of other selves and of their parts, is direct, since we perceive both the selves and their parts. In this respect the knowledge which each self, in absolute reality, has of other selves, resembles, to some degree, the knowledge which he has, in present experience, of himself. For a self can perceive himself and he can perceive his parts. But in present experience a self's knowledge of himself and of his parts need not be perception, though it can be perception. It can also be judgment. But in absolute reality there is no knowledge except perception, and we cannot judge of other selves and of their parts, but only perceive them.

Thus, while my present knowledge of myself and my parts does to some degree serve as a type of my knowledge, in absolute reality, of other selves and their parts, it is, after all, an imperfect type. My present knowledge of other selves differs more from my knowledge of other selves in absolute reality, than it does from my present knowledge of my own self.

The great difference between the knowledge of other selves in our present experience and in absolute reality renders it unsafe to argue from the emotional qualities of the one to the emotional qualities of the other. And thus the fact that our knowledge of selves in present experience is often without any emotional quality, and often presents a great variety of emotions, does not entitle us to conclude that the same will be the case in absolute reality.

459. I believe that in absolute reality the knowledge of other selves will always have one emotional quality (whether it will always, or ever, have others also, is a question which will be considered later). And the emotion which I believe will always be present is love.

What is meant by love? I propose to use the word for a species of liking. Liking, as was said above (p. 144, footnote 1), is an emotion which can only be felt towards substances. In confining the name of love to an emotion which is only felt towards substances,

I think that I am in accordance with usage. It is true that, if a man admires courage or benevolence with a certain intensity, it is not unusual to say that he loves courage or benevolence. But this, I think, would generally be admitted to be a metaphor.

But how is love to be distinguished from other sorts of liking? I propose to confine the word, in the first place, to a liking which is felt towards persons. Here, perhaps, it is more doubtful if common usage supports the restriction. It is not so clear that we are speaking metaphorically when we say that a man loves the Alps, as when we say that he loves justice. Still less is it clear that we are speaking metaphorically when we say that he loves his school or his country. But it is important to have a separate name for the liking which is felt only towards persons, and there is, I think, no question that, however far the common use of the word may extend, the central and typical use of it is for an emotion felt towards persons. And thus, in using it exclusively for that emotion, we shall not depart much from the common use, if we depart at all.

Again, I propose to use the word only of a liking which is intense and passionate[1]. This is in accordance with the general usage of the present, though not of the past[2].

460. Love then is a liking which is felt towards persons, and which is intense and passionate. It is clear that love must be carefully distinguished both from benevolence and from sympathy. The difference from benevolence is fundamental, since benevolence is not an emotion at all, but a desire—a desire to do good to some person, or to all persons. Nevertheless it has sometimes been confounded with love. It is true that we shall generally desire to do good to any person whom we love. But the emotion and the desire are quite separate. And we often desire to do good to people whom we do not love, and even to people whom we hate.

[1] The word liking is often used to exclude any emotion which is intense and passionate. It would not be unusual for a man to say " I do not like *A*. I love him." But some general word is needed to include emotions of this kind without reference to their intensity. And no better word than liking seems available.

[2] Both Spinoza and Hume, for example, use "love" of every emotion of liking towards another person who has any quality which gives me pleasure. (Cp. *Ethics*, Book III, Prop. 13, note. *Treatise of Human Nature*, Book II, Part II.) But so wide a use of the word seems to have dropped out early in the nineteenth century.

Sympathy is more closely connected with love, for it is an emotion. But it is a different emotion—the emotion which affects us pleasurably in the pleasure of others, and painfully in their pain. If we love a person, we shall generally sympathize with him. But we can sympathize with people whom we do not love. It is even possible to sympathize with people whom we hate—at any rate, if the hatred is not very intense.

461. We are confining the use of the word to emotion which is intense and passionate. This must not mislead us into exaggerating the closeness of its relation to sexual desire. It is often found in connection with that desire. But it is also found in connection with other bonds of union—kindred, early intimacy, similarity of disposition or of opinions, gratitude, and so forth. And it is also found without any such connection in instances where it can only be said that two people belong to one another —such love as is recorded in the *Vita Nuova* and *In Memoriam*.

462. Can we discover any characteristic which, in our present experience, is always present when B loves C (either in B, or in C, or as a relation between them)? It has been maintained that, when B loves C, the love is always dependent on the fact that the action or existence of C has given or is giving pleasure to B. But this is inconsistent with the facts. Love often arises without any such pleasure. And, when there is such pleasure, it often happens that the pleasure B owes to C, whom he loves, is much less than the pleasure which he owes to D, whom he does not love.

Love then is not always caused by pleasure. Nor, when love has arisen, does it always cause pleasure. There are many cases where it produces far more pain than pleasure, and it does not seem impossible that cases could arise where it produced only pain. A love which leads to jealousy may produce a great balance of pain over pleasure, and even if it were said that love which leads to jealousy was not the highest sort of love, it would be preposterous to maintain that it was not love at all. And a love which is unreturned may produce much more pain than pleasure, even if it is free from jealousy. The view that love must be pleasurable is, I believe, due to people who accepted or assumed the validity of psychological hedonism, and then argued that, if

a lover was unwilling to cease to love, it could only be because
he found love pleasant.

463. Nor is it the case that *B*'s love of *C* involves *B*'s moral
approbation of *C*. It has been maintained—not exactly that there
can be no love without moral approbation, but that this only
happens when love has been led astray by sexual desire, or by
some other influence which is regarded as distorting it, and that
love when left to itself "needs must love the highest when we see
it." This seems to me to be utterly mistaken. I cannot see that
moral approbation stands in any special relation to love. It may,
of course, be found with it, and may cause it. *B* may have come
to love *C* because he was virtuous, or he may have come to love
him because he was beautiful. But it is possible that *B* should
love *C*, though he knows him to be ugly, and it is possible that
he should love him, though he knows him to be wicked. And,
while virtue is more important than beauty, it seems to me that
love towards a person known to be wicked is just as truly love
(and, for that matter, just as good) as love towards a person known
to be virtuous.

Nor can it be said that benevolence and sympathy are always
found together with love. There are cases where men have rejoiced
in, and desired to promote, the ill-being of those whom they
really loved. Such cases are probably rare, they are certainly evil,
and perhaps they are always caused by influences which may be
called morbid. But they do occur. And, whatever may be said of
such exceptional cases, it is clear that benevolence and sympathy,
even if they were never absent when love is present, are often
present when love is absent. Indeed, as was said above, they are
sometimes present together with hatred.

464. Neither pleasure, then, nor approbation, nor benevolence,
nor sympathy, is always found with love. Is there anything that
is? I think that there is one thing. When *B* loves *C*, he feels
that he is connected with him by a bond of peculiar strength and
intimacy—a bond stronger and more intimate than any other
by which two selves can be joined. In present experience, as
was said above, our knowledge of any other self is never perception, and is reached through a double mediation. Yet there
are times when the intimacy of the relation in love is felt to be

scarcely less than the intimacy of a man's relation with his own self[1].

And this seems to me to be the essence of love. Love is an emotion which springs from a sense of union with another self. The sense of union is essential—without it there is no love. And it is sufficient—whenever there is a sense of a sufficiently close union, then there is love, whatever may be the qualities of lover and beloved, and whatever may be the other relations between them.

465. This leads us to another consideration about love—that it is more independent than any other emotion of the qualities of the substance towards which it is felt. I do not mean that love is not reached, in our present experience, by means of the qualities which the beloved has, or is believed to have. If B loves C and does not love D, it can often be explained by the fact that C possesses some quality which D does not possess. And in some of the cases where neither B nor anyone else can explain why he loves C and not D, there may be such an explanation, though it has not been discovered. What I mean is that, while the love may be *because* of those qualities, it is not in *respect* of them.

The difference between an emotion occurring because of a quality and in respect of a quality may be seen more clearly if we take the case of approval of another man, which, as we said on p. 144, footnote 1, is always in respect of a quality. And it is also always because of a quality. But the quality in respect of which I approve of him may be different from the quality because of which I approve of him. I approve of Cromwell, let us say, in respect of his courage. But what causes my approval? Its immediate cause is my belief that he was courageous. If we state this in terms of Cromwell's qualities, the cause is that he has the quality of being believed by me to be courageous. My approval is then in respect of one quality, and is because of quite a different quality. For to be courageous, and to be believed by me to be courageous, are quite different qualities. Of course, my belief that he was courageous may be determined by the fact that he was

---

[1] I doubt if, in present experience, we can go further than "scarcely less." Some difference in the degree of intimacy appears always, in present experience, to remain. Cp. p. 156.

courageous, and then this second fact—his courage—is the
remote cause of my approval. But my approval is in respect of
his courage directly, and without any intermediate stage. Again,
I might have believed him to be courageous when he was really
not courageous. Or I might have believed him to be courageous
because I believed that he led the Guards at Waterloo. In these
cases his courage would not have determined my approval at all,
but my approval would be in respect of his courage.

Nor could I have approved of Cromwell if I had never heard
of him. The facts that I did not die before he was born, and that
I have read some history, are therefore factors in the cause of my
approval of him, though I certainly do not approve him in respect
of his having been born before my death, or of having been read
of by me in history.

This, then, is the difference between an emotion being because
of a quality and in respect of a quality. And my contention is
that while love may be because of qualities, it is never in respect
of qualities.

466. There are three characteristics of love, as we find it in
present experience, which support this view. The first is that love
is not necessarily proportional to the dignity or adequacy of the
qualities which determine it. A trivial cause may determine the
direction of intense love. It may be determined by birth in the
same family, or by childhood in the same house. It may be
determined by physical beauty, or by purely sexual desire. And
yet it may be all that love can be.

Other emotions, no doubt, may be determined by causes not
proportioned to them in dignity or adequacy. I may admire a
man passionately because he plays football well. I may be proud
of myself because of the virtues of my great-grandfather. And so
also with acquiescence. I may acquiesce in a state of civil war
because it makes the life of a spectator more exciting. But the
difference is that, in the case of the other emotions, and the
acquiescence, we condemn the result if the cause is trivial and
inadequate[1]. The admiration, the pride, and the acquiescence

---

[1] It might be said that we should not condemn sympathy felt on an inadequate
ground, as when a man sympathizes only with members of his own social class.
But in such a case, I think, we approve of the sympathy because we hold that it

which we have just mentioned would all be condemned because they would be held to be unjustified. But with love, it seems to me, we judge differently. If the love does arise, it justifies itself, regardless of what causes produce it. To love one person above all the world for all one's life because her eyes are beautiful when she is young, is to be determined to a very great thing by a very small cause. But if what is caused is really love—and this is sometimes the case—it is not condemned on that ground. It is there, and that is enough. This would seem to indicate that the emotion is directed to the person, independently of his qualities, and that the determining qualities are not the justification of that emotion, but only the means by which it arises. If this is so, it is natural that their value should sometimes bear no greater relation to the value of the emotion than the intrinsic value of the key of a safe bears to the value of the gold to which it gives us access.

467. The second characteristic is to be found in our attitude in those cases in which we are unable to find any quality in the object of love which determines the love to arise. In such a case, if the emotion were other than love, we should condemn the emotion. For since we do not know what the cause is, we cannot know if the cause is adequate. And without an adequate cause, the emotion is to be condemned. But we do not condemn love because it is not known why it is C, and not D, whom B loves. No cause can be inadequate, if it produces such a result.

468. The third characteristic becomes evident in those cases in which a man discovers that a person, whom he has loved because he believed him to have a certain quality, has ceased to have it, or never had it at all. With other emotions, such a discovery would at once condemn the emotion, and in many cases, though not in all, would soon destroy it. Continued admiration or fear of anything because of some quality which it had ceased to possess, or which it had erroneously been believed to possess, would be admitted to be absurd, and would seldom last for long.

is good to feel sympathy with *any* being who can feel pleasure and pain. We do, however, condemn his *selection* of these people, as the only class for whom he feels sympathy, because the ground of that selection is inadequate. On the other hand, we do not, I think, condemn B for being determined to love C rather than D by the fact that C is beautiful and that D is not.

But with love it is different. If love has once arisen, there is no reason why it ought to cease, because the belief has ceased which was its cause. And this is true, however important the quality believed in may be. If a man whom I have come to love because I believed him virtuous or brave proves to be vicious or cowardly, this may make me miserable. It may make me judge him to be evil. But that I should be miserable, or that he should be evil, is irrelevant to my love.

It often happens, of course, that such a strain is too hard for love, and destroys it. But while such a result would be accepted as the only reasonable course with any other emotion, it is felt here as a failure. Admiration, hope, trust, ought to yield. But love, if it were strong enough, could have resisted, and ought to have resisted[1].

We come, then, to the conclusion that love, as we see it in our present experience, involves a connection between the lover and the beloved which is of peculiar strength and intimacy, and which is stronger and more intimate than any other bond by which two selves can be joined. And we must hold, also, that whenever one of these selves is conscious of this unity, then he loves the other. And this is regardless of the qualities of the two persons, or of the other relations between them. The fact that the union is there, or that the sense of it is there, may depend on the qualities and relations of the two persons. But if there is the union and the sense of it, then there is love, whether the qualities and relations which determine it are known or unknown, vital or trivial. Qualities and relations can only prevent love by preventing the union, or the sense of it, and can only destroy love by destroying the union, or the sense of it. Love is for the person, and not for his qualities, nor is it for him in respect of his qualities. It is for him.

---

[1] Although hatred is specially connected with love, as its polar opposite, hatred does not share these characteristics of love. It would be admitted that, if hate can be justified at all, it can only be when it is grounded on qualities in the person hated, and on qualities which afford an adequate ground for hate. If B hated F on no grounds at all, or because F's great-grandfather had killed B's great-grandfather, B would certainly be condemned. And again, while B might perhaps be excused for hating F, if he believed that F himself was a murderer, he certainly would not be excused for continuing to hate him after he had discovered that his belief was erroneous.

469. Such, then, is the love which one person bears another. But is it possible that anyone should feel love for himself? I think that it is not. It is true that love brings us, more than anything else in our present experience does, into a relation with other selves resembling that in which each of us stands to himself. A man's relation to himself is very close—even omitting the fundamental relation of identity—because he can perceive himself. And, since he can perceive himself, his knowledge of himself is more independent of his knowledge of his qualities than is the case, in present experience, with his knowledge of other selves. The intensity of his interest in himself, again, is independent of the qualities which he believes himself to have.

But the emotion which a man feels towards himself is never the same emotion which, when felt towards others, is called love. While it is essential to love that it should be felt towards a person, it is also essential that it should be felt towards another person. Common usage is not inconsistent with this, for what is called self-love is, I think, generally recognized as not being love at all. The phrase is often used as equivalent to selfishness. Even when—usually qualified as "reasonable" self-love—it is used of a feeling which does not deserve condemnation, it seems only to mean an interest in my own well-being, which errs neither by excess nor by defect. Now love of another person is very much more than an interest in his well-being—indeed it is not such an interest at all, though the interest will generally follow from it.

470. We have thus determined, as far as we can, what is the nature of love, as we see it in our present experience. And now we return to the original question—the place which will be held by love in absolute reality. I believe that in absolute reality every self will love every other self whom he directly perceives. (The question of his relation to selves whom he perceives indirectly will be considered on p. 162.)

We came to the conclusion that no condition was necessary for love except that the lover should be conscious of his unity with the beloved. Now every man who knows any other is in some degree conscious of his unity with him. But in present experience this consciousness of unity is not always strong enough to be love, since we do not love all the people we know. On the other hand, it is sometimes strong enough, since we do love some of the people we know. As we do not know the people we love in any other way than that in which we know the people we do not love, the consciousness of unity in the case of love must derive some of its strength from characteristics other than cognitional.

The more intense the consciousness of unity, the greater is the love. If, therefore, the consciousness of unity with a self is always more intense in perceiving that self than it can be when the self is otherwise cognized, then all such perception of selves will be love. For then the consciousness of unity will be more intense than it is ever in present experience, in which no self perceives another. And yet even in present experience, the intensity is sometimes great enough for love.

But is it the case that the consciousness of unity must always be more intense when a self is perceived, than when it is otherwise cognized? There is no doubt that the consciousness of the unity will be more intense, in so far as the intensity is determined by the cognitional characteristics of the unity. For perception is direct, while other cognition of selves is mediated by the sensa of the knowing self, and by the qualities of the self who is known. But then, in present experience, the consciousness of the unity, in the cases in which love does occur, derives part of its intensity from characteristics other than cognitional. And might it not be possible that no consciousness of unity would be intense enough to produce love unless it derived some of its strength from characteristics other than cognitional? If this were so, it would be possible that some or all of the direct perceptions of other selves in absolute reality might not be states of love.

Now we have, of course, no present experience of the perception of other selves. But we have experience of perception—the perception of sensa, of our own parts, and of our own selves. And we know, though we do not perceive, other selves. And thus it is possible, I think, to image fairly adequately what a perception of another self would be like. And I think we may learn from that imaging that a perception of another self would unite the knower with the known more closely than he could ever be united with any self, however beloved, known to him in any other way.

471. Love, as we now experience it, has often been described as an essentially restful state, and also as essentially a state of unrest. The incompatibility of these statements is only apparent. It is essentially restful because it presents itself as something which is sufficient in itself, which needs no justification, which is good unconditionally, whenever it does arise, whatever may be the circumstances in which it has arisen. And it is essentially unrestful because, in proportion as it becomes intense, we desire, more and more intensely, not indeed anything else but love, but love more intense and more absorbing. Of all true lovers it is true in this world:

> The wind's is their doom and their blessing:
> To desire, and have always above
> A possession beyond their possessing,
> A love beyond reach of their love[1].

What is it that they want? Is it just a quantitative increase in the intensity of love? If so, the desire must remain for ever unsatiated, since beyond any intensity of love which was reached there would be a greater intensity which could be desired. But most people who have endeavoured to interpret it, have interpreted it, and I think rightly, as a desire for a state whose greater intensity of love will flow from a qualitative difference in the nature of the union, a difference which brings the perfect rest which love here only longs for.

Some thinkers, especially Oriental mystics, have concluded that love could only reach its goal when the lover and the beloved became identical. But then the attainment would be suicidal. Love would be destroyed by it, since love depends on a relation between two persons. And does love seek for its fruition in anything but love? Surely the truer interpretation is that which looks for attainment when we shall no more see through a glass darkly, but face to face—when the lover knows the beloved as he knows himself.

This desire for more direct knowledge of the beloved, this conviction that only by the removal of all mediation can our longing be satisfied, is found in many men, scattered over many countries and many ages. The fact that many men have this desire is, of

---

[1] Swinburne, *By the North Sea.*

course, no evidence that the desire is likely to be gratified. But the fact that they have it does prove that they hold that the perception of selves gives a unity of selves which can never be attained without perception.

472. Thus a self will love every other self whom he perceives directly. And since every self perceives at least one other self directly, every self will love. But, as we have seen, it is not necessary that every self should be perceived, and therefore it is possible that there are selves who are not loved, though it is also possible that all selves are loved. Again, it is possible that every self should be a member of the differentiating group of every other self, in which case every self would be loved by every other self. But this is not necessary. And, again, even if every self did not love every other self, it would be possible that all love should be reciprocal—that, if B loves C, C always loves B. But this again is not necessary.

The conclusion that in absolute reality each of us loves every person that he knows may appear to be paradoxical because it maintains that every person known must be loved, regardless of his qualities. But any appearance of paradox is illusory. The qualities of C either prevent B from perceiving him, or they do not. If they do prevent the perception, then they prevent the love, since B can only love the persons he perceives. But if they do not prevent the perception, then they do not prevent the unity, which lies in the perception. Nor do they prevent B's consciousness of the unity. And present experience is sufficient to show us that it is possible to love a man, whatever his qualities are, provided that the unity and the consciousness of the unity are sufficiently intense[1].

473. Every self, then, will love every other self whom he directly perceives. And the intensity of this love, we must also conclude,

---

[1] When we go further we shall see that all selves have in absolute reality such a nature that it would be difficult, if not impossible, that they should also have any of those qualities which, in present experience, tend to check love. But the demonstration of this nature of the selves is dependent on the fact that each self loves all the others whom he knows. It would therefore involve a vicious circle if we appealed to the nature of the selves to show that it would be possible to love them. Nor is such an appeal wanted. The possibility follows from the grounds given in the text.

will be much greater than that of any love which occurs in present experience.

The chief ground for this conclusion is to be found in the same fact which we have already considered as a proof that love must be there in every case—the greater closeness of unity between the two selves when one of them directly perceives the other. If I perceive another self, I know him with the same directness, the same immediacy, the same intimacy, with which I know myself. There is no longer any of that separation which weakens love. Separation—or rather distinction—of course remains, for if there were not two distinct selves, there could be no love. But there is no barrier between the selves. The unity is unhampered. Love is no longer held back by the inadequacy of knowledge. Must it not reach an intensity which we can only estimate dimly by considering that in it all the longings of our present love are satisfied? *"Quam bonus te petentibus, sed quid invenientibus!"*

This is the chief reason for holding that the intensity will be much greater in absolute reality. But there are also others. A second reason is that when love does exist in present experience, it is often weakened by the recognition of qualities in the beloved which are uncongenial to the lover. But many of these qualities, as we shall see later, are certainly incompatible with the nature of absolute reality, and the others are such that it seems almost impossible that they should be compatible with it, and in any case they would be relatively insignificant. And thus love will be free from checks found in present experience, and must therefore be more intense[1].

In the third place, the effect of such uncongenial qualities in checking love is often increased, in present experience, by a volition that the quality disapproved should have been different. But this cannot occur in absolute reality, since in absolute reality there are no ungratified volitions.

Fourthly, in absolute reality all the life of every self is, or is dependent on, love. The self has no parts except his perceptions

---

[1] The fact that absolute reality has such a nature is dependent on the fact that every self loves, but not on the fact that every self loves intensely. And therefore, though there would be a vicious circle in using it to prove that every self loves, there is nothing vicious in using it to prove, of selves already proved to love, that they love intensely.

of himself, of other selves, and of parts of selves. All perceptions
of other selves are states of love. His perceptions of the parts of
other selves are parts of states of love, and, as we shall see later,
derive their emotional and volitional qualities from this. And his
perceptions of himself and of his parts, as we shall also see, derive
their emotional and volitional qualities from the fact that he
loves others. In absolute reality, then, love is supreme, not only
in value—for that we have not to wait for absolute reality—but
supreme in power. Nothing is alien to love, everything is dependent on it. The harmony and the absence of distraction which
this involves must increase the intensity of love—all the more
because this supremacy of love will not only be real, but will be
known as real.

Absolute reality is timeless. We shall see later (Chap. LXVII,
p. 461) that this makes the value of absolute reality infinite in
amount. But the fact that love in absolute reality has infinite
value does not, as might perhaps be supposed at first sight,
involve that its intensity is infinite. The infinity comes in a
different dimension from the intensity—the dimension of the
C series, which in present experience appears as time. The timelessness of absolute reality has thus no direct effect on the
intensity of love in that reality. But it has an indirect effect.

For, fifthly, love in present experience can never keep permanently, or even for long together, at the highest intensity
which it occasionally reaches. It cannot be permanently on that
level, because the necessities of life compel us to turn our
attention to other things besides loving our friends. Nor can it
be on that level for long together, because the strain of intense
love—of love which has to fight its way through its cognitional
inadequacy—is such that it cannot endure for more than a brief
period. But in timeless reality there is no change, and no weari-
ness, and that which is highest can exist without ceasing. What
this would mean, even if the highest were no higher than it is
now, it is useless to try to say, except to those who do not need
to be reminded of it.

All these causes, then, will operate to make the intensity of
love in absolute reality greater, much greater, than in present
experience. But the first—the greater unity which comes with

perception—seems to me to be much the most important of all of them.

474. Besides perceiving other selves, we perceive their parts. What can we say about the emotional quality of our perceptions of parts of selves? In the first place, is it true, as it is with our perceptions of selves, that the closeness of the cognitive union with these parts of selves determines the emotional value in that cognition? I do not think that it is so. To be closely united to a self involves love of that self, but I can see nothing analogous in the case of parts of selves. Our emotions towards them, it would seem, must depend on the qualities which we perceive them as having.

All these parts will be parts of selves whom we perceive, and will be perceived as being such parts. (Chap. XXXVII, pp. 98–104.) And the selves which are perceived are loved. Now when we consider how, in present experience, we regard states and events in the life of a person whom we love, we find that we tend to regard them with a special sort of liking. This sort of liking would not be called love, since the name of love is reserved for an emotion towards persons. We might perhaps call it complacency, though the name suggests a milder emotion than that which is often felt.

With regard to present experience it can only be said, as was said above, that we *tend* to regard it in this way. The tendency comes from our recognition of the state as the state of a person whom we love. But if the state has other qualities which tend to render it repugnant to us, the complacency with which we regard it may be weakened, or even completely prevented.

But in absolute reality could such states have any qualities which would tend to render them repugnant to us? I do not see that they could. The states of C which are perceived by B will all be perceptions of selves other than C, or of the parts of these selves, or of C himself, or of parts of C. None of them can be ungratified volitions. The perceptions of other selves will be states of love. C's perceptions of the states of other selves will themselves be states of complacency, unless this should be interfered with by some quality of the states of which they are perceptions. And as the same argument will apply to these latter states, and so on infinitely, there will be no place for any such quality to be introduced.

It might be said that C's state might be a state of pain, and that this might make it repugnant to B. The question of pleasure and pain will be treated in Book VII. We shall see then (Chap. LXVII, pp. 471–472) that in absolute reality the pleasure of any state would infinitely exceed the pain of that state. It would be impossible, therefore, that the state as a whole could be an object of repulsion on hedonic grounds.

All C's perceptions of other selves and their parts will be emotions of love or of complacency. And we shall see later that C's perceptions of himself and of his parts will be emotions of self-reverence and complacency respectively. Can such states have any quality which should cause repugnance in B—who, it must be remembered, loves C himself[1]? I cannot conceive that they can. If this is so, there will be nothing to check the tendency which B has to regard them with complacency, because they are parts of C, whom he loves. And so he will regard them with complacency.

475. It is possible that C should be in the differentiating group of B, and D in the differentiating group of C, but not in that of B. In that case B will directly perceive C, and C will directly perceive D, but B will not directly perceive D. He will, however, perceive C's perception of D, and we have seen (Chap. XXXIX, p. 126) that this may properly be called an indirect perception of D. Will this indirect perception give any emotional relation towards D? I think that it will. B will directly perceive $C! D$—that is, C's perception of D, which will also be a state of love of C to D. This perception by B (symbolically $B! C! D$) will, as we have seen, have the quality of being an emotion of complacency towards $C! D$. And I think that it will also have the quality of being an emotion towards D—that emotion which we feel, in present experience, towards those whom we do not love,

---

[1] It must be remembered that we are speaking here of repugnance, which is an emotion towards a substance as a whole, not in respect of its qualities, though it may be caused by its qualities. The possibility, in absolute reality, of the disapproval of a substance in respect of its qualities, will be considered later (p. 167). But disapproval is not incompatible with love or complacency, which are forms of liking.

but who are loved by those whom we do love. I do not know
that this sort of emotion has any special name, but we are not
at a loss to know its nature, for there are few men to whom it
does not form a part of present experience. We might perhaps
use the word affection to denote that sort of liking which is felt
for persons, as distinct from that sort which is felt for other
substances, and in that case both love and this emotion will be
instances of affection[1].

It is, of course, the same perception $B! C! D$ which is both
the emotion towards $C! D$ and the emotion towards D. I do not
think that this causes any difficulties. We often find that a cognition may have two emotional qualities towards the same object.
For I may simultaneously love C and admire him, and there
seems no reason to suppose it necessary that I should have, in
that case, two simultaneous cognitions of C. Nor is there any
difficulty in the supposition that the same cognition should
have the two qualities of being an emotion of love towards C,
and an emotion of admiration towards C. In this case, no doubt,
both the emotions have the same object. But if a perception
can have both a direct object and an indirect object, there seems
no reason why it should not be an emotion towards both those
objects.

In the same way $B! C! D! E$ will be an emotion felt by B towards E, arising from the fact that B loves C, who loves D, who
loves E. And so on with lower grades.

476. We have now spoken of emotions towards selves, and towards parts of selves. Can we say anything positive as to our
emotions towards groups of selves? I do not think that we can.
If I know a group of selves directly, I shall love all its members.
If I know it indirectly, I shall regard all its members with
affection. And I can have no emotions towards the group which
are incompatible with these facts. But it does not follow that I
shall have any emotion towards the group at all. A man may
love each of his school friends, each of his college friends, and each

---

[1] In present experience, indeed, I may not regard the friend of my friend with
affection, because of some qualities in him which excite sufficient repugnance in
me to prevent the affection arising. But an argument analogous to that in pp. 161–
162 will show that no such repugnance could arise in absolute reality.

of his children. But it does not follow that he will have any emotion towards a group consisting of one of his school friends, one of his college friends, and one of his children—a group of which, perhaps, no one member has ever met either of the other members.

477. So much for emotions towards other selves and their parts. But some selves, at any rate, perceive themselves, and, if they do so, they perceive their own parts. What can we say about emotions here?

I do not think that the closeness of relation between a self and himself determines any emotion in the self towards himself. But such an emotion must exist in absolute reality. Every person who perceives himself directly must also perceive directly at least one other person. He will therefore love at least one person. Now love induces in the lover an emotion towards himself which we may call self-reverence. Since I love, I have value—supreme value, since I am possessing the highest good. And since I have value I shall regard myself with reverence. And if I reverence myself I shall regard my parts with a feeling of complacency analogous to that with which I regard the parts of the persons whom I love[1].

Thus it will be seen that our whole emotional attitude in absolute reality, so far as we can now determine it, depends on love. It is because B loves C, that he feels complacency towards the parts of C, affection for D, whom C loves, reverence for himself, and complacency towards his own parts. And we shall see (p. 165) that it is his love of C which determines his acquiescence in the existence, not only of C and of the parts of C, but of D, and of himself, and of his own parts. Both his emotions and his volitions towards himself depend on his love of someone else. But if we consider life, even as we find it here and now, we shall find nothing surprising in the view that a man finds himself worthy of reverence, and his existence desirable, only for the sake of the love he bears his friends.

---

[1] In present experience we cannot say more than that there is a tendency towards complacency, and perhaps not more than that there is a tendency towards self-reverence. But, by an argument analogous to that in pp. 161–162, it can be shown that in absolute reality the tendency could not be thwarted.

478. We are now able to return to a question which we left unsolved at the end of the last chapter. We saw there that in absolute reality there could be no volitions which were not perceptions, and that there could be no ungratified volitions. But it was not then possible to determine whether all the perceptions were volitions, or even to determine whether any of them were so.

But now we can answer this question. I shall love all the other selves which I directly perceive. And acquiescence is a necessary consequence of love. I may not get happiness from my beloved, or from my love of him. I may not approve of him morally. I may desire that many of his qualities should have been otherwise. But there is one thing I must desire if I love him. I desire his existence. I want him to be there.

In present experience, indeed, my desire that he should exist may be accompanied with a simultaneous desire that he should have different qualities, or even that he should not exist. For my own sake I desire his existence. But for his own sake, if his life were miserable, or even for the sake of others, I might also desire that he should not exist. And the latter desire might be the stronger.

But in absolute reality there are no ungratified volitions, since all volitions are perceptions. And therefore, since he exists and is as he is, I could have no desire that he should not exist, or that he should be different. Thus there would be no desire to conflict with the desire for his existence, and that desire, which, as we have said, exists whenever love exists, will have undisputed sway.

My direct perceptions of other selves, then, will all be gratified volitions. And the same will be the case with my indirect perceptions of other selves, and with my perception of myself, and with my perceptions of parts of selves. For I shall regard these with affection, or with self-reverence, or with complacency, and whatever I regard with any of these emotions I shall desire to exist.

In present experience these desires, like desires for the existence of selves whom I love, may be accompanied by desires that the same object should exist with different qualities, or should not exist at all. But, since there can be no ungratified volitions in absolute reality, there can, in absolute reality, be no such opposing desires, and I shall have no desires except the gratified desires for existence.

Thus all my perceptions will be, in a volitional aspect, states of acquiescence in the objects perceived. And, in an emotional aspect, all my perceptions will be states either of love, or of other affection, or of self-reverence, or of complacency.

479. What other emotions, we must now enquire, are possible in absolute reality besides these four?

It is clear that some emotions cannot occur, because they are incompatible with those which, as we have seen, must occur. There can be no hatred, since I regard every person whom I know, directly or indirectly, either with love or with another sort of affection. Nor can there be any other sort of repugnance. For the only things which I know, besides selves, are parts of selves or groups of selves. Parts of selves, as we have seen, are all regarded with complacency, which excludes repugnance. Nor can we regard a group of selves with repugnance, when we regard each member of it with affection.

Malignancy presents a more difficult problem, for, as we have seen (p. 150), it is not incompatible with love. But it would seem that, when malignancy is found together with love, it is always found in connection with, and dependent on, some ungratified volition (usually, though not always, of a sexual nature). In this case it could not occur in absolute reality, since in absolute reality there is no ungratified volition.

On this ground also we must reject various other emotions which involve ungratified volitions. Anger is one of these. Jealousy and envy, also, however they may be analyzed, clearly depend on ungratified volitions. And so do those special varieties of sadness and humility which are known as regret and remorse. None of these, then, can find any place in absolute reality.

Thirdly, absolute reality is timeless. And this will exclude hope and fear, which relate only to the future. It will also exclude wonder, in the sense in which it signifies an emotion excited by what is new or surprising. (Wonder, in the sense of an emotion excited by what is great or sublime, is a variety of approval, which will be considered below.)

Finally, in absolute reality there can be no assumptions, since all our cogitations are perceptions. And every question contains an assumption. Consequently in absolute reality there can be no questions, and therefore no unanswered questions. And this is incompatible with the existence of curiosity.

480. Courage and cowardice do not seem entirely impossible. It is true that their sphere would be considerably lessened. For they occur at present almost always in connection with a volition that the evil, with respect to which the courage or cowardice is shown, should not be taking place, or should not take place in the future. Volitions of this sort are cogitatively assumptions, and therefore cannot occur in absolute reality. Still in absolute reality there is some evil, and therefore, I suppose, a place for courage or cowardice. But as this evil, as we shall see later, must be infinitely small in proportion to the good, the importance of courage and cowardice is infinitesimal.

There remain sympathy, approval, disapproval, pride, humility, gladness and sadness. I do not know that any proof can be given for the assertion that we shall, in absolute reality, sympathize with those whom we love, or for whom we feel affection—the only persons whom we shall know at all. But everything that we can gather from our present experience gives a presumption that we shall do so. It can scarcely be supposed that approval will be absent. If it is good to love, we shall, in respect of their love, approve of those persons whom we perceive as loving. And self-reverence, which we have seen to exist, is a form of pride.

But it is possible that disapproval and humility may also occur. Approval is not inconsistent with disapproval, nor pride with humility, in spite of their polar opposition. For all four are felt for persons in respect of their possession of certain qualities. Thus the same substance may possess qualities which excite approval, and others which excite disapproval; and the same is the case with pride and humility.

It would seem certainly that there could be little, if anything, in the condition of selves in absolute reality which could excite either humility or disapproval. Each self will have a set of parts, each of which is a state either of love or of self-reverence, and all the parts of those parts will be states of complacency. Still there seems a place at any rate for disapproval. Disapproval, of course, can be excited by any quality which is bad, and not only by those which are morally bad. What evil there is in absolute reality, will, as we shall see later, take the form of pain. And when a state is painful, it will so far excite disapproval.

Again, so far as there is pain, there will presumably be the emotion of sadness, while gladness cannot be absent from a universe in which each person acquiesces—and often passionately acquiesces—in everything which he knows to exist. More specifically, love must bring gladness, unless it raises ungratified volitions. And this cannot be the case in absolute reality.

Loyalty—in the sense of an emotion towards a community of which I am part—is so important in our present experience that we must ask whether we can know anything about its place in absolute reality. It consists, I should say, in an emotion of devotion—which is one of the more intense forms of liking— towards the community, combined with an emotion of self-reverence towards myself as part of that community. Thus it partly consists in, and partly depends on, an emotion towards a group of selves. And we saw above (p. 163) that we have no means of deciding whether, in absolute reality, we feel any emotions towards groups of selves.

481. We have so far said nothing about pleasure and pain. They are not emotions, but the class of which pleasure and pain are members—sometimes called the class of feelings—is analogous to emotions and to volitions. To be pleasurable or to be painful are qualities which can belong to states of cogitation, and only to states of cogitation. When a state of cogitation has the quality of being pleasurable, it is a state of pleasure; when it has the quality of being painful, it is a state of pain. In our present experience the most usual and typical pleasures and pains are perceptions—especially perceptions of sensa. But judgments, assumptions, and imagings, can also be pleasures or pains.

Can we say anything about the position of pleasure and pain in absolute reality? It seems evident that there will be some pleasure. For everyone will love. It might perhaps be maintained that love always involves some pleasure, even if it involves a balance of pain. But it cannot be doubted that love involves pleasure, when it is not connected with some ungratified volition. And this is the case here.

But we cannot conclude at once that there will be no pain, or even that the pain will not exceed the pleasure. For the perception of a loved person may give pain at the same time as it gives pleasure, and the pain may be the greater. Nor can we argue that there can be no pain because there can be no ungratified volition. The only reason that there can be no ungratified volition in absolute reality, is that, where there are no cogitations but perceptions, we can only desire what exists. And this obviously can give no guarantee of any quality of the existent—except the quality that it will never be wished to be otherwise. We have not removed the possibility that there should be pain, and much pain, although it is certain that, whatever pain there may be, we shall be spared the secondary pain of ineffectual protest and revolt.

The question of pleasure and pain will be considered again towards the end of Book VII. We shall then see that there is reason to believe that in absolute reality the pain will be infinitely less than the pleasure.

CHAPTER XLII

DISSIMILARITY OF SELVES

482. In order that determining correspondence may produce a series of parts within parts to infinity, it is, of course, necessary that we should start with primary parts which are dissimilar. And we have assumed, up to this point, that the selves, which we have found reason to believe are the primary parts of the universe, are dissimilar to one another. But it is now time to enquire in what way they can be dissimilar.

Three ways present themselves as possible. They might differ in their original qualities, or in the quantities of their original qualities, or in their relations[1]. All these three, or any two of them, might be combined.

483. It will be convenient to begin with relations. We know that the selves stand to one another in the relations required by determining correspondence. Each self has two or more selves as its determining group. And, if there are more than two selves in the universe—which, on empirical grounds, seems probable—then however large the number of selves, and however small the differentiating groups, it would be possible that no two selves should have the same differentiating group. And in that case it would be an exclusive description of any self that it was the self which had such and such selves as its differentiating group.

But if we endeavour to adopt this as the only manner in which the selves are to be dissimilar, we shall find that our attempt involves a vicious infinite. We find the dissimilarity between B and C, let us say, in the fact that the differentiating group of B is EF, and the differentiating group of C is FG. But this fails to make B and C dissimilar, unless the dissimilarity of E, F, and G has been already established. If we attempt to do this by relying on their differentiating groups being respectively HJ, KL, and

---

[1] Different quantities of the same original quality are themselves different original qualities, but it will be more convenient to treat them separately.

Difference of relations implies, of course, difference in *derivative* qualities.

MN, this requires that we should have previously established the dissimilarity of these six selves. And so we shall go on in an infinite series which is vicious, since the dissimilarity of B and C can only be established by establishing the dissimilarity of the last members of a series which has no last members.

Nor should we be in a better position if the terms concerned formed a closed circle, so that, to take a simple example, the differentiating groups of B, C, and D were respectively CD, DB, and BC. For if the series returns on itself, so that B is a member of the differentiating group of a member of its own differentiating group, then to establish the dissimilarity of B will require the previous establishment of the dissimilarity of B, and, instead of a vicious infinite series, there will be a vicious circle.

This result follows from the conclusion reached in Section 105. We saw there that every substance must have at least one exclusive description which is a sufficient description, and that the attempt to differentiate substance by means of exclusive descriptions which were not sufficient descriptions would lead to a vicious infinite. And if we try to differentiate substances by the difference of the terms to which they stand in a certain relation, we are trying to differentiate them by exclusive descriptions which are not sufficient descriptions.

484. We saw, however, in Section 104, that a description of a substance by means of its relations to one or more other substances may be a sufficient description under certain circumstances. In the first place, if B is exclusively described as having the relation M to C, and C can be sufficiently described as having the original qualities XYZ, then we have a sufficient description of B as the only substance which has the relation M to the only substance which has the qualities XYZ. But this requires, as we have just seen, that C, unlike B, should be capable of discrimination otherwise than by the difference of the terms to which it is related. And therefore this would not enable us to differentiate all selves exclusively by their relations. But there remains a second alternative. By a combination of relations a compound relation may be formed which is so rare that only one substance stands in that relation to any substance. In that case it would be an exclusive description of B to say that it was the only substance which stands in the relation PQR to any substance. And this exclusive description will be a sufficient description, since it consists entirely of general characteristics.

Now these possible variations in the relation of a self to its differentiating group which can be expressed entirely in general characteristics would be sufficient to form dissimilar descriptions for any number of selves. It is possible, for example, that a self, B, might be discriminated from many others by the fact that he had a differentiating group of seven members. From those other selves who had also differentiating groups of seven members he might perhaps be distinguished by the fact that his group contained one member who had himself a differentiating group of six members, one member who had himself a differentiating group of ten members, and so on. If this were still insufficient to distinguish B from all other selves, he might perhaps be distinguished by similar variations in the differentiating groups of the next grade, and so on until a description had been reached which applied only to B.

It is not impossible that the numbers of members in the differentiating groups should vary in this way, and therefore it is not impossible that the dissimilarity of the selves should arise exclusively in this way. But although not impossible, it seems very improbable. If it were the fact, then all differentiations of selves, on which all other differentiations of substance depend, would not themselves depend on any internal differences in the selves, nor on any internal differences in those other selves with which he was in relation, nor in the sort of relation in which he stood to them. It would depend entirely on such an external characteristic as the number of selves in the groups with which he was connected directly or indirectly.

485. The appearance of the world in our present experience suggests very strongly that ultimate differentiation does not depend only on relations, but, at any rate to some degree, on original qualities. It may be said that we have already seen that absolute reality is very different from what it appears to be in our present experience, and that this may be another point of difference. But then what we have seen about absolute reality has all tended to emphasize the importance of the selves. For it is the selves who are primary parts, and we saw in Section 256 that the primary parts must be regarded as more fundamentally important than the primary wholes or the universe. It seems therefore to be very improbable that each of the primary parts should derive his individuality from characteristics of the groups to which he is related.

Nor is it necessary that it should do so, for the necessary differentiation of the selves could arise in other ways. It could arise in respect of their original qualities. It will be sufficient if each self has some quality—simple, compound, or complex—which no other self has.

We must remember, however, that we have reached the conclusion that the different selves are qualitatively very much alike. Each of them perceives selves, and the parts of selves, and has no other content but such perceptions. All these perceptions are also volitions of acquiescence in what is perceived. And all such perceptions are states either of love, of self-reverence, of affection, or of complacency.

Now this uniformity seems to leave very little possibility of much variation in other qualities. We saw, indeed, in the last chapter, that it was at any rate possible that there were other emotions which the self might or might not have. But the number of these—at any rate of those now known to us—is so small that their combinations would not afford sufficient differentiations even for the number of selves which we have empirical grounds for believing to exist—and the total number of selves may be much greater.

The dissimilarity might, however, be quantitative. It might consist in a variation of the intensity of the qualities which are possessed in common by all selves. Perception can be observed to vary in intensity. And so can acquiescence and all the emotions of which we have spoken. Here, then, is another source of dissimilarity. For example, it might be the case that there was one self who could be sufficiently described as the only self perceived by all other selves, and that all the other selves could be sufficiently described by the varying intensity of the love which they felt for him. There is nothing to suggest that this is the case, but I can see nothing which makes it specially improbable. And of course there are many other ways in which selves could be differentiated quantitatively.

A quantitative series could afford as many differentiations as could be required. For the series of degrees of intensity might be compact, so that an infinite number of degrees could be found. And even if this were not the case, the number of degrees might be so large that it might equal the number of selves. For, although the number of selves may be infinite, it is possible that it is not infinite.

486. There is also another possibility. Could there be qualitative differences in the way in which different selves perceive the same percepta? The same words can be pronounced in different tones. The same design can be drawn in different colours. Would it not be possible for B to differ from C in some quality or qualities analogous to tone or colour? Then we might have the necessary differentiation by means of these differences. Suppose that B and C had the same differentiating group, and that they did not vary in the qualities which they perceived the members of that group, or the parts of those members, as having. Then B and C would have no other nature than to perceive the same things, and to perceive them as having the same qualities. They would acquiesce in all that they perceived. And they might regard them with the same emotions. But their perceptions could be qualitatively different in the same way in which two repetitions of the same words in different tones might be different. And this qualitative difference of the perceptions might affect their volitional and emotional qualities. Thus there might be a qualitative difference between B's love for D and C's love for D.

However many selves there might be in the universe, they might all be differentiated from one another by such means, since there is no reason why the variations in such qualities should have any limit in minuteness, and therefore any number of them might occur in different selves. And the qualities might be of more than one sort.

I do not suggest that there is any proof of the existence of such tone-differences. In some way or other each self must be differentiated from all the rest. But we have seen earlier in the chapter that they could be differentiated by other means than tone-differences, and therefore it is possible that tone-differences do not exist.

487. But it is possible that they do. And it may be well to pursue this excursion into possibilities rather further. It is, as we have seen, possible that every self should be included in the differentiating group of every self, but it is also possible that this is not the case, and that different selves have different differentiating groups. D, for example, may be in the differentiating group of B, and not in the differentiating group of C. Now, if there are such differences, they may, of course, be ultimate. Something must be ultimate, and these differences might very well be so. But, on the other hand, they need not be ultimate. And, if there were such tone-differences, it might be that the differentiating group of each self should consist of those other selves whose tone-qualities were in certain relations to his own. (These relations need not, of course, be relations of special similarity. They might, for example, be relations of complementary difference.)

If this should be so, then, if B loves C, it is caused by the fact that the tone-qualities of C have certain qualities—the qualities which place them in a certain relation to the tone-qualities of B. There is nothing in this which is inconsistent with the conclusions which we have reached with regard to love. For we saw in the last chapter (p. 151) that while love is never in respect of the qualities of the beloved, it is often because of his qualities.

We have thus shown that the dissimilarity of selves, which is essential to our theory, could be realized in several different ways. And therefore the fact that such dissimilarity is necessary is no objection to our theory. But there seems no way of deciding in which of the possible ways it actually is realized.

CHAPTER XLIII

GOD AND IMMORTALITY

488. Do the conclusions which we have reached as to the nature of the existent throw any light on the questions whether God exists, and whether men are immortal?

I shall take the word God to mean a being who is personal, supreme, and good. Personality is the quality of being a self, and we have already discussed what is meant by a self. In including supremacy in the definition of the quality of deity, I do not mean that a being should not be called a God unless he is omnipotent, but that he must be, at the least, much more powerful than any other self, and so powerful that his volition can affect profoundly all else that exists. In including goodness, I do not mean that a being should not be called a God unless he is morally perfect, but that he must be, at the least, more good than evil.

All these three qualities—personality, supremacy, and goodness—are, I think, included in the definition of the quality of deity, in the theology of the western world at the present day. Personality, in the first place, always seems to be regarded as essential. Cases can be quoted, no doubt, in which an impersonal reality has been called God. But I think that such statements do not mean more than that the reality spoken of is a worthy substitute for a God, or that the belief in it is a worthy substitute for the idea of a God. They do not mean that the name can be used in a strict sense of an impersonal reality.

It is sufficiently clear that a person would not be called God unless he possessed such supremacy as is spoken of above. And it is also clear that modern usage would not permit any person, however powerful, to be called God, if he were held to be more evil than good.

489. This is the usage of theology and of common language. But in philosophy we have high authority—including Spinoza and Hegel—for a much wider definition. God, it is said, is all that truly exists, provided only that it possesses some sort of unity and is not a mere aggregate, or a mere chaos. If the word is used in this sense, every one, except absolute sceptics or the most extreme pluralists, must be said to believe that a God exists. The question of the existence of God becomes, on this definition, very trivial. The important question is not whether there is a God, but what sort of nature he, or it, possesses.

If the usages of theology and philosophy differ in such a matter, it is surely philosophy which ought to give way. A deliberate effort may possibly change the meaning of terms which are used by a comparatively small number of students. But no such effort could change the popular usage of such a word as God. Now popular usage is distinctly in favour of the narrower definition, and philosophy ought to accommodate itself to this, to avoid a dangerous ambiguity.

Again, while the conception of the whole of the truly existent is of fundamental importance for philosophy, the conception of a supreme and good person is also of great philosophical importance. It is desirable that each of these substances should have a separate name in philosophical terminology. For the second substance no name but God has ever been proposed, while the first is often called the Absolute, or, as we have called it, the Universe. If God were used as another synonym, we should have more than one philosophical name for one important idea, and no name for the other.

Finally, philosophical usage is by no means uniform. Against Spinoza and Hegel we may put Kant, who uses the word in the theological sense. The balance of convenience, then, seems in favour of confining the name of God to a being who is personal, supreme, and good.

490. The definition which I have proposed, then, is not too narrow. Nor can it be condemned as too wide. It is true that most theists would go much further in what they said about God. They would believe that he was absolutely perfect; they would believe that he was the creator of all else that exists; and they would profess to believe in his omnipotence[1]. But they would not, I think,

---

[1] It is scarcely ever the case that God is really believed to be omnipotent, so that there would be nothing whatever which he could not do. If we consider the views of any writer on theism, we almost always find that there are various things

include these qualities in the definition of the quality of deity. If they came across a man who denied that the supreme person was creative and omnipotent, or that his goodness was absolutely perfect, they would not, I think, call such a man an atheist. They would allow that he believed in God, though they would regard his conception of God as inadequate.

491. God, conceived according to this definition of deity, may be regarded as having one of three relations to the universe. He may be believed to be identical with all that exists. He may be believed to create all that exists except himself. Or he may be believed, not to create the universe, but to guide and control it.

On the first of these suppositions God will be the universe, or, to put it the other way round, the universe will be a person. But we have come to the conclusion that no self can be a part of another self (Chap. XXXVI, pp. 82–86). If, therefore, the universe is a self, it follows that no part of the universe is a self, and that there are no selves but God. Now we have found that every primary part is a self, so that not only are some selves parts of the universe, but selves form a complete set of parts of the universe. And, as the existence of even a single self within the universe would show that the universe is not a self, we must reject the view that God is the universe.

If there is a God, then, he must be part of the universe, and there are other parts of the universe which are not God, or part of him. Let us now consider the second supposition—that God creates all that exists, except himself.

492. This supposition must, I think, be rejected on several grounds. If it were true, God would create all other selves. This would make God more fundamental in the universe than all other selves. But we have seen that all selves are primary parts. And all primary parts are fundamental. From the natures of primary

which he does not believe that God can do. But to *call* God omnipotent is a piece of theological etiquette from which few theists seem capable of escaping.

It seems to me that, if the word omnipotent is taken strictly, it is impossible that any person should be omnipotent. I shall not give here the arguments for this conclusion, as they have no special connection with our theory as to the nature of the existent. I have stated them in *Some Dogmas of Religion* (Section 166, and Sections 171–176. To the arguments discussed in Sections 167–170, I should now attach even less weight than I did when those sections were first published).

parts follow the natures of all other substances—secondary parts, primary wholes, and the universe. But the natures of primary parts áre ultimate facts. It would not, I think, be possible to combine this co-equal primacy of the selves with such pre-dominance of one self as would be involved in creation.

But a more definite objection arises from the unreality of time. Some reference to time is essential in creation. It has sometimes been held that the creator is timeless, and even that the creator's volition is timeless, but the thing created is always taken as being in time. It is essential, not only that something should be caused to exist, but that something should be caused to exist which did not exist before. And this involves time. If there is no time, there can be no creation.

And the unreality of time is fatal to creation in another way. Creation is a causal relation, and a causal relation which is not reciprocal. If I am created by God, it is impossible that God should also be created by me. God must be the cause, and I the effect. Now we saw in Section 212 that we cannot say that one thing is the cause of another unless it is prior to it in time. When two terms are timeless or simultaneous, we can only say that they are in a causal relation, without distinguishing either of them as cause. And this is not sufficient for creation[1].

493. But, it might be said, even though creation must be a temporal relation, it is possible that there should be a God who was in a timeless causal relation to other substances, and this relation would be so like creation that the result would not be gravely misrepresented by saying that God was the creator of those other substances.

But this would be wrong. For in such cases the other selves would be co-equal with God, and God would only be the cause

[1] If time were real, this argument would, of course, be invalid. And it might then be difficult to prove that there was not a creative God. But it follows from another of our previous results that the ordinary reason given for believing that there is a creative God is invalid. This reason is that, without a creative God, either what exists must begin without a cause, or else there must be an infinite causal regress. And it is asserted that both of these are impossible. But we saw in Section 214 that there is no ground to suppose that an infinite causal regress involves a contradiction. Thus we are not justified in assuming the existence of a creator to avoid such a regress. Nor does there seem any reason why we should reject as impossible a causal series with an uncaused beginning.

of other substances in the sense in which they were his cause. And this differs so completely from what is ordinarily meant by creation that it would be gravely misrepresented if it were spoken of as creation.

We have found reason, indeed, to hold that every self in absolute reality desires the existence of everything which he knows. And so a God who knew the whole universe would desire the existence of the whole universe. But in this there is nothing at all analogous to creation. For the only desires in absolute reality are those which are, in their cogitative aspect, perceptions. And in absolute reality the perception of anything, and of anything as having a certain nature, depends on that thing existing and having that nature[1]. The acquiescence in anything, then, depends on the perception of it, and so on its existence. And in such a desire as this there can be nothing analogous to a process of creation in which a creator should call into existence that which, independently of its existence, he wills to exist.

Moreover, God's acquiescence in the existence of himself and his own parts is of exactly the same sort as his acquiescence in the existence of other selves and their parts. And a relation which God has to himself cannot be analogous to the relation of creation.

494. There can, then, be no relation between God and other substances which would be so like the relation of creation that we should not gravely misrepresent the truth by saying that God created those substances. But one more possibility remains. Granting that there is nothing in the nature of absolute reality which is either creation, or anything analogous to creation, yet, it may be said, things may appear as other than they really are, and appearance may be a *phenomenon bene fundatum*. Suppose then that there were a self, of such a nature as to be called God, who appeared, *sub specie temporis*, to exist before all other parts of the universe, and whose existence was really causally related to their existence. Since he appeared to precede them, would not he appear as their cause? And in that case would not the state-

---

[1] The acquiescence in the existence of a primary part would itself be a secondary part of the first grade, the acquiescence in the existence of a secondary part of the first grade would itself be a secondary part of the second grade, and so on.

ment that he was their creator possess, not indeed absolute truth, but as much truth as any statement can have which deals with time? And would not this be all that is required by advocates of the doctrine of a creative God?

But, *sub specie temporis*, no self appears to exist before any other self. The grounds for this assertion must be postponed to the next Book, in which the relations of present experience to absolute reality will be considered. We shall there find reason to conclude that the relation of all selves to the C series is such that, *sub specie temporis*, the first moments of the existence of all selves, must be taken as simultaneous (Chap. LI, p. 275). It would be impossible, therefore, that God could appear as temporally prior to the rest of the universe.

495. We have now to consider the possibility of a God, who, while he does not create anything, yet controls and governs the universe. It is clear that, if there is such a God, he must do more than exercise *some* influence on the universe. Each of us does as much as that, and it would certainly not be said that this made each of us a God. On the other hand a being who was not omnipotent, and who therefore had not unrestricted control over events, could be called a God. Certain quantitative considerations appear to enter into the question. In the first place, he must have more power than any non-divine person, and, it would seem, much more power. And, in the second place, though not controlling the course of events completely, it would be held to be essential that his volition was sufficient to change it materially. The analogies which suggest themselves are those of a statesman who is ruling a country, or of a general who is directing an army. And, as with the statesman and the general, we should conceive that God's power, though not unlimited, would be such as might well make all the difference in the value of that which he ruled.

Can there be any person who is really a controlling God? The existence of such a God is, like the existence of a creative God, rendered impossible by the unreality of time. For a controlling God is also a cause—not of the existence of other selves, nor, perhaps, of the existence of their parts, but certainly of the occurrence of certain qualities of selves and their parts. And divine control is, like creation, a relation which cannot be reciprocal. If a divine volition causes the occurrence of a quality in me, then the occurrence of that quality in me cannot cause the divine volition. But, as we have said, it is only in time that the terms of a causal relation can be discriminated into a cause which is not an effect, and an effect which is not a cause. And so, since time is unreal, there can be no divine control.

Nor could there be any timeless causal relation between one self and the rest of the universe which should be such that it would not be gravely misrepresented by saying that that self was a controlling God. For in such a relation, as we have just seen, neither of the terms could be discriminated as being the cause and not the effect. And any acquiescence by that self in the possession of certain qualities by particular things would depend on the existence of those things with those qualities, and would have therefore no analogy with a process of control in which a quality occurs because it has been previously and independently willed. It would therefore gravely misrepresent the position of such a self to speak of the self as a controlling God.

496. So far the effect of the unreality of time on the doctrines of a creating God, and of a controlling God, has been similar. But when we come to the possibility of the appearance of a self as a God being a *phenomenon bene fundatum,* the position is different. We saw that the possibility of any self appearing as a creative God was destroyed by the fact that, *sub specie temporis,* the first moments of the existence of all selves must be taken as simultaneous, and that therefore no self could appear as prior to any other self, and consequently could not appear as his cause. But for a self to appear as a controlling God, it would only be necessary that his volitions should appear as earlier than the events which fulfilled them. And this can happen, and does happen.

From the point of view of our present experience there are persons whose volitions are viewed as being the causes of events. And, from this point of view, we may say that some of them influence events more than others—that Napoleon, for example, influenced the history of Europe more than one of his grenadiers influenced it. Now it is not impossible that there may be a person of whom, from this standpoint, we might say that his influence was so great that it affected the whole course of the whole universe, as much as, or more than, Napoleon's influence affected the action of the French army at Austerlitz. And if a person of this sort were also good, he would be a person who, from the standpoint of which we speak, would be regarded as a controlling God.

I see no reason why there should not be such a person—a person who was not a God, but who, *sub specie temporis* and from the standpoint of our present experience, appeared as a controlling God. And such an appearance would be a *phenomenon bene fundatum.* The statement that there was a God would not be true, but it would have as close a relation to the truth as the statements that there are mountains in Switzerland, and that thunder follows lightning.

497. Would such a belief as this be of much religious value? This is a difficult question. Such a being would be really a person, and really good. But he would not really be God, for his supreme power would not be real but apparent. It might be said that if we regard him as supremely powerful in the same way in which we regard Napoleon as more powerful than the grenadier, we have got all that is practically wanted. But it may be doubted whether a religious emotion does not require to be based on something which is believed to be absolutely true. My feelings of fear or admiration towards Napoleon would probably not be diminished by my conclusion, based on philosophical arguments, that he was not powerful in absolute reality, but only appeared to be so, as there appear to be mountains in Switzerland. But then these emotions are not religious. And would the emotions with which I might regard a person whom I believed to be really God remain at all the same if I came to believe that he only appeared to be God? But, whether this be the case or no, there is no doubt that the unique position in the universe occupied by such a being would render the question of his existence one of very great interest.

498. There seems, as I have said, no reason why there should not be such a person. But on the other hand there seems no reason why there should be such a person. The conclusions at which we have arrived in our theory of the nature of absolute

reality give us no such reason. And, further, our conclusions invalidate the usual, and the strongest, argument for the existence
of such a being.

The usual argument for the existence of a God who is at any
rate a controller, whether he is a creator or not, is the argument
from design. It is asserted that we can see in the universe such
marks of its fitness as means to carry out certain ends, that we
must suppose that it was constructed in order to carry out those
ends. Any other supposition, it is said, would be as wild as the
supposition that a watch had come into existence in any other
way than as the result of a volition to produce something which
would tell the time. In that case, it is said, the universe, as it
exists at present, must at any rate have been arranged, if not
created, by a person. To be able to do this, he must possess far
more power than any other person. And the character of the ends
to which we judge that the universe is a means involves that the
person who acts for such ends must be good. Then he must be
God.

I have put the argument in its ordinary form, in which it
reaches the conclusion that there is a controlling God. This conclusion we have already rejected. But the argument, in so far as
it is valid at all, could be used to support the view which we are
now considering—that there is a self who appears as God, and
whose appearance is a *phenomenon bene fundatum*.

I do not propose to consider here any general criticisms which
might be made on this argument[1], but only to discuss how far it
is affected by the conclusions which we have reached as to the
nature of absolute reality.

499. The argument from design has two stages, which attempt
to prove respectively that the universe must be controlled by
a person, and that the controlling person must be good. The first
asserts that there is some order and system in the universe. And
this by itself, it is urged, is a reason to suppose that the universe
is controlled by some person. For the universe might have been
a chaos, and that it should have been some sort of chaos is ante-
cedently more probable than that there should have been order
and system without a controlling person.

[1] I have considered them in *Some Dogmas of Religion*, Sections 196–207.

There seems, certainly, no reason to hold that a chaotic universe is an *ultimate* impossibility, such as the inequality of two things which are equal to the same thing. But we have seen that it is an *à priori* impossibility, though not ultimate. Its impossibility, that is, is not self-evident, but, as we saw in Sections 258–262, it is necessary, as a consequence of various propositions whose truth is self-evident, that every substance should be connected by general causal laws with some other substances. This involves that the universe is not completely chaotic. And thus it is necessary that the universe should be more or less of an ordered system, if it is to exist at all, and we have no more right to say that a God is required to keep the universe from being chaotic, than we should have to say that a God was required to keep the universe from being a non-universe.

500. The second stage of the argument asserts that the order and system of the universe form appropriate means to a good end, and infers from this that the person who controls the universe must be good. Now I think it can be proved—the question will be discussed in Book VII—that it is necessary, as a consequence of various propositions whose truth is self-evident, that the universe must be more good than evil, and that the good and evil are so distributed that, *sub specie temporis*, the universe improves as time goes on. If this is so, it will not be dependent on any control or guidance exercised by one of those selves on the others. In that case a result which is more good than evil cannot be used as a ground for inferring the control of the universe by a person whose will is directed towards the good.

Of course these considerations do not disprove the existence of a person who would appear as a controlling God. And the existence of such a person remains possible. But they do, as it seems to me, remove the only serious argument for holding that his existence is necessary or probable, since they account for the facts, which are held to indicate his existence, in a way which shows that they would be what they are now, even if he did not exist.

Thus—to sum up—there can be no being who is a God, or who is anything so resembling a God that the name would not be very deceptive. Nor can there be any being who could even appear, as a *phenomenon bene fundatum*, to be a creative God. But there might be a being who appeared, as a *phenomenon bene fundatum*, to be a controlling God. We have found no reason, however, to believe that such a person does exist. And we have found a reason for rejecting the contention that his existence would be required to account for what order and goodness we can discern in the universe.

501. We now pass to the second subject to be considered in this chapter. Do the results which we have reached as to the nature of the existent give us any light on the question whether men are immortal?

This question is not free from ambiguity. "Am I immortal?" may mean "Is there any future moment at which I shall cease to exist?" Or it may mean "Shall I have an endless existence in future time?" Affirmative answers to these questions are not contradictories, for they may both be false. If time is unreal there can be no future moment at which I shall cease to exist, and there can be no future in which I can have an endless existence. Now this, as we have seen, is actually the case, for time is unreal. And so, if immortality is taken in the first sense, we must affirm that all men are immortal. If, on the other hand, it is to be taken in the second sense, we must affirm that no man can be immortal.

In which sense shall we use the word? It is certainly desirable that it should only be used of something of great importance and significance. But, in whichever sense it is used, this would be the case. If we call ourselves immortal because we are timeless, and so cannot cease in time, we are using immortality of a quality which has the greatest importance and significance for our moral and religious life. For it is because of our timelessness that it is true that we shall see ourselves as we really are, in proportion as we are able to regard ourselves as eternal, and in proportion as we realize, vividly and continuously, that our existence does not end. To regard ourselves as substances which are, but which will cease to be, is erroneous. To regard ourselves as substances which are, and which do not cease to be, is correct. And this is very much[1].

---

[1] A view rather resembling this can, no doubt, be held, and rightly held, even by those who hold that time is real. I cannot express this better than by quoting

502. The position which we have reached here is much the same as Spinoza's. He held that all that is real is really time-less. And he held that this fact made death insignificant, and freed those who realized it from the fear of death. "The free man thinks of death least of all things, and his wisdom is a medita-tion, not of death, but of life."[1]

Spinoza, however, appears to have reached the conclusion that whatever exists is timeless by means of a confusion between a fact and a proposition about that fact. If propositions are to be accepted as separate realities, then the proposition "Waterloo is the scene of a battle" is timelessly true. And so is the pro-position "the date of the battle of Waterloo is 1815." But it does not follow from the timelessness of these propositions that the fact which they deal with—the battle itself—is also timeless. And yet this seems to be the only reason on which Spinoza relies in asserting the timelessness of the fact. This criticism, however, does not apply to our own view, since we reached the conclusion that all existence was timeless by a line of argument which involved no confusion between a fact and a proposition about it.

the words of Mr Russell (*Our Knowledge of the External World*, p. 166): "The contention that time is unreal and that the world of sense is illusory must, I think, be regarded as based upon fallacious reasoning. Nevertheless, there is some sense—easier to feel than to describe—in which time is an unimportant and superficial character of reality. Past and future must be acknowledged to be as real as the present, and a certain emancipation from slavery to time is essential to philosophical thought. The importance of time is practical rather than theoretical, rather in relation to our desires than in relation to truth. A truer image of the world, I think, is obtained by picturing things as entering into the stream of time from an eternal world outside than from a view which regards time as the devouring tyrant of all that is. Both in thought and in feeling, to realize the unimportance of time is the gate of wisdom. But unimportance is not unreality."

This seems to me profoundly true. But the importance of time will be still less, if, as I have maintained, nothing is really in time, and the temporal is merely an appearance. And, as the importance of time diminishes, so also diminishes the importance of the cessation of our lives in time.

[1] *Ethics*, IV, 67. At the point at which these words occur they have no relation to the eternity of existence, which Spinoza only reaches in the Fifth Book. But the eternity of existence, and the attitude towards reality which follows from the apprehension of that existence, give them a deeper and fuller meaning than they had previously. It seems probable, as Sir Frederick Pollock has suggested, that Spinoza, in placing them where he did, intended to foreshadow the later position which they express so perfectly.

503. Thus, if we did use the word immortality in the first sense we should be using it of something important enough. But I do not think that we should be justified in doing so. For I think that the second sense is that which is generally adopted, and which therefore philosophy ought to adopt. The timelessness which we have affirmed of selves must also be affirmed of every other substance—that is, of every combination of selves, and of every part of a self. Every nation, every bridge-party, every perception is as eternal as our selves. We should not, indeed, call them immortal, even if we adopted the first meaning of immortality, since they are not selves. But would it be in accordance with usage to call a self immortal on the ground of a timelessness which is shared with every event in its own life, and which would be shared with soap-bubbles, if soap-bubbles existed? It seems clear that this is not the common usage, and that what, in common usage, is meant by immortality is endless existence in future time.

If we are to say that no one is immortal unless he really exists endlessly in future time, we must say of course that no one can be immortal, since there is no time, and so no future time, and therefore no one can exist endlessly in the future. But there is another possibility. There is no time, but there is an appearance of time. And we may ask whether the nature of the eternal self is such that it will appear to persist through all future time, or is such that, like a bridge-party or a perception, it will appear to cease at some moment within future time. And if the answer should be that it will appear to persist through all future time, we may further ask whether this involves that it will appear to have an endless existence in future time.

These questions will be discussed in Books VI and VII. I shall there endeavour to show that the self will appear to persist through all future time, and that it will appear to have an endless existence in future time.

In consequence of this, I think we may properly say that the self is immortal. It is as true that it exists endlessly in future time as that it exists at the present. Neither sort of existence is real, but there is an appearance of each which is a *phenomenon bene fundatum*. And I think that it would be generally admitted that if it is as true that I shall live endlessly in the future as that my body will die at some future time, then I may properly be called immortal.

504. We have now completed all that we are in a position to say about the general structure of the existent as it is in absolute reality—with the exception of certain considerations relating to value, which will be discussed in Book VII. Different as is the nature of absolute reality from that of present experience, we have seen that it is not impossible to form a fairly adequate picture of it by means of the materials given to us in present experience. I know what it is to love others, and to reverence myself for loving them, to regard the actions of myself and my friends with complacency, and to regard the friends of my friends with affection. And I know what it is to acquiesce in what I cognize as existing. If I imagine that this should comprise the whole content of my mind, and that my knowledge of my friends should be as direct as my knowledge of myself, I should be contemplating in imagination a state which resembled absolute reality in many ways—though it would not be, like absolute reality, timeless, nor, like absolute reality, differentiated into perceptions with perceptions as their parts in an infinite series.

BOOK VI

ERROR

CHAPTER XLIV

ERROR

505. It is obvious that, if the conclusions which we have reached in the last Book are correct, the real nature of the universe differs very much from the nature which it appears *primâ facie* to possess. In the present Book we must consider whether it is possible that such an appearance is compatible with such a reality.

The difference between the apparent and real natures may be brought under five heads. In the first place, the apparent nature of the universe includes the fact that it contains matter and sensa, while, according to our view, nothing exists but spirit.

In the second place, according to the *primâ facie* view, I perceive myself[1], I perceive parts of myself, and I also perceive sensa. And, according to this view, I perceive nothing else. All my knowledge of anything else existent is based on my perception, not of that thing, but of sensa connected with it in a certain way. But in reality, if our theory is right, I do perceive other selves, and their parts, and I do not perceive sensa.

In the third place, the content of myself comprises, according to the *primâ facie* view, not only perceptions but also other awarenesses, judgments, assumptions, and imagings, while, according to our theory, it comprises in reality nothing but perceptions.

In the fourth place, from the fact that the self contains nothing but perceptions, it follows that all volitions and emotions must be perceptions, which *primâ facie* is not the case.

In the fifth place, according to the *primâ facie* view, all that I perceive, and much, if not all, of what I believe to exist without

---

[1] It might possibly be doubted whether it is part of the *primâ facie* view that a man perceives himself. But, even if it is not part of the *primâ facie* view, it does not involve a departure from that view. The perception of a self by itself is not contrary to what appears at first sight to be the nature of the universe, while the perception of one self by another is contrary to that appearance.

perceiving it, exists in time. But according to our view, nothing does really exist in time.

Of these five differences from the *primâ facie* point of view, the first and the fifth are found in many philosophies. The second is found explicitly in Leibniz[1], and is perhaps implied in Hegel. I do not know that the third and fourth are found explicitly anywhere, though it might perhaps be argued that they are implied both in Leibniz and in Hegel.

506. We have to consider whether this appearance is compatible with the real nature of the universe being what we have determined it to be. That there is such an appearance is beyond doubt. And, if this should prove to be inconsistent with our conclusions, it would be necessary to abandon the latter. Nor is this all. If the compatibility of such an appearance with such a reality should prove to involve, not an absolute impossibility, but something which was very highly improbable, we ought perhaps to suspend our belief in our conclusions, or even abandon them altogether. Our arguments in the first four Books professed, indeed, to give absolute demonstration. But, in the face of a very improbable conclusion, we ought not to ignore the possibility of some error which has passed undetected. And the argument of the fifth Book, as we saw, makes no claim to be an absolute demonstration.

But, on the other hand, we must be careful not to doubt our conclusions for inadequate reasons. It is not, for example, right to doubt them because they surprise us very greatly, or because they appear to us paradoxical, or because they are incompatible with what had previously been regarded as certain. Nor, again, must we regard a conclusion with suspicion only because it makes the real state of the universe very different from its *primâ facie* state. The presumption, no doubt, is for the reality of what appears

[1] It might be objected that, since Leibniz holds that our knowledge of other substances is not *caused* by the substances we know, he should rather be said to deny that there is perception at all. But Leibniz holds that we have cognitions of other substances, and that, while to some extent they are confused and false, to some extent they are true knowledge. And the inferred nature of these cognitions is, at any rate in many cases, that of perceptions and not of judgments. I think therefore that they are properly called perceptions of other substances. And all these other substances, according to Leibniz, are either selves or parts of selves.

to be real, and we must not hold that the reality is different from the appearance, until the difference has been proved. But there is no reason to doubt a line of argument because it reaches such a conclusion.

Indeed, it is impossible both to avoid paradox and to treat the universe as being what it appears to be, because the views which are generally accepted as true, are views which treat the universe as being very different from what it *primâ facie* appears to be. The ordinary believer in matter, for example, holds matter to be very different from what it appears to be. And when an attempt is made to hold that matter is really what it appears to be[1], it has to reject so unsparingly the conclusions which are generally accepted, that the theory, whether true or false, is as paradoxical as Hume and far more paradoxical than either Leibniz or Kant.

507. We have said that there are five divergences which, according to our theory, exist between the reality and the appearance. As to the first of them, which arises from the fact that nothing exists but spirit, no difficulty will arise from rejecting the existence of matter. It was pointed out in Chapter XXXIV that matter is not perceived, but only inferred. And therefore, if matter does not exist, all that is involved is that certain judgments, which are very generally made, and which all men have a tendency to make, will be erroneous. Now many judgments which are very generally made, and which all men have a tendency to make, are admittedly erroneous. And the fact that our conclusions about matter involve that some more judgments of this sort are also erroneous, will be no ground for doubting the arguments which lead to those conclusions[2].

508. But the fact that nothing exists but spirit, involves the non-existence, not only of matter, but of sensa. And here, and also with the remaining points of divergence, it does not seem possible to avoid the conclusion that there is something erroneous in perception, that is, that some percepta are perceived as having characteristics which they do not possess. It is true that nothing

---

[1] As by Mr Russell in *Our Knowledge of the External World.*

[2] It is true, no doubt, that on our theory there are no judgments, and therefore no erroneous judgments. But this does not arise from our denial of matter, but from our theory of the nature of spirit, and is considered in the following section.

can be logically incompatible with a judgment except another judgment. Anything incompatible, for example, with the judgment "selves contain no parts but perceptions," must be another judgment such as "selves contain some parts which are assumptions and not perceptions." And thus it might seem that here too, as with matter, our theory only involves the falsity of certain judgments which are frequently made.

But this would not be correct. Even if our theory admitted that there were any judgments at all, it would be difficult to suppose that the error here was in all cases in the judgment. My judgments that I do perceive sensa, that I perceive judgments and assumptions as occurring in me, and that I perceive percepta as occurring in time, seem based with so much certainty on the perceptions that the error could hardly fall in the judgments. And this is still more the case with the wider and more negative judgment—equally incompatible with our theory—that I perceive things which, whatever they are, are not perceptions. And if the error is not in the judgment about the perception, it must be in the perception itself.

But, whether this is so or not, it is quite certain that on our theory the error will have to be within perception, because there is nowhere else for it to be. It cannot be in judgments, because there are really no judgments. Whatever falls within the mind— and there can be error nowhere else—is either a perception or a group of perceptions. And so the error must be in some way in perception. In the last paragraph we spoke of my judgment that I perceive assumptions as occurring in me. But, if our theory is right, this apparent judgment is not really a judgment, and is really a perception. And so, whether the error is to be found in what appears as judgment, or in what appears as perception, it must really be found somewhere in perception.

We must therefore find a theory which allows for both knowledge and error in perception. For if there is any knowledge it can be nowhere but in perception, and if there is any error it also can be nowhere else. And there is certainly both knowledge and error.

509. It is tempting to say that the proposition that there is no knowledge contradicts itself. But this is not strictly correct.

For the proposition that no knowledge exists is not false unless some true proposition is asserted by someone. And the proposition "no knowledge exists" does not involve the proposition "it is asserted that no knowledge exists," nor does it involve the proposition that any proposition is asserted. Since it does not involve the assertion of any proposition, it does not involve the assertion of any true proposition, and therefore does not involve its own falsity.

But it is as impossible to maintain it as if it were self-contradictory. For, although its truth would not involve its falsehood, its assertion involves its falsehood. If I assert that no knowledge exists then I am maintaining that my belief in the proposition "no knowledge exists" is false. For if it were true, it would be knowledge. At the same time, since I am asserting the proposition, I am maintaining it to be true. There is therefore a contradiction involved in the assertion of the proposition, although not in the proposition itself. And any system which involves the proposition cannot be believed without a contradiction[1].

510. In the second place, no theory can be true which does not allow for the existence of error. The proposition that there is no error is not self-contradictory, nor does its assertion involve a contradiction. For no contradiction is involved either in the proposition itself or in the assertion of it. But it must be false, since it is incompatible with facts which cannot be denied.

I have believed on various occasions that other people have fallen into error, or that I myself have fallen into it in the past. And I have good reason to believe that other people have, on various occasions, entertained similar beliefs. Now if a single belief of this sort has been held by anyone, it is certain that error exists. Such a belief must be either true or false. If it is true, then the error, which the belief asserts to exist, does exist. If it is false, it is itself an error.

[1] I am taking the words "no knowledge exists" in the sense in which they would probably be used by a sceptic—*i.e.* that the characteristics of being knowledge and of being existent are such that they can never be possessed by the same thing. In this sense the words have a meaning. They express a proposition which, as said above, is proved to be false by the fact that it is asserted. If, on the other hand, they were taken in the sense that, in point of fact, no existent thing *was* knowledge, they would be unmeaning. Cf. my article "Propositions applicable to themselves," *Mind*, 1923, p. 462.

It might be answered that the universe is, in many respects, not what it appears to be, and that perhaps what appears to be error is not really so. But then, behind Y, which appears to be, but is not, the belief that Z is an error, there must be a mental state, X, which is the appearance that Y is such a belief. And then X will be an error. And if it is said that it may only appear to be an appearance that Y is a belief that Z is an error, then there must be another mental state, W, which is the appearance of X as what it is not. And then W will be an error. Thus the attempt to remove one error produces another through an infinite series. For whenever we say that anything appears to be what it is not, we assert the existence of a mental state, which is the appearance, and which is erroneous.

511. We may contrast the way in which the denial of error involves a vicious infinite with the way in which the denial of time does not involve one. If, in present experience, I deny of Z, which appears in time, that it is really in time, it is true, no doubt, that my denial Y, if I perceive it, will appear to me as being in time. And if I deny Y to be in time, then this denial, X, if I perceive it, will also appear to me as being in time, and so on infinitely. But no vicious infinite will arise. For my denial that Z is really in time does not involve that Y, that denial, is really in time, and therefore it does not require, if all existence in time is to be denied, a fresh step to deal with Y. Thus no infinite series is necessary, and, if in fact it did occur, it would not be vicious. For it would not be necessary that we should reach the end of the endless series before we were entitled to assert that Z was not really in time.

But with error it is different. For here, if we get rid of the *primâ facie* error in Z, by asserting that it is not really erroneous, though it appears to be so, that implies that Z has been contemplated, and that its contemplation, Y, is erroneous. Thus in denying the reality of one error, we are asserting the reality of another, whose reality can only be denied by asserting the reality of a third. If we stop anywhere, we stop with a real error, and our attempt to get rid of error has failed. And if we go on without end, the infinite series is vicious, for the proposition for whose sake it was begun—the proposition that

there is no real error—will not be justified as long as we have
not reached the end of that endless series.

Thus we must allow both for knowledge and for error. About
the reality of knowledge there is, on our theory, no difficulty.
For every self has perceptions, and perceptions, if correct, are
knowledge. Thus knowledge can exist.

512. Of course this does not dispose of all the problems which
will, on our theory, arise about knowledge. There is much reality
which appears to us in the form, not of perceptions, but of judgments. And some of this must be knowledge. Everything which
appears as a judgment cannot be false. There are judgments
which assert what is self-evidently true. And it is obvious that
it is suicidal to assert that all which appears to us as judgments
is false. For that very assertion appears to us as a judgment.

Yet, if our theory is correct, what appears as judgment is
not really judgment, but perception. And we shall have to
consider how it is possible that knowledge, which is really not
judgment but perception, can appear as judgment, and yet be
true. This will be discussed in Chapter LIV. At present, all that
we are considering is the general requirement that some knowledge should exist, which is satisfied by the fact that perception
does exist.

513. But what of the other requirement—that error must
also exist? All cognition must be in perception—there is nowhere
else for it to be. Then error, like other forms of cognition, must
be in perception. Now can perception be erroneous? Is it not an
essential and self-evident characteristic of perception that it
cannot be erroneous—that, when I perceive A as being X, then
A must exist and be X? If we remove this characteristic from
anything, do we not thereby declare that it is not perception?
And so if, as we have seen, it is essential that error should exist,
would it not follow, contrary to our theory, that selves have
some parts which are not perceptions?

It is not, however, universally accepted as certain that, when
I perceive A as being X, then A must exist *and* be X[1]. It is not

---

[1] Dr Moore, in his *Philosophical Studies* (pp. 245–247), says he is not sure that
this is unquestionable, though he says "I am not sure that I may not be talking
sheer nonsense in suggesting that it can be questioned."

doubted, as far as I know, that, when I perceive A as being X, A must exist, but it is thought possible that it should not be X. If we could accept this view, the difficulty would disappear.

I do not think, however, that this view can be accepted. When I contemplate any case in which I perceive any perceptum A as having a quality X, it seems to me self-evident, not only that A then exists, but that it then has the quality X[1]. And, when in general I contemplate what is the nature of perception, and what is the nature of the relation of a perception to its perceptum, it seems to me self-evident that such a self-evident correctness belongs to all perceptions.

I may, of course, be mistaken in this view, but I do not think that I am. And my contention may be indirectly supported by considering what would happen if perceptions had not this self-evident correctness. For the correctness of a perception can certainly not be proved, and if it is not self-evident we have no right to believe that any perception of anything as having any particular quality is correct at all. We should therefore know nothing about the percepta. Nor could we say that, although we were not entitled to say that A was X, we were at any rate sure that our perception of A was a perception of it as being X. For if the perception of A gives knowledge at all, it can only be about A. It cannot give knowledge of itself—the perception of A. That could only be given by another perception—the perception of the perception of A, which would have the perception of A as its perceptum. And we could know no more about this perceptum than about any other.

We should therefore have no right to believe any existential proposition. For, although many existential propositions are not about percepta, yet even in these cases our only justification for believing in the proposition is that its truth is implied in something we perceive. One exception, indeed, there would be. For it is not doubted that, when A is perceived, it must exist. The person, therefore, who perceived A, could assert "this exists,'

---

[1] The self-evident certainty only relates to the perceptum, and not to anything else. If I perceive a sensum as being round and yellow, then it is self-evidently round and yellow. But the conclusion that there is, besides the perceptum, a round and yellow piece of matter, however naturally it springs from the perception, is not self-evidently true, even if it is true at all.

when "this" denoted the A he was then perceiving. But he would know nothing else about A—not even that he was perceiving it. Nor would he know that he himself was an existent self, or that anything had the quality of being one of his perceptions. Thus the denial of the self-evident correctness of perceptions would reduce us to almost complete scepticism—a result which indirectly supports the view that they are self-evidently correct.

514. It would seem then that all perceptions must be correct. But we see that there is one limit on the self-evident correctness of the perceptions which occur in our present experience. And this is, or appears as, a limit of duration. What is self-evident is, as we said above, that *when* I perceive A as being X, then A must exist and be X. But it is not self-evident that A exists, or is X, when I am not perceiving it. A *may* exist and be X at some time at which I do not perceive it. And it is in some cases a legitimate inference that, since it exists and is X when I am perceiving it, it exists and is X at some other time. But all that is immediately and self-evidently certain is that it exists and is X when I am perceiving it.

But "when I am perceiving it" requires to be stated more definitely. It does not mean here "at the moment at which I perceive it." For all presents are what are sometimes called specious presents. And so at any moment, p, I perceive not only what is happening at that moment, but also what happens at the earlier moments between m and p. Thus if A existed and was X at the moment o, I may perceive it at the moment p, when perhaps it has ceased to exist or to be X. And thus A need not exist or be X at the moment at which I perceive it. What is meant is that, if at the moment p I perceive A as X, then it is self-evidently certain that A exists and is X at some moment or moments which I am then perceiving as present.

515. Now it seems to me that two propositions are self-evidently certain. The first of these is that all perceptions of anything as in time must be subject to this limitation on their self-evident correctness. The second is that there can be no limitation to the self-evident correctness of any perception other than this[1].

---

[1] From this it follows that a perception which did not perceive its perceptum as in time would be self-evidently correct without any limitation. (Cp. Chap. XLVII, p. 232, and Chap. XLIX, p. 255.)

The limitation, then, of the self-evident correctness of a perception is a limitation to a period in time measured from a point in time. But nothing is really in time at all. And so the limitation must be restated "A exists and is X at a point or points in the C series which appear to be present at the point in the C series at which the perception exists."

The limitation of the self-evident correctness is thus to a certain position in the C series. And therefore we cannot know what perception does guarantee until we know more precisely what the terms of the C series are, and what the generating relation of the series is. And it is possible that we may find that the answers to these questions are such as to make the limitation of the correctness into a qualification of the correctness, in such a manner as to allow for perceptions being in some degree erroneous, while allowing them at the same time to give in some degree true knowledge. I shall endeavour to show in Chapter L[1] that this is actually the case. At present I confine myself to pointing out that, until the questions are answered, we cannot be certain that erroneous perception is impossible.

516. If it should thus prove possible for perception to be erroneous, how great must that error be? In the first place, with regard to that perception which at present appears to us as perception, some error will be found in all of it, since in none of my present perceptions do I perceive things as they really are. In reality, nothing exists except selves, parts of selves and groups of selves. Now the greater part of my present perceptions are of things as being sensa, which are neither selves, parts of selves, nor groups of selves. In my present experience I never perceive anything as a group of selves. I perceive nothing but myself as a self, and nothing as parts of selves except my own parts. And even my perceptions of myself and my parts are more or less erroneous. I perceive parts of myself as judgments, assumptions, ungratified volitions, and so on, which is erroneous. Finally, I perceive everything, sensa, parts of myself, and myself, as in time, while in reality none of them are in time.

Of these errors, it is only the last which is common to all perceptions which appear as perceptions. For when I perceive

---

[1] pp. 256–257.

myself, there need not, it would seem, be any other error but
this. But this error is always present[1].

It should be noticed that a perception which is erroneous is
not necessarily entirely erroneous. I may perceive A as having
the qualities Y and Z, and it may really have Y, though it does
not really have Z.

517. In the second place, there are the perceptions which
appear to us in the form of judgments. Here some are certainly
erroneous—for, as we saw above, some of our present judgments
must be false. But, on the other hand, as we also saw, it is
certain that some of our present judgments are true, and an indefinite number of them may be so. Thus those perceptions
which appear to us as judgments appear to be less universally
infected with error than those which appear to us as perceptions.
But although there can be, in some cases, less error *in* them,
there is always more error *about* them. For those perceptions
which appear to us as perceptions, appear in that respect what
they really are, while those which appear, not as perceptions
but as judgments, appear in that respect what they really are
not.

Erroneous judgments are called false, but it does not seem
convenient to apply the terms true and false to perceptions.
Thus we want two terms, one to include perceptions which give
knowledge, together with true judgments, and the other to include those perceptions which are erroneous, together with false
judgments. I propose to use the terms correct and erroneous.
I do not know of any convenient term to designate the correctness of perceptions, as distinguished from the truth of judgments,
but the erroneousness of perceptions may be distinguished as
misperception.

A misperception is, of course, as real as a correct perception,
in the same way as a false judgment is as real as a true judgment.
And if it is further asked whether what is misperceived is real,
we must realize that the question is ambiguous. When I misperceive anything, I perceive something which is real, as having
characteristics which it really has not. I perceive a self as being

---

[1] The view that this error is never absent in apparent perceptions will be defended
against some objections in the next chapter, pp. 208–211.

in time, which it is not, or a group of spiritual substances as being a chain, and so as extended in space, which they are not. In one sense of the word, the misperception is of the self or group of selves, and in that sense what is misperceived is real. But in another sense what is misperceived is in one case the temporal duration and in the other case the spatial extension, and in this sense what is misperceived is not real.

518. The theory that there is misperception has not, as far as I know, been definitely put forward by any philosophers in the past, excepting by Leibniz, who holds that what we perceive as material is really spiritual, the misperception being due to the confusion which belongs to each created percipient. But we may briefly consider whether any other of the great philosophers ought logically to have accepted misperception.

In the majority of cases, I think, there was no absolute necessity for them to do so. They all, indeed, as was pointed out earlier in the chapter, depart more or less from the *primâ facie* view of reality, and have to recognize therefore that, besides the errors which exist in particular persons from particular causes, there are others which naturally arise in the whole human race, and which are only to be extirpated by philosophical reflection, if, indeed, they can be extirpated then. But in most cases it would be logically possible to explain these errors simply as false judgments, which various causes rendered natural, general, and very persistent, and so to avoid the necessity of admitting misperception.

But, it may be said, how about the systems which deny the reality of time? Do we not perceive things in time, and if they are not really in time does not this mean that we misperceive them? Still, it might be said it is only a judgment that we do perceive them in time, and the error may be only in the judgment. This answer could not be given on our theory, which rejects the existence of judgments. But if the existence of judgments is admitted, there would be no contradiction in taking this view, though, as was said above (p. 196), it seems contrary to the weight of the evidence.

Kant's denial of the reality of time, however, presents special features. For he says that time is not a category, but a form of intuition. And if that is so, it seems clear that we do not only judge the manifold of intuition to be in time, but also perceive it as being in time (using "perceive" in the sense in which we have been using it). Now, according to Kant, nothing is really in time, and thus that which we perceive as being in time is misperceived[1].

Hegel's system, again, involves an enormous difference between the real world and the world of appearance, not merely as to time, but as to many other points. Does it necessarily involve misperception, or could the difference be explained by erroneous judgments? It is difficult to say what answer Hegel himself would have given to this question, since his exclusive devotion to ontology leaves his epistemology very obscure. But it certainly seems as if the difference could not be explained entirely by errors of judgment, and would require misperception.

519. There are, then, misperception and error. Can we say anything about their cause? Why does what we perceive appear to us to be so different from what it really is?

The error in question can be reduced, as we have seen, under a few general heads—those enumerated on p. 193. There is therefore a strong presumption that, if causes can be found for the error, they will also be limited in number. For it is not probable that similar errors should be produced by dissimilar causes in different cases, and still less probable that different causes should work so uniformly in detail as to produce a system showing as much order as we find in the world as it appears to us. This improbability does not afford a ground for asserting beforehand that there can only be a few causes, but it does make it worth

---

[1] It may be objected that the manifold of intuition only exists in phenomena, and that phenomena are in time, so that perception in time is correct. But this is a confusion. To say that the phenomenal world is in time only means that the world as we observe it is in time. And this only means that we observe the world in time. But the world itself, which we observe, exists independently of our observation of it, and this, according to Kant, cannot be in time, because time is only, for Kant, a way in which we observe the timeless. This is perhaps clearest when we consider introspection. According to Kant, I can observe, among other things, my own perceptions. Now the perceptions I observe can clearly not be reduced to the further perceptions by which I observe them, since this would start a vicious infinite series. The observed perceptions must really exist, independent of our observations of them. And so, for Kant, they cannot be in time, though they observe their objects as in time, and may in their turn be observed as in time.

while to begin by looking for a few, and, first of all, to consider whether a single cause can be found for all the error. It is true that our erroneous cognition is highly differentiated, and that differentiation in what is determined can only be accounted for by differentiation in the determinant. But it might turn out that the element of differentiation was entirely supplied by the true nature of what is perceived (a nature which we have already found to be infinitely differentiated), and that the element of error was introduced by a single cause, which produces a differentiated result by acting uniformly on the differentiated. We shall begin, therefore, by enquiring whether the error can be accounted for by a single cause.

520. It is well to remind ourselves at this point that, wherever the cause of the error can be found, the error itself can only be found in one place—in the observing subject. It is well to remind ourselves of this, because some philosophers have found it very easy to forget it. It would be universally admitted that knowledge is only in the mind of the observer, though the thing known may be outside that mind, and generally is outside it. And when an error is confined to one person, or to certain people, and when it is one which tends, until removed, to bring disorder and confusion into the experience of the persons who hold it, it would be admitted that the error is only in the minds of those persons. But when the error is one which is believed to be shared by all thinking beings in the universe, or, at any rate, by all human beings, and when the effects of the error are not such as to prevent the formation of an orderly and uniform system of experience, it often happens that the error is called phenomenal truth. In this there are considerable advantages, since such errors require to be distinguished from the other errors previously mentioned. But if we are to speak of phenomenal truth it is essential to remember that what is phenomenally true is not really true, but really false. This, however, is often forgotten. It is supposed that what is said to be phenomenally true is really true in some sense or another. But since a belief which is really true has an object outside itself, it is supposed that what is phenomenally true must have an object outside itself. And then the content of the error—that which is erroneously asserted—is hypostatized and set up as an object which has some sort of bastard reality, though not real reality. This, for example, is the case in Kant's philosophy, with the phenomenal objects in space and time. He does not hold that these exist independently in their own right, as the observing selves and the things in themselves do. Yet it seems that he holds that there is a phenomenal table, which is not in the minds of those who observe it, and which is such that the same table can be observed by two different people.

But such a view as this is untenable. If nothing really exists in space, then no tables exist, and any perception or judgment which perceives anything as a table, or asserts that a table exists, is erroneous—however inevitable and however useful the error may be. There are no tables, but only erroneous perceptions of tables, and erroneous judgments that tables exist. And these perceptions and judgments, like all other perceptions and judgments, are only within the observing self. A phenomenal object of phenomenally true cognitions is nothing but an objectified error detached from the self who has the erroneous cognition. And this is impossible. But although the error is only within the erring subject, the cause of the error—or, at least, a part-cause of the error—may be outside the subject.

CHAPTER XLV

ERROR AND THE C SERIES

521. It is clear that, if we are to find a single cause for error, we must find it in close connection with the appearance of time, and with the reality on which that series is based. For that appearance, and the reality behind it, must in any case play a very important part in the explanation of error. We are, as we have seen, faced with the difficulty that all error must be in perception, since we have no mental states except perceptions, and yet that the correctness of perceptions is self-evidently certain. There was only one possibility of meeting this. We found[1] that whenever anything is perceived as being in time, there is one limit to its self-evident correctness, which appears as a limit of duration. If I perceive A as being X, all that is certain is that A exists, and is X, when I am perceiving it. And if what appears as a limitation of the correctness in time should turn out to be really a qualification of the correctness, then erroneous perception might turn out to be possible. Thus the nature of that series, the C series, which appears as the time-series, must in any case be essential to the explanation of error, even if it should not prove to be the whole cause of error.

522. But is the appearance of time to be found in all the erroneous perceptions of our present experience? It would appear at first sight that it is not. For what appear as judgments are really perceptions, and we can certainly judge things to be timeless. And errors will be found among these judgments as well as among other cognitions. This question must be postponed for the present. We shall see later[2] that there is reason to think that the perceptions which appear as such judgments are in reality perceptions which perceive their percepta as being in time.

523. But, again, is it the case that even all those perceptions in present experience, which appear as being perceptions, perceive

---

[1] Chap. XLIV, pp. 201–202.　　　　　[2] Chap. LIV.

their percepta as in time? It was mentioned in Chap. XLIV, p. 203 foot-note, that this view is not universally accepted, and we must now proceed to discuss the objections which have been made to it.

These objections, I think, relate exclusively to certain mystical perceptions of God, of the universe, or of one's own self, which are said to be experienced by some people at certain times. With regard to ordinary perception, there is, as far as I know, a general agreement that it always perceives things as being in time. In some cases the characteristic of being in time may not be very prominent in the perceptum, but if the attention of the observer is directed to the point, so that he asks himself whether he perceives this perceptum as in time, the answer, it is agreed, would always be affirmative.

But how about perceptions of God or of the universe? And, in the first place, are there any such perceptions at all in our present experience? It has been asserted by various people that they have had such perceptions. I doubt, however, whether this is a correct account of what they have experienced. No doubt they have experienced something, and, if our theory of the nature of spirit is correct, that experience must really be perceptions. But when they say that they have perceived God, or the universe, they mean that they have had perceptions of them, as distinct from judgments about them, or imagings of them. In the language we have adopted, they are speaking not only of perceptions, but of perceptions which are apparent perceptions.

In such cases, I am inclined to think, a judgment has been mistaken for a perception[1]. It is not difficult to mistake an immediate judgment that something exists for a perception of that thing. Nor is it difficult to make the mistake, even when the judgment is not immediate, but has been based on reasons, provided that the judgment is firmly held, and is one of sufficient interest to excite a strong emotion in the person who makes it[2].

---

[1] For the sake of brevity, I speak of judgments and perceptions, instead of perceptions appearing as judgments, and perceptions appearing as perceptions.

[2] Spinoza speaks of a Third Knowledge, which is intuitive in its character, and which, starting from knowledge of God's nature, proceeds to trace how the existence and nature of each mode flow from the existence and nature of God. This knowledge, he tells us, can be attained by men, and it would seem that he thought that he had himself attained it. But it is very difficult to see how he

To the possibility of these mistakes, and, indeed, to the great difficulty of avoiding them, I believe that most people would bear witness who have any experience of mystical states of mind, and who have the power and the resolution to analyze the states they experience. And it does not seem improbable that the cases in which people have supposed themselves to have perceptions of God, or of the universe, are cases in which a judgment has in this manner been mistaken for a perception.

But we cannot absolutely exclude the possibility that there may occur, in present experience, a perception of God or of the universe. And we have seen that selves do have perceptions of themselves. The universe and the self are eternal, and so is God, should God exist. Would it be possible that, in our present system of perception, we should perceive them as eternal?

I do not think that this is possible, and, in the cases in which it is supposed to occur, I believe that, where there is a perception at all, there is a perception of the object as in time, together with a judgment that the object is eternal, and that these two are confused together, and mistaken for a perception of the object as eternal[1].

524. The view that the perception is always of the object as being in time can be proved, I think, by the following considerations. When a man perceives an object—himself or anything else—he does not, of course, always perceive his perception of the object, but he frequently does so, and can always do so if his attention is called to it. And this is as much the case with perceptions which are supposed to perceive their objects as being eternal, as with any other perceptions.

could have regarded such knowledge as attainable within the limitations of a single life. I am inclined to think that these difficulties can only be removed by supposing that Spinoza made the mistake referred to in the text, and mistook certain philosophical judgments as to the relation of God and the modes, together with the perception of some particular mode, for a perception of the particular relation which existed between God and the mode in question. For, if the relation had really been perceived, then the perception of God and the mode in that relation would have been Third Knowledge. (I owe the first suggestion of this view to a remark of my friend and pupil, Mr J. C. Chatterji.)

[1] This mistake, it will be seen, is closely analogous to the mistake mentioned on p. 209. How it is possible, on our theory, to have both a perception of anything as in time, and also a judgment that it is eternal, will be discussed in Chap. LIV, p. 306.

Now a perception, when perceived, is always perceived as being in time. And when we perceive an object, and also its perception, and perceive the latter as in time, then the former also—the original object—must be perceived as in time. For with all apparent perception—and it is only apparent perception of which we are now speaking—the object and its perception will be perceived—if the perception is perceived at all—as in a time-relation to one another (cp. Chap. XLIV, p. 201).

The view that such perceptions must appear as being in time has been denied. It has been said by persons who have believed themselves to have perceived God, or the universe, or themselves, as eternal, that the perception itself appears as eternal, and not as temporal. But the supporters of this view would not deny—it would be impossible to deny of anything in our present system of experience—that there was a point in the time-series at which it might be said of any such perception that it had not begun, and another point at which it might be said of it that it had ceased. And then they are inconsistent in denying it to appear as in time. Whatever is between two points in the time-series, and not beyond either of them, is itself in time.

525. The reasons why such perceptions—or judgments supposed to be perceptions—have been supposed to be timeless, are, I think, two. In the first place, it has been held (erroneously, as I have maintained above) that what is perceived in such perceptions is timeless, and then a confusion has arisen between the characteristics of what is perceived and the characteristics of the perception—a confusion by no means uncommon. In the second place, such states are usually states of high excitement, and always of intense contemplation, and, while they continue, the lapse of time is not noticed, nor, when they are over, is it always easy to judge how much time has elapsed. And so it is supposed that the state is not temporal at all.

We are therefore, I submit, justified in concluding that all perceptions which appear to us as being perceptions, misperceive their percepta as being in time, and that therefore a solution of the problem of error which depended on time would apply all such apparent perceptions. The question whether the same would be true of perceptions which are not apparent perceptions must, as was said above, be postponed for the present. When we consider it, we shall find reasons for answering it in the affirmative.

526. But if we can explain error by means of time, it can only be if we are able to determine what is the nature of that reality which we misperceive as the time-series. Now when anything is perceived as in time, a plurality of existent states appear to follow one another in what we have called the *B* series—the series of earlier and later. To constitute such a series there is required a transitive asymmetrical relation, and a collection of terms such that, of any two of them, either the first is in this relation to the second, or the second is in this relation to the first. We may take for this purpose either the relation of "earlier than" or the relation of "later than." Taking the first, then the terms have to be such that, of any two of them, either the first is earlier than the second, or the second is earlier than the first[1].

Since nothing is really in time, there is really no *B* series. Things are misperceived as being in such a series, but they are not really in it. But, as was said in Chap. XXXIII, p. 30, the fact that things appear as being in this series forces us to conclude that they really do form another series. For the *B* series depends on the *A* series. We begin with the misperception of certain terms as being present, terms on one side of such a term being misperceived as future, and terms on the other side as past. If they did not thus appear as being in the *A* series, they could not appear as being in the *B* series. For the *B* series is temporal, and there can be no time without change, and change can only take place by the passage of a term from future to present, and from present to past—that is, by being in the *A* series.

But the misperception which gives us the *A* series clearly implies that the terms which are misperceived as forming it, do really form a series. When we misperceive one term as present, we misperceive those on one side of it as future, and those on

---

[1] Cp. Chap. XXXIII, p. 10. All events are not separate terms in this series, since two events can be simultaneous, in which case neither is earlier than the other. Each term in the series will consist of a group of events, simultaneous with one another, and not simultaneous with any event outside the group. Such a group is itself an event.

the other side as past, and, among future and past terms, those
which are further away from the present as further in the future
or the past[1]. But this misperception gives us no way of deciding
which terms are on the same side of any one term, and which
are on different sides, or which, of those which are on the same
side of a term, are further away from it. In order that the misperception should produce an *A* series, it must be a misperception
of something which is a series already, though not a time-series.
The fact that the terms are in such a series involves that each
term has a definite position on one side or the other of any given
term, and is either nearer to it or further from it than any other
term on the same side of it. And such a series can appear, when
misperceived in the way we have mentioned, as an *A* series, and,
in consequence of this, as a *B* series.

We decided in Chapter XXXIII to call this real series by the name
of the *C* series. The name suggests that it is some way posterior
to the *A* and *B* series. In a sense this is true, but only as respects
the order of our knowledge. In present experience we can never
perceive the *C* series as such. We only infer its existence from
the fact that we do misperceive things as in the *A* and *B* series.
But in the order of existence the *C* series does not depend on
the *A* and *B* series, which, on the contrary, depend on it. If the
misperception did not arise the series which we have called the
C series would still exist, while the *A* and *B* series would not
appear to exist.

It is clear that the relation which really connects any two
terms, and which appears as the relation of "earlier than," must,
like the relation of "earlier than," be transitive and asymmetrical.
The nature of this relation will be discussed in Chapter XLVIII.

527. When a self *G* perceives an object *H* (which will, of
course, be a self, or a part of a self, or a group of selves or parts
of selves), what different series in *G* and in *H* have to be taken
into account?

With regard to *A* and *B* series, there are no such series at all.
Nothing *is* past, present, or future. Nothing *is* earlier or later

---

[1] In present experience our misperceptions of terms as future or past are never
apparent perceptions. They are apparent judgments, or, in the case of some
misperceptions of terms as past, apparent memories.

than anything else. All that exists is the erroneous perceptions of H as being in those series. And those perceptions, of course, are in the percipient G, and not in the perceptum H.

From this there follows the important consequence that there are as many time-series as there are selves who perceive things as in time, since the only real series in the matter is the series of misperceptions in the percipient. It is possible that two or more time-series should have a certain resemblance or congruence which should enable us to speak of two moments, one in F's time and one in G's time, as being in a sense the same moment, and so to reach the conception of a general and objective time. And we shall see in Chap. LI, pp. 273–275, that this is actually the case. But, strictly speaking, no time can be common to two selves.

This result would also follow from any other theory which accepted time as unreal. It would follow, for example, from the theories of Spinoza, of Kant, and of Hegel. I do not think that any of them would have been willing to accept it. But, as was said in the last chapter, few philosophers have distinctly realized that, if cognition is not strictly correct, it must be partially erroneous, and that what is erroneous has no place outside the person who is in error.

528. The C series, however, are real. And the C series which appears to G as successive states of H is a series in H. If, for example, H appears to G as successively black, red, and white, then H has three states, which appear to G as being black, red, and white respectively, and which are terms in a series in which the term appearing as red is between the terms which appear as black and white. And this series is a C series. It is in H, but it is called a C series, because it is the series which G, and other selves observing H, perceive as a B series. (Of course, H, if it is a self, may be one of the selves who observe H.)

The C series, then, which is misperceived by G as the time-series in H, must be in H, and not in G. But this C series cannot account for the fact of misrepresentation. Since G perceives H, the fact that there is a C series in H can account for G perceiving the C series in H. But it could not account for his misperceiving the C series as a series in time, which it is not. Nor would it account for his misperceiving the terms of the

C series as material objects, sensa, judgments, assumptions, and so forth, because in reality none of them are anything of the sort.

529. But, on the other hand, it is possible that the C series in G should account for these misperceptions. For every such misperception is a term in the C series in G, since, when it is itself perceived by introspection, it is misperceived as being in time. And it is possible that the nature of the C series might turn out to be such that it would allow for misperceptions in spite of the apparent self-evident correctness of all perceptions. (The manner in which this could happen was explained in Chap. XLIV, p. 202.)

I believe that this is not only possible, but actual, and that we can account for G's misperception of H by the consideration of that series in G in which the misperceptions fall, and which is a C series. This view will be considered in the rest of the present Book. If it is justified, the explanation of all error will have been found in close connection with time, since that depends on the C series. But it is with time in the observer, and not time in the object observed. One of the errors to be explained is G's misperception of the C series of H as in time. But the explanation of that, and of all other errors, is to be found in connection with the C series of G, and with its misperception as temporal.

CHAPTER XLVI

THE *C* SERIES—CONDITIONS OF THE PROBLEM

530. We have now to consider in more detail the nature of the *C* series. We must consider, firstly, what nature is possessed in common by all the terms of the *C* series in any substance, and, secondly, what is the nature of the relation between the terms which, when the series appears as a *B* series, appears as the relation "earlier than." In the present chapter we shall enumerate the conditions which must be fulfilled by any theory of the *C* series, if that theory is to enable us to find in that series the explanation of error.

To begin with, whatever view we adopt about the nature of the terms of the *C* series must be consistent with the results already reached as to the nature of substance. The terms must be such as can be parts of a substance which is spiritual, and which is divided into parts within parts to infinity by determining correspondence. This is the *First* condition.

In the next place, the *Second* condition is that the terms of this series shall be such as shall allow both of correct and of erroneous cognitions. We have seen that some of the cognitions which occur in the time-series must be correct, and some erroneous. As the reality which appears in the time-series is really the *C* series, it is in the *C* series that both the correct and the erroneous cognitions must be found.

The *Third* condition is that our theory shall allow not only for the existence of some erroneous cognitions, but for the different sorts of erroneous cognition which do actually exist. We perceive certain objects as being matter, sensa, judgments, assumptions, and so forth. And these perceptions are erroneous, since nothing exists but spirit, and spirit has no content except perception. Our theory must be consistent with this.

The *Fourth* condition is that the *C* series shall be a series of one dimension, and that the relation which constitutes it shall be transitive and asymmetrical. For the relation of "earlier than" is transitive and asymmetrical, and a series of terms connected by it has only one dimension. And the real series which appears as the B series, and the relation which really connects its terms, must also have these characteristics.

The *Fifth* condition is that the series should have a number of terms sufficient to account for the number of terms in the time-series. There must therefore be at least as many terms in the C series as can be distinguished from each other in the B series. But there may be more. The C series must, by its definition, have a term for each term in the B series. And the B series has at least as many terms as we can distinguish in it, but may have more. For our observation may not be sufficiently minute to distinguish terms which nevertheless are there.

The C series, therefore, must have at least a large number of terms. And, it may have an infinite number of terms. For, as far as we have yet seen, it may be unbounded in one or both directions, or it may have no next terms, or it may have no simple terms—any of which possibilities would make its terms infinite in number. But it is also possible, as far as we have yet seen, that the time-series is bounded in both directions, and that it has simple terms which are next to each other. In that case the C series might have a finite number of terms. Nor is the infinite divisibility of all substances incompatible with this. For, as we have seen in Section 102, they need not be divisible in every dimension. And if the C series is (as will be shown in the next chapter) in a different dimension from those of determining correspondence, substances need not be infinitely divisible in the dimension of the C series.

The *Sixth* condition is that our theory must allow for the fact that, while in absolute reality my knowledge of any substance is differentiated into parts of parts to infinity, in present experience there is no such infinite differentiation of knowledge. Any substance is made up of parts of parts to infinity. And determining correspondence involves that, when I perceive a substance, I perceive all its parts in the determining correspondence system— my perceptions of the parts being parts of my perception of the whole. In absolute reality, then, all my perceptions will be differentiated into parts of parts to infinity. But this is certainly not the case in present experience. There, when I perceive anything, I may perhaps perceive some sets of its parts, but I certainly do not perceive an infinite number of such sets.

The *Seventh* condition is that we must be able to allow for the persistence and recurrence of certain contents in the time-series of our present experience. I suppose that it is not absolutely certain that any content which appears in time as the content of an apparent perception does persist or recur. With regard to what appears as perception of sensa, it is not certain that the same sensum is ever perceived as existing at several consecutive moments of time, so as to persist, or as existing at several separated moments of time, so as to recur. The events which are spoken of as the persistence or the recurrence of the same sensum in perception might possibly be explained—I do not say that they could be—as the perception of different but similar sensa. And even of a man's perception of himself, it might possibly be maintained that all that he perceived was the state of himself at the time of the perception, which would not be persistent or recurrent.

But there can be no doubt that various contents which appear as the contents of judgments and assumptions do persist and recur in the time-series. During the whole of an appreciable time I can keep my mind on the fact that Caesar died in Rome. And then I can pass to the fact that Lincoln died in Washington, and return to the fact that Caesar died in Rome. And thus the content of my judgment that Caesar died in Rome both persists and recurs.

Judgments and assumptions, indeed, do not exist any more than the time-series does. What appear as judgments or assumptions, persistent or recurrent in time, must in reality be timeless perceptions, occupying certain positions in the *C* series. But our account of the *C* series will have to be such as to allow for these perceptions appearing as persistent and recurrent in the time-series.

531. We now come to three conditions which arise from the fact that the content of experience appears to change from time to time, and also appears to oscillate in its nature, having a characteristic at two separate times, while it does not have it in the intermediate time.

The first of these, which is the *Eighth* condition, deals with changes and oscillations in the apparent extent of our experience, and in its clearness as a whole. As to the extent, we certainly appear to have more objects before our consciousness at one time than at another. Here, then, is a change in the extent of our experience, and, since this extent often diminishes after it has increased, and subsequently increases again, there is oscillation in the extent. I am not asserting that there really are more objects before consciousness at one time than at another (*i.e.*, in reality, at one position in the C series rather than another). But there certainly appear to be more, and our theory of the C series will have to be compatible with the fact of this appearance.

So, also, with the clearness of our experiences. That also appears to be different at different times. Such changes take place often, and for many reasons, but the most obvious examples are to be found in the gradual increase or diminution of clearness which takes place as we gradually wake or fall asleep.

We must also consider that the continuity of consciousness appears to be broken altogether by any sleep which is—or at any rate appears to us in recollection to be—completely free from dreams. What meaning, if any, can we attach to the statement "I have slept dreamlessly for an hour," if the time-series is what we have held it to be?

The *Ninth* condition relates to changes and oscillations in the clearness of our knowledge of particular objects. Even while an object remains in consciousness continuously, the clearness with which I am conscious of it often varies, and varies not only independently of the variations in general clearness spoken of above, but also in direct opposition to them. It is possible for an object, while remaining continuously in consciousness, to be apprehended less and less clearly, while the general clearness increases, provided that there is some circumstance which turns the attention of the self to a greater degree upon other objects. And here, too, there is oscillation.

But there is a still stronger case—when something which appeared not to have been in consciousness at all, appears to come into consciousness, or *vice versa*. That this does happen was already implied when we spoke earlier of the recurrence of the content of experience, since such a recurrence involves that it was, or appeared to be, the case that what was in consciousness ceased to be so, and then, after having been out of consciousness, came into it again. But in speaking of recurrence before, our attention was directed to the characteristic which it shares with persistence—the presence of the same content at different points of time. Here we are considering the characteristic which recurrence of a content shares with the first appearance of a content—the characteristic of presence at a point immediately before which it was not present.

The *Tenth* condition is that our theory must be compatible with the fact that there appears to be change and oscillation in the accuracy of our knowledge, as well as in its extent and clearness. It might be difficult to prove that the whole of a man's experience is more accurate at one time than at another, though the probability that it is so seems overwhelming. But with regard to particular questions both the change and the oscillation can be proved. A man often believes A to be X, and subsequently believes it to be not-X. One of these beliefs must be true, and the other must be false; and, whichever is true and whichever is false, there must be change in the accuracy. And in cases of doubt and difficulty, it often happens that a man successively believes A to be X, not-X, X again, and not-X again. In this case, whichever is the truth, a true belief has been succeeded by a false belief, and a false belief by a true belief. And thus there is not only change in accuracy, but oscillation in accuracy.

And it is also clear that such a change and oscillation can take place with regard to one problem, while it does not take place with regard to another. For if I successively believe A to be X, not-X, X, and not-X, while all the time I believe B to be Y, and C to be Z, it is clear that there has been a variation of accuracy in the one case, and that there has not been in the others.

532. With regard to all the changes to which the eighth, ninth, and tenth conditions relate, it is obvious that we cannot find any explanation by ascribing the changes in question to the simple passage of time, or, rather, to that difference of position in the C series which appears as the passage of time. For we have seen that in each case the change is sometimes an oscillation. And therefore it is impossible that the passage of time should by itself account for change in either direction, since it is found in connection with changes in both directions.

But on the other hand we could not accept an explanation in which there was no relation of the content of the series to its place in the apparent time-series. And the *Eleventh* condition is that our theory should allow for some such relation.

533. It is possible, of course, to have terms arranged in a series in such a way that their position in it has no relation to some of their characteristics. If, for example, a number of books were to be arranged in the order in which each of them had been lifted up by a child who could not read, their position in the series would have no connection with the subjects of which they treated. We could not argue that, because E and F were near together in the series, their subjects were similar, or that, because they were far from one another in the series, their subjects were unlike.

The C series cannot be indifferent in this way to the nature of its terms. It is true that we have not found, and that there is no reason to expect that we shall ever find, any connection between the position and the nature which can be known *à priori*, so that, from the facts that E had a certain nature, and that it stood in a certain position in the C series in relation to F, it would follow *à priori* that F had a certain nature. But although there is no connection which can be known *à priori*, there are many connections which can be known empirically. If a man's head is cut off, his death follows at once. The condition in this proposition is never fulfilled, for it is impossible to cut off a man's head, since neither matter nor time exist. But behind this there is a proposition whose condition has been often fulfilled. And this is the proposition that if any term in the C series appears as the cutting off of a man's head, it will have, in close conjunction with it in the C series, a term which appears as the death of that man. And thus there is a connection between the natures of the two terms and their relative places in the C series.

534. Such a connection is indicated by every causal law which connects together any two things which appear as events in time. And most of the causal laws which are recognized in ordinary life are of this nature. But, it may be objected, we have not proved that any such causal laws are valid. The only case in which we have shown that any causal laws were valid was that of the connections between determining correspondence parts treated in Sections 262 and 263. These have not been shown to be terms in the *C* series, and we shall see in the next chapter that they cannot be terms in it.

It must, however, be remembered that all that is necessary, in order that there should be some connection between the position of the terms and their nature, is to prove that causation should occur in some cases among those terms. It is not necessary that it should occur in every case. If, for example, the cutting off of a man's head does involve his death, there is some connection between the position of some terms in the *C* series and their nature, even if in other cases a man dies without any cause at all, or if the cutting off of an elephant's head at the North Pole should produce no effect whatever. Therefore we must accept the eleventh condition, unless we are prepared to say that all laws which profess to be laws of causal sequence in time could be safely dismissed as not indicating laws of causal connection between terms in the *C* series. In that case we should have to dismiss the assertion that a man dies if his head is cut off, as having no closer relation to the truth than the assertion that he dies if his hair is cut off. And a theory which depended on our being able to do this would not have much claim to be considered satisfactory.

535. Causality itself, however, cannot be the relation for which we are looking—the relation by which the terms of the *C* series are connected with each other. Such a relation must be asymmetrical; for if it were not so it could not determine the order of the terms. Now causality is not necessarily asymmetrical, though it can be so. It is possible for causal determination to be reciprocal, so that each of the two terms determines the other. And this prevents it from being the relation required.

Still, some causation is unreciprocal. Could this special sort of causation be the relation wanted? This also is impossible. For the C series is that which corresponds to the time-series in such a way that all those terms which appear as later than a given term are on one side of that term in the C series, while those that appear as earlier are on the other side of it in the C series. If any sort of causation, then, was the relation which generated the C series, it would be necessary that the term which causally determines the other should either be always earlier than the other, or always later than the other.

Now we saw in Section 210 that the determining term in an asymmetrical causal relation is sometimes earlier and sometimes later than the determined term; and, further, that of any two substances, either may be said to determine the other, according to the descriptions of the substances which we have taken[1]. And thus causation could never determine the order of terms in the C series, which are substances.

536. With these eleven conditions, then, our theory must comply. In order to get a complete solution we shall have to answer two separate questions. The first is, what is the nature of the substances which are terms of the C series? The second is, what is the nature of the relations between them—those two relations in one of which each term in the series stands to every other term in the series, and which, when the series appears as a B series, appear as the relations of "earlier than" and "later than"?

---

[1] This is due to the fact that causality is a relation between qualities, though only between the qualities of existent substances. Cp. Section 208.

CHAPTER XLVII

THE C SERIES—NATURE OF THE TERMS

537. What then is the nature of the terms of the C series? Are they, in the first place, the terms of the system of determining correspondence? Every C series falls within a self. And every self is a primary part, and is divided into secondary parts of the first grade, secondary parts of the second grade, and so on infinitely. Is it not possible that these parts, taken in a certain order, are the terms of the C series?

The C series, as we have seen, must be a series of one dimension. The determining correspondence system has two dimensions. There is the dimension in which the terms are the different grades of parts. And each of these grades has a dimension in which the terms are the parts in that grade.

It would, however, be possible to arrange the parts as a series of one dimension. Let us suppose that each secondary part of the first grade, for example, $B!C$, corresponded to a certain stretch, M, in the C series, and that the parts of the second grade within it (for example, $B!C!B$, $B!C!C$, $B!C!D$) which formed a set of its parts corresponded to parts of M which formed a set of parts of M. And let us also suppose that there was a similar correspondence with all lower grades. There would then be a series of determining correspondence parts, which would be a series of one dimension.

538. The C series is one in which each term has a definite place. Now there is nothing in the determining correspondence series as such which determines any definite place for the parts of any part—which determines, for example, whether $B!E$ comes between $B!C$ and $B!F$, or whether $B!F$ comes between $B!C$ and $B!E$. But it is possible that there might be something which determined such definite places. There might be something which determined the primary parts as having a certain order. And it might be the case that the C series was one in which the parts of each part were arranged within that part in the order of their determinants[1]. We do not know that there is any such order, but we cannot say that it is impossible that there should be, and we cannot say, therefore, that the condition cannot be satisfied.

539. But there is an objection which renders it impossible that the parts of any determining correspondence system should be the terms of the *C* series. We saw in Section 229 that only one part within any primary part can have the same direct determinant. Since all primary parts are now known to be selves, and all parts within them to be perceptions, this means that, in the determining correspondence system, a self can only have one perception of any one perceptum.

Now this makes it impossible that the perceptions in the determining correspondence system can be those which appear to be terms in the time-series. For, as we have said above (Chap. XLVI, p. 218), whatever may be said about those perceptions which appear as perceptions, the contents of many of these perceptions which appear as judgments undoubtedly recur after an interval. I may judge that Caesar was killed at Rome. I may then never think of Caesar's death for a year, during which I shall have a good number of other perceptions—some appearing as perceptions, and some as judgments and in other forms. At the end of the time I may again judge that Caesar was killed at Rome.

Now if the two perceptions of this content are to be considered as really separate—that is, as separated by an interval of time in which the perceptum was not perceived—then it is clear that they really are two perceptions. And as they are perceptions of the same perceptum, they cannot be two perceptions in the determining correspondence system.

If, on the other hand, we try to avoid this by suggesting that the perception which appears as "Caesar was killed at Rome" had existed all through the interval by the side of other perceptions, but was so faint that it could not be recognized by introspection—this would be fatal to the theory in another way. For in that case different perceptions would be present at what

---

[1] For example, if the primary parts were arranged in the alphabetical order of the symbols we have given them, so that *E* came between *C* and *F*, then *B ! E* would come between *B ! C* and *B ! F*, and *B ! C ! E* would come between *B ! C ! C* and *B ! C ! F*.

appears as the same moment of time, while the same perception would be present at what appear to be different moments of time. This would involve that, in reality, different perceptions would be present at the same point in the *C* series, while the same perception would be present at different points of the *C* series. And thus it is impossible that the perceptions of which we are speaking—the perceptions of the determining correspondence system—could be the terms of the *C* series.

540. We cannot, therefore, find the *C* series in the system of two dimensions which forms the determining correspondence system. We must find it in some other dimension of existence. From this it follows that there must be simple and indivisible terms in the *C* series. For infinite divisibility, as we have seen, involves a contradiction unless the parts are determined by determining correspondence.

541. I believe that it is possible to find the *C* series in another dimension, and I shall now explain how I believe it can be done. The proof of this theory will rest on its satisfaction of the eleven conditions laid down in the preceding chapter, and of a twelfth condition which will be found in the next chapter. This proof will occupy the remainder of the present Book. It will be in one respect negative, for if any other theory should be put forward which would equally satisfy the twelve conditions it would be uncertain which of the two theories was correct. And therefore the proof of our theory will rest in part on the fact that no other theory yet suggested will satisfy the conditions. Proofs of this nature, however, are the only proofs available when the problem is—as it often is in all fields of knowledge—to find a solution which shall satisfy given conditions. The amount of certainty which can be gained from these varies with the circumstances of the particular case, but may be such as not to fall much short of absolute demonstration. In the present case, I am, of course, making no claim to absolute demonstration. I hope, however, to show that there are reasonable grounds for a strong and confident belief.

My view is, then, that whenever a self (or a part of a self determined by determining correspondence) appears as being in time, it is divided in another dimension besides those of its determining correspondence parts, and that the terms in this fresh dimension form the C series.

542. In order to give a sufficient description of any term in this series, it will be enough to find a description which will distinguish it from all the other terms in the same series which are within the same self, or the same determining correspondence part of a self. (We may refer to these descriptions as c_1, c_2, etc.) For every self, and every determining correspondence part of a self, has, as we have seen, a sufficient description, and by combining this with the other, we shall get, for example, $c_2\,G!\,H$, which will be a sufficient description of this term in $G!\,H$.

543. And I shall maintain that all parts of $G!\,H$ in the C series are states of misperception of H, of which $G!\,H$, of which they are parts, is a correct perception[1].

Each of these parts of $G!\,H$ in the C series of G will be a misperception of the terms of H's C series, c_1H, c_2H, and so on. But part of the erroneous element of G's perception of H will be to regard this C series as a B series, and consequently they will be misperceived as being in time. (G, of course, can have a perception $G!\,G$, and will then perceive himself in the same way as he perceives other selves, and so perceive himself as being in time.)

At any stage in the C series G will perceive as present whatever in H is at that stage in the C series. He will perceive as future or as past whatever is at a different stage in the C series. This involves that different selves have correspondent C series. The question of correspondent C series will be discussed later (Chap. LI, pp. 274–275).

544. The only perceptions which are apparent perceptions— that is, which appear to be, as they are, perceptions—are *some* of those which are at the same stage in the C series as their percepta. For an apparent perception always perceives its perceptum as present, and therefore it can never be at a different stage of the C series from its perceptum. All perceptions which

---

[1] It is only the *parts* of $G!\,H$ in the C series which are states of misperception. We shall see later (Chap. XLIX, p. 247) that the whole, $G!\,H$, which is a correct perception, is itself a term in the same C series.

I say "states of misperception" and not "misperceptions" to allow for cases where many such parts form only a single misperception of a perceptum misperceived as being undifferentiated. Cp. Chap. L, p. 260.

are not at the same stage as their percepta appear, not as perceptions, but as judgments or as cogitations of some other sort. But even perceptions which are at the same stage as their percepta do not in every case appear as perceptions. In some cases they appear as judgments or as cogitations of some other sort.

545. The parts in this dimension include our present experience, all of which appears to be in time. And the fact that they are states of misperception will account for the error in that experience. The fact that parts in this dimension are states of misperception must be taken as ultimate. We cannot explain it further, though we may have good reason to believe it, if we find that this theory satisfies the conditions required for a solution, and that no other theory does so.

But while the acceptance of error as ultimate is necessary, it is not sufficient. For, as we saw in Chap. XLIV, p. 202, it is necessary to hold, not only that there are erroneous perceptions, but that those perceptions, although erroneous, are correct, except for whatever qualification is introduced owing to their position in the *C* series.

I shall endeavour later on to show that this is possible. Meanwhile we see that on this theory error is closely connected with the *C* series, and, through the *C* series, with the time-series.

546. We have said that, while the parts of $G!H$ in this dimension are states of misperception of H, $G!H$ itself is a correct perception of H. But what reason have we for asserting that $G!H$ is a correct perception of H—that is, that it does not perceive H as being anything which it is not? $G!H$, no doubt, is a perception in the determining correspondence system. And we have seen in Chapter XXXVII that all the perceptions in that system must be correct in some respects, since they must perceive selves as selves, and perceptions as perceptions, and must also perceive perceptions as determined by the determinants which do in fact determine them. But why is it impossible that they might be erroneous in other respects? The discussion of this point will occupy us for the rest of this chapter.

547. We saw above (Chap. XLIV, p. 193) that our present experience misperceives its percepta in five respects. In the first place, it perceives various existents as being matter and sensa, while in reality nothing exists but spirit. In the second place, it does not perceive selves as perceiving other selves and their parts, though in reality they do perceive them. In the third place, it perceives certain parts of selves as judgments, assumptions, imagings, and awarenesses of characteristics, though all such parts are in reality perceptions. In the fourth place, it perceives certain volitions and emotions as being judgments, assumptions, and imagings, though in reality they are all perceptions. In the fifth place, it perceives various existents as in time, though in reality nothing is in time.

It is thus clear that in our present experience perception is largely misperception. But when we consider how the parts of the determining correspondence system perceive other such parts as wholes—how, for example, $G!H!K$ is a perception of $H!K$ as a whole, we shall see that it cannot misperceive it in any of the four ways first mentioned. For it must perceive $H!K$ as being a perception, and as being a perception of a self. To perceive it as being a perception excludes the first, third, and fourth errors, and to perceive it as being a perception of a self excludes the second error.

548. But $G!H!K$, besides being a perception of $H!K$ as a whole, may also perceive the states of misperception which are parts of $H!K$ in the dimension of its C series. And we shall see later (Chap. LXIII, p. 388) that this is actually the case. (We may call the parts in this dimension Fragmentary Parts, to distinguish them from the parts of $H!K$ in the determining correspondence system, such as $H!K!L$, $H!K!M!P$, and so on.) Now is it possible that $G!H!K$ should misperceive these fragmentary parts in any of the four ways just mentioned, though it cannot misperceive $H!K$ as a whole in any of these ways?

The answer must be that this is not possible. For if the fragmentary parts are perceived in the perception $G!H!K$, they must be perceived as parts of $H!K$. In the first place, since the perception $G!H!K$ is the perception of $H!K$, whatever is part of $G!H!K$ must be part of the perception of $H!K$. But, secondly, we have seen (Chap. XXXVII, p. 97) that the perception of a part cannot be part of the perception of its whole, unless the part is perceived as being a part of that whole. And therefore the fragmentary parts of $H!K$ can only be perceived in $G!H!K$ as being parts of $H!K$.

This will exclude the first four errors. For if the fragmentary parts of $H!K$ are perceived as being parts of the determining correspondence parts, which are themselves perceived as being perceptions, then the fragmentary parts cannot be perceived as being matter, or sensa, or judgments, or assumptions, or imagings, or awarenesses of characteristics. For nothing which is perceived as being part of a perception could be perceived as being any of those things. This result excludes the first, third, and fourth errors. Nor can the second error occur, since the fragmentary parts are perceived as being parts of selves who are perceived as perceiving other selves and their parts.

549. There remains the fifth error in our present experience—that various existents are perceived as being in time. Can this error occur in the determining correspondence perceptions?

It will be convenient to reverse here the order which we adopted in the last section, and to enquire, firstly, as to the fragmentary parts, whether a determining correspondence perception can perceive them as being in time. Can the fragmentary parts of H be perceived by G in $G!H$ as being in time, or the fragmentary parts of $H!K$ be perceived in $G!H!K$ as being in time?

This cannot be the case. For, in $G!H!K$, $H!K$ as a whole must be perceived either as being in time or not as being in time. If it is not perceived as being in time, then, in $G!H!K$, the fragmentary parts of $H!K$ cannot be perceived as being in time. If any whole, P, is perceived, and not perceived as being in time, and there is a perception of Q, a part of P, which perception is part of the perception of P, it is impossible that the perception of Q should be a perception of it as being in time. The qualities of being in time, and of being a part of a perceived whole, are qualities which nothing can be perceived as having together, unless the whole is perceived as being in time.

550. Is it possible, then, that in $G!H!K$ both the fragmentary parts of $H!K$, and also $H!K$ as a whole, are perceived as being in time? Here there are two alternatives. The first of them is that $H!K$ as a whole should be perceived as occupying a period of time which was the aggregate of the periods occupied by the fragmentary parts of $H!K$.

This, however, is impossible. For $H!K$ as a whole is, as we have seen (Chap. XXXVII, pp. 102–104), a perception of K as a self, and so as spiritual. And, again, $G!H!K$ is, by the same argument, a perception of $H!K$ as having a perception of K as a self, and so as spiritual. But the fragmentary parts of $H!K$, or some of them, are, in some cases, misperceptions of K as material, since we have experience of existents as matter. And these fragmentary parts must be perceived in $G!H!K$ as being such misperceptions. For otherwise there would be nothing by which they could be distinguished from $H!K$ as a whole.

Thus in $G!H!K$, while $H!K$ was perceived as the perception of K as spiritual, some of the fragmentary parts of $H!K$ would be perceived in some cases as perceptions of K as material. And then, on the alternative we are considering, $H!K$ would be perceived as the perception of K as in time, and as spiritual, and as containing a set of parts, some of which were perceptions of K as in time and as material. And this is impossible[1].

551. The second alternative is that $H!K$ should be perceived as occupying a different position in the time-series from those occupied by its fragmentary parts. And it is clearly impossible that when parts of a whole are perceived as being parts of that whole, the parts and the whole should be perceived as being at different positions in time.

The fragmentary parts, then, cannot be perceived by the determining correspondence parts as being in time. For we have now seen that this is equally impossible, whether the wholes, in which the fragmentary parts fall, are perceived as being in time, or not perceived as being in time.

But if the fragmentary parts are not perceived as being in time, then the wholes cannot be perceived as being in time. For the dimension, whose terms appear as the time-series, is, by our original hypothesis, a dimension in which the plurality of terms comes only from the plurality of fragmentary parts. The whole, therefore, taken by itself, could not be more than one term in this dimension. And one term by itself could not appear

---

[1] The general question how it is possible for incorrect perceptions to be parts of correct perceptions will be considered in the next chapter. All that we require here is the impossibility of the particular case mentioned in the text.

as in time, since time requires that the relation of earlier and later should hold between different terms.

Thus none of the five errors which are found to belong to the fragmentary perceptions can be found in the determining correspondence perceptions, whether they are perceiving terms of the determining correspondence system, or fragmentary parts of those terms. And thus we have no reason to suppose that any error is to be found in the determining correspondence perceptions.

552. But, it may be said, it remains possible that there should be error in them. We saw that they must be correct in certain respects, and we have just seen that they cannot be incorrect in some others. But is it not possible that there may yet be other respects in which they might be incorrect?

This also, however, is impossible. We saw (Chap. XLIV, p. 202) that it is self-evident that what I perceive exists, and exists as I perceive it, subject to one condition only. That condition, in the form in which it appears to us in present experience, is "when I perceive it." And we saw that the whole possibility of erroneous perception depended on the question whether that condition, when restated in terms of the *C* series, allowed for erroneous perception.

Now, as we have seen, no determining correspondence part, taken as a whole, can perceive itself, or anything else, as being in time. Consequently there can be no such condition, and the self-evident correctness of the perception is without any possible limitation. And so there can be no error in determining correspondence perceptions. They need not give complete knowledge. They need not perceive their percepta as having all the characteristics which they actually do have. But they cannot perceive them as having any characteristics which they have not. And so they cannot be erroneous.

553. I have now stated what seems to me to be the true theory as to the nature of the substances which are terms of the *C* series. Before proceeding to explain and defend it, it is necessary to consider the second question mentioned in Chap. XLVI, p. 223—what are the relations between the terms of the *C* series, which, when that series appears as a *B* series, appear as the relations of "earlier than" and "later than"?

CHAPTER XLVIII

THE C SERIES—NATURE OF THE RELATIONS

554. What are the relations in the C series which appear in the B series as "earlier" and "later"? They must be transitive and asymmetrical relations, since "earlier" and "later" are so, and they must be such as to make the terms of the C series into a series of one dimension. Further, they must, like "earlier" and "later," be converse relations. And, of any two terms in the C series, one must stand in one of these relations to the other, while the other term will stand in the other relation to the first term.

555. We have seen (Chap. XLVI, p. 220) that our perceptions in the C series vary in accuracy. Could it be possible that the relations which appear as "earlier" and "later" are the relations "more accurate" and "less accurate" (reserving the question which member of the second pair appears as which member of the first pair)?

If this were the case, it would follow that, as time went on (in reality, as time appeared to go on), our cognitions would become steadily more accurate or less accurate. But this is not the case. To begin with, it is clearly not the case with every particular C series within each self. As we also saw in Chap. XLVI, p. 220, the accuracy in such series frequently oscillates. If a man successively believes A to be X, not-X, X again, and not-X again, then, whatever is the truth, a true belief has been succeeded by a false belief, and a false belief by a true belief.

556. But, it may be replied, it is possible that, while particular C series oscillate like this, the whole cognition of each self shows a steady progress towards accuracy or inaccuracy. I do not know that there is any way of disproving this suggestion, though the supposition that every man is less in error at every moment of his life than at the moment before, and the supposition that he is more in error at every moment of his life than at the moment before, both seem to be wildly improbable. But even if one of them were true, it would not help us. For what we want is a relation which orders the terms in every C series—not merely those in the C series of each self, taken as a whole, but those in the C series of every determining correspondence perception within every self. And we have just seen that it is certain, when we come to these, that the passage of time involves neither a continuous increase, nor a continuous decrease, in accuracy.

557. Similar reasons prevent us from holding that the relations for which we are searching could be "more extensive" and "less extensive," or, again, "more clear" and "less clear." (Cp. Chap. XLVI, p. 219.) And in these cases, we may add, the oscillation over the whole content of a self is as obvious as that in any part of a self. This is shown whenever a man first concentrates his attention, and then relaxes it again, and whenever a man gradually becomes drowsy and then again wakeful.

558. We have now rejected three possible relations, and they seem to be the only alternatives which present themselves at first sight as suitable for our purpose. We must go deeper, and look for something which is not so immediately obvious. And, before doing so, we must discuss a preliminary question. The terms of the C series are all parts of the self in which they fall. But do they form a set of parts of that self?

In Section 124 we defined a set of parts of any whole as any collection of its parts which together make up the whole, and do not more than make it up, so that the whole would not be made up, if any of those parts, or of their parts, should be subtracted. We saw, too, that in any substance no set of parts can contain a member twice over, or contain any two members which have a part in common, and that, in any substance, the content is expressed in each set of parts.

Now the terms of the C series exhaust the dimension in which they are terms. And so there is no content in $G!H$ which does not fall within the terms of the C series of parts of $G!H$. But it does not follow from this that the C series must be a set of parts of that substance, all of whose content is included in it. There is another alternative. If we take a line a foot long, we can divide it into twelve parts, each an inch long. And these will contain all its content, and be a set of parts of it. But we can also find a series of parts in a foot which should be respectively one inch, two inches, three inches, and so on, up to eleven inches; and going on with the series, we should get a twelfth member, which was not a part, but the whole foot. Now here the parts would be members of a series which contained the whole content of the foot, and contained nothing else besides that content, but yet was not a set of parts of the foot, since all the terms of the series, except the last, might be subtracted, and yet the whole would be made up. Part of the content, in this series, would be taken twelve times over, part eleven times over, and so on.

559. Now of which sort is the C series? It might seem natural to conclude that it was a set of parts of the self. But against this view there are objections, and objections which, I think, are insuperable.

The perception $G!H$ is by our theory a correct perception. If those parts of $G!H$ which are the terms of its C series are a set of its parts, then, when the different terms $c_1 G!H$, $c_2 G!H$, and so on, are all added together, they will by themselves, without any addition or subtraction, form $G!H$, which is a correct perception. But they themselves are all misperceptions. And so, by adding together misperceptions, we get a correct perception. Or again, by subtracting misperceptions from a correct perception, we get other misperceptions[1].

Now this, I submit, is impossible. The defect in accuracy of a misperception cannot be removed by the addition of other misperceptions. If, indeed, a misperception were merely an incomplete perception—if its only defect were that it did not perceive all the parts of the perceptum, or did not perceive it as having all its qualities—then the matter would be different. In that case it might be true, under certain conditions, that the sum of these perceptions might be the correct perception of the whole.

[1] It is true, no doubt, that we called the terms "states of misperception" and not "misperceptions" to allow for the case where many such parts form only a single misperception of a perceptum which is perceived as undifferentiated. (Cp. Chap. xlvii, p. 227 footnote.) But the argument will not be affected. For what would then happen, on the hypothesis discussed in the text, is that by adding misperceptions we should get a group of correct perceptions. And this would be as impossible as it would be to get a single correct perception.

But the perceptions in the *C* series cannot be merely incomplete. We have to account for real error, since real error certainly exists. And if we are to find that error in the perceptions in the *C* series, those perceptions must be definitely and positively erroneous. And it seems clear that the difference between an erroneous cognition and a correct cognition cannot possibly consist in one or more additional erroneous cognitions.

And, further, it is impossible that two or more misperceptions should be members of the same set of parts of a correct perception, even in company with other parts which are not misperceptions. They can no more be some members of such a set of parts, than they can be all the members of it. For on either view the difference between a misperception and a correct perception would contain a misperception. And it is this which is impossible.

560. Two misperceptions, then, cannot be members of the same set of parts of a correct perception. Nor can we avoid this result by saying that the terms of the *C* series are themselves not perceived as they really are, and that it is perhaps the erroneous element in our perception of them which leads us to the conclusion that they cannot be combined in this way. It is, of course, true that they are not perceived altogether as they really are. (For example, they are perceived as being in time while they are really timeless.) But this does not affect the question. For the difficulty arises from the fact that the terms in question are misperceptions. And this is not appearance but reality. The terms must really be misperceptions. If they were not really misperceptions, it would be impossible that the error, for which we are trying to find a place, could be found in them.

561. Nor could we escape—as might perhaps be suggested—by saying that the objection looks at the question too mechanically. A whole, it might be said, has characteristics other than those possessed by a mere aggregate of its parts, and though it might be impossible that a mere aggregate of parts, some of which were misperceptions, could form a correct perception, yet it might be possible that such parts, arranged in a certain way, might form a correct perception. The school of thinkers from whom such an argument might be expected to come would probably regard it as having special force in a case in which the whole could fairly be held to be logically prior to its parts. And this might be said to be the case here, since any proof which could be given of the existence of the parts in question—the parts which are terms of the C series—would have to start by first proving, as we have proved in Book V, the existence of the whole—the correct perception—which is part of the determining correspondence system.

I do not think, however, that their position is tenable. No doubt the whole of which we are speaking consists of terms arranged in a definite order. And no doubt we must distinguish between what is true of the terms as arranged in a certain order and what is true of the terms in whatever order they are arranged. It does not follow, because the terms would not make up a correct perception if arranged in some order, that they would not make up a correct perception if arranged in another order. But the objection that I have put forward is that they could not make up a correct perception in any circumstances—that there is no order in which they can be arranged in which this would be possible[1]. And if this is true, then it is useless to appeal to the fact that the whole is more than a mere aggregate. For in the whole the parts must be arranged in some particular order, and if they cannot make up a correct perception in any possible order, they cannot make up a correct perception in the order in which they are.

562. Two misperceptions, then, cannot be members of the same set of parts of a correct perception. But this leaves it possible that every misperception should be a member of some set of parts of a correct perception, provided that no two of them are members of the same set of parts. For if there is only one of them in a set of parts, then the difference between it and the whole which is the correct perception will not consist of, or contain, other misperceptions. And it is only this which is the difficulty.

563. Let us consider in passing whether our conclusion would

[1] It will be remembered that the order in question must in any case be an order which forms them into a series of one dimension. For the question is of the terms as arranged so as to form a C series. And that is a series of one dimension.

have any effect on the possibility of such a dialectic as Hegel's. It is true that categories in Hegel's logic are not perceptions. He does not regard all cognitions as perceptions, though it might perhaps be argued that he ought, on his own premises, to do so. And the categories—or rather the assertion of them as valid, which is the essence of the dialectic—are not perceptions, but judgments. But they are cognitions. And if the difference between an erroneous cognition and a correct cognition cannot be another erroneous cognition, then, if Hegel's system does involve that such a difference is an erroneous cognition, Hegel's system must be wrong.

None of the conclusions at which we have arrived involve that Hegel's dialectic is valid. And therefore we should not be in any difficulty if we decided that Hegel's dialectic broke down in this manner. But, whatever other objections there may be to its validity, no such objection arises from the results we have just reached. For Hegel's system does not involve that the difference in any such case is an erroneous cognition.

No doubt Hegel considers that all the categories except the Absolute Idea are erroneous cognitions, in the sense in which we have used that phrase, that is, cognitions which are not completely true, and which are partially erroneous. But then the lower categories do not form a set of parts of the Absolute Idea.

In the first place, the only way in which the lower categories can be said to be parts of the Absolute Idea is that they are synthesized in it. But it is just as much the case that some of the lower categories are synthesized in others. And thus they could not all be members of the same set of parts of the Absolute Idea. Being, for example, is synthesized in Becoming. And, if to be synthesized implies to be part of, Being is part of Becoming. They cannot then be members of the same set of parts of the Absolute Idea. For if Being, as a separate term, were taken away, no part of the Absolute Idea would be taken away, if Becoming were left, since Being is a part of Becoming.

564. And, in the second place, it is not correct to say that the erroneous cognitions which are the thesis and antithesis of a triad are parts of the synthesis. In a certain sense they may be said to be absorbed into it. But in order to be synthesized they must, according to Hegel, be transcended. And when they are transcended, they are no longer what they were before. Indeed, if they had remained as they were before, it is obvious that they could not be united in the synthesis, since the whole spring of the dialectic lies in the fact that, as they were before synthesis, they were incompatible. Now it is only as they were before synthesis that they were erroneous cognitions. In the synthesis they have been purged of their error, except in so far as the synthesis is itself erroneous. But in the final synthesis, which, according to Hegel, is free from all error, all the categories have been successively synthesized, and are so altered and transcended that they are no longer erroneous cognitions.

565. We have now added a twelfth condition to those which, as we saw in Chapter XLVI, must be fulfilled by any satisfactory theory as to the nature of the C series. It may be convenient to recapitulate the other eleven. (1) The series must be one which can be found in a substance which is spiritual, and which is divided into parts within parts to infinity by determining correspondence. (2) It must allow for the occurrence both of correct and of erroneous cognitions. (3) It must allow for the different sorts of erroneous cognition which do actually exist. (4) It must be a series of one dimension, and the relation which constitutes it must be transitive and asymmetrical. (5) It must have at least as many terms as can be distinguished from each other in the B series. (6) It must allow for the fact that, while in absolute reality my knowledge of any substance is differentiated into parts of parts to infinity, in present experience there is no such infinite differentiation of knowledge. (7) It must allow for the persistence and recurrence of certain contents in the time-series of our present experience. (8) It must allow for changes and oscillations in the apparent extent of the content of our experience, and in its clearness as a whole. (9) It must also allow for changes and oscillations in the clearness of our knowledge of particular objects. (10) It must also allow for changes and oscillations in the accuracy of our knowledge. (11) It must allow for some relation of the content of experience to its place in the apparent time-series.

We have put forward the theory that terms of the C series are states of misperception by some self of some determining correspondence part of the universe, and that all this series falls within the correct perception which the self has of that object. And we have now the *Twelfth* condition that, although these terms are all parts of this correct perception, no two of them can be in the same set of its parts, because the difference between a state of misperception and a correct perception cannot be or include another state of misperception.

566. From this twelfth condition it follows that no two terms of the *C* series which are misperceptions can be mutually outside one another, or, in other words, that there can be no such terms which have no content in common. For if there were, then the difference between one of them and the whole, $G!H$, would include the other of them. There remains only one alternative— that, of any two terms in the *C* series, one must include the other. In this case the difference between $G!H$ and any part of $G!H$ would not be, or contain, a state of misperception. For the terms of the *C* series which are intermediate between the given term, $c_x G!H$, and the whole $G!H$, will each include the term $c_x G!H$. And thus, while they are states of misperception, they do not form the difference between $c_x G!H$ and the whole, since they include $c_x G!H$.

The terms, then, which appear as terms of the *B* series must, in reality, be terms each of which includes or is included by each of the others. And it is in this fact, I think, that we must find the clue to the relations which appear as "earlier than" and "later than." They are the relations "included in" and "inclusive of." Of any two terms in the *B* series, one is earlier than the other, which is later than the first, and by means of these relations all the terms can be arranged in one definite order. And of any two terms in the *C* series, one is included in the other, which includes the first, and by means of these relations all the terms can be arranged in one definite order. And it seems to me possible, as I shall explain in detail in the remainder of the present Book, that it is the relations of "included in" and "inclusive of" which appear as the relations of "earlier than" and "later than," while I cannot see that there are any other relations which the *C* series could possess, and which could appear as "earlier than" and "later than." There is thus good reason to believe that it is "included in" and "inclusive of" which do appear in this way, though that reason, of course, depends on the success of the explanations in detail which will follow.

567. One point must remain for the present unsettled. The pair of relations "included in" and "inclusive of" appear as the pair of relations "earlier than" and "later than." But which of the first pair appears as which of the second pair? We must postpone this question for the present. In Chapter LX we shall find reason to believe that it is "included in" which appears as "earlier than," and "inclusive of" which appears as "later than."

[EDITORIAL NOTE. *Here ends Draft C. The rest of the book is printed from Draft B.*]

568. Since each term of the C series either includes or is included in any other, it follows that each term is either greater or less than any other. The terms, therefore, have magnitude. Magnitudes are either extensive or intensive. An extensive magnitude is one in which the difference between two magnitudes is another magnitude of the same sort. Thus the difference between a length of a foot and a length of seven inches is also a length. The difference between a duration of an hour and the duration of a minute is also a duration. On the other hand, the difference in magnitude between one state of pleasure and another is not a third state of pleasure. And the difference between a temperature of a hundred degrees and a temperature of eighty degrees is not a temperature of twenty degrees. Such magnitudes as these are not extensive, but intensive.

The magnitudes of the C series are intensive. $G!H$ is a state of perception, and so is any part of $G!H$ in the C series. But, as we have seen, the difference between the two cannot be a state of perception.

569. The existence, however, of this series of intensive magnitudes involves the existence of another series of magnitudes which is extensive. Of any two terms in the C series, one includes the other. Inclusion is not the same as identity. And there must, therefore, be more in the inclusive term than in the included term. Whenever any term, M, is included in another term, N,

there must be some increment added to *M*, which, with the content of *M*, forms the content of *N*. And so of each other term in
the intensive series. Now these increments, if taken in the order
of the terms in which they are added, will themselves form a
series. None of the members of this series, of course, will be included in or include any other. And they will have an extensive
magnitude. If, starting from the same point in the complete
series, we take two lesser series, one of twelve increments and
one of seven increments, the difference between them will be a
series of five increments. Or, in the event of the series having no
next terms, if, starting from the same point, we take two lesser
series of increments, one of which is half the complete series,
while the other is a third of the complete series, then the difference
between them is a sixth of the complete series. Thus the difference
between one series of increments and another series of increments
is a third series of increments, and the magnitude of the series
is extensive.

This series is less obvious, and of less immediate interest to us,
than the intensive series of states of perception. And it is somewhat difficult to get names for the increments which shall be
different from the names of the terms in the intensive series.
But—to take the examples given above of intensive series—I
suppose we should say that the increments in that series were
amounts of heat, though not states of heat, or temperatures.
And we should also say, I think, that the increments in the
pleasure series were amounts of pleasure, though not states of
pleasure.

570. The question has been much disputed whether states
which have intensive quantity, such, for example, as pleasures,
can be summed in respect of their dimension of intensity[1]. From
what was said in the last section, we can now see that this
question is ambiguous. Two states of pleasure cannot be added
together so as to make a stronger state of pleasure. But the increment between the strength of the state of pleasure *M* and
the greater strength of the state of pleasure *N* can be added to

[1] It is sometimes asserted that pleasures cannot be summed at all. But it is
perfectly obvious that they can be summed in respect of their dimension of
duration. The pleasure I have from sunrise to noon to-day, added to the pleasure
I have from noon to sunset, make up the pleasure I have from sunrise to sunset.

the increment between the strength of N and the greater strength
of the state of pleasure O. And the sum of these will be the increment between the strength of M and the strength of O. If,
therefore, these increments are to be spoken of as amounts of
pleasure—and it seems difficult to see what else they could be—
we must admit that amounts of pleasure can be added in this
way, though states of pleasure cannot.

Thus it is theoretically possible to measure the *differences*
between states of pleasure by some common unit. And, if the
series of such states has a first term, or initial boundary, it is
theoretically possible to measure, not the intensive states themselves, but the total amount of increment in each of them, which
may be considered as an indirect way of measuring the intensive
states themselves. If we have a series of rich men's fortunes, the
terms in themselves have intensive magnitude, and cannot be
added. For the fortunes of two rich men cannot also be the
fortune of a richer man, in the way that two durations can be
also a longer duration. But the intensive series of fortunes is
connected with an extensive series of increments, and we can
measure the fortunes indirectly by comparing the number of
equal increments (say of farthings) which are required to reach
each from zero. In the same way it is theoretically possible to
measure states of pleasure. How far it is possible to do this in
practice, and to what degree of accuracy, is, of course, a different
question. That, with some degree of accuracy, it can be done,
and that we do it every day, seems to me to be certain[1].

571. The C series, then, will have an extensive series of increments corresponding to it. We will call this the D series. It is
clear that, if an intensive series has a first term, that term will
be identical with the first term of the extensive series of increments, counting from zero. No other term in the intensive series,
however, will be a term in the series of increments, since each
of them will contain a plurality of increments. Thus, if the C
series has a first term, that first term will be identical with the
first term of the D series. (Such a term will be a state of misperception, and, being also an increment of the amount of perception, will be a member of a set of parts of the correct perception.

[1] Cp. Chap. LXVI, pp. 448-449.

In this, however, there is no difficulty. Indeed, every state of misperception in the *C* series is a member of a set of parts of the correct perception. What we have found to be impossible is only that two or more misperceptions should be members of the same set of parts of the correct perception.) If, on the other hand, the *C* series has no first term, it will have no term which is identical with any term of the *D* series.

572. What, then, is the nature of the increments in the *D* series? It is clear that they cannot be increases or decreases in the extent, the clearness, or the accuracy of the perceptions. For, as we have seen, all these characteristics oscillate in the time-series, sometimes increasing and sometimes diminishing. They cannot, therefore, in the *C* series, change uniformly in the same direction.

I believe that it is possible that the nature of the increments might be additional perception of the perceptum in a sense which we shall proceed to discuss. And I believe that uniform increase in the amount of perception, taken in this sense, would be compatible both with increases and with decreases in the extent, clearness, and accuracy of the perception. (This point will be considered in Chapter L.) I believe that it could be additional perception of this sort, and I do not see anything else that it could be.

Before deciding what this additional perception can be, we must see what it cannot be. It cannot, taking the case of *G! H*, be a perception of fresh stages in *H* corresponding to the increases in *G*—stages which would therefore form the *D* series of *H*. For if the perception of these stages in *H* gave us the *D* series in *G*, it could only be because the stages in *H* were independently a series. The *D* series in *G*, therefore, would depend on the *D* series in *H*. But the same question would arise with regard to the *D* series in *H*, and if we tried to answer it by reference to the stages of *J*, and so on, we should be led into a vicious infinite series.

And, more generally, *G*'s additional perception of *H* cannot be a perception of more parts of *H*, of any sort. For then any of the misperceptions in the series of fragmentary parts within *G! H* would only differ from the perception *G! H* as a whole by being incomplete—by being perceptions only of a part, while

$G!H$ was a perception of the whole. But this is not the case. For $G!H$ as a whole is, as we have seen, a correct perception of H, while its fragmentary parts are incorrect perceptions of H. The difference is not one of relative completeness compared with relative incompleteness, but of correctness compared with incorrectness.

Neither could G's additional perception of H be an increase in the number of characteristics which H was perceived as having. For then the same difficulty would recur. The difference between $G!H$ as a whole and any of its fragmentary parts would be the difference, not between correctness and incorrectness, but between relative completeness and relative incompleteness.

The increased perception of H, then, cannot be any of these things. It must be increased perception of H as a whole. Nothing more must be perceived, but everything must be perceived more. And the difference between the different stages of it must be due to the nature of the percipient G, and not to the nature of the perceptum H.

573. Can we conceive this quantitative increase of perception of the same perceptum? The conception is no doubt difficult, for it is one which we have no occasion to employ in ordinary life. Some analogy may be found in our present experience, interpreted as we ordinarily interpret it, by considering what happens as we slowly wake after sleep, or again by considering what happens as we see an object through a mist which gradually diminishes. But the analogies are by no means close. For, in such cases, the change often, though not always, consists in an increase in the number of parts of the object, or in the number of its characteristics, which are known. And we have seen that this cannot be the case here. And, again, though in such cases the external object is held to be observed imperfectly and erroneously, it is not *perceived* imperfectly and erroneously. For it is not perceived at all. All that is perceived are the sensa, which are held to be produced in us by the joint action of the external object, on the one hand, and of our sleepiness or the mist on the other hand. And the sensa, according to our ordinary interpretation of experience, are not perceived erroneously, but correctly. On this point, indeed, we can never find any analogy to help us

in our ordinary experience as ordinarily interpreted. For in that
we find no recognition of erroneous perception.

Still, I think that such analogies will help us in conceiving
the nature of such increases of perception. And, in any case, I
think that it can be conceived—though the conception is, as has
been admitted, difficult. It will, I think, become plainer when,
in Chapter L, we discuss in detail the adequacy of the conception
to satisfy the twelve conditions which we have already laid
down.

574. The terms of the *D* series are not known to us by perception, nor are the terms of the *C* series known to us by perception, as being such terms, though we do misperceive the latter
as the *B* series. We have thus no empirical evidence for the
existence either of the *C* series or of the *D* series. Nor is there,
so far as I can see, any *à priori* reason for believing in their
existence, except the line of argument which I have given. If
our general theory is right, there must be a timeless *C* series,
which is the reality which is misperceived as the *B* series. As to
the nature of the *C* series, various alternatives, which seemed at
first sight possible, have had to be rejected on closer examination.
No alternative, so far as we have been able to see, remains open
except the one we have chosen, which involves the existence of
a *D* series. And so, since we have found reason to accept our
general theory, we must accept this, as something which necessarily follows from our general theory. There is no reason why
we should consider this alternative as impossible, or even as so
improbable as to throw any doubts on the theory which requires
it. If it does appear strange and improbable at first sight, it is
due, I think, to the belief that perception cannot be erroneous—
a belief which, as I have pointed out, cannot be accepted as
absolutely true until we have considered the real meaning of
the qualification that the perception is correct at the time the
perception is made.

CHAPTER XLIX

THE RELATIONS OF THE THREE SERIES

575. We have now investigated the nature of that third dimension which we have decided exists in all our perception, in addition to the two dimensions of the determining correspondence system. It is, we have seen, the terms of this dimension which are the terms of our present experience. We have found that the terms of this dimension form a series whose members are connected by the relations "inclusive of" and "included in," so that of any two terms one will be inclusive of the other, and the other will be included in it. We may call this the Inclusion Series. We have also found that some, at least, of the terms in this dimension are states of misperception, and we may call them the Misperception Series. We have also found that some, at least, of the terms in this dimension appear as a B series to certain percipients, and are therefore a C series for these percipients. In this chapter we shall consider the relations of these three series to one another.

576. It is evident, to begin with, that the inclusion series runs right through any substance, H, in that dimension in which any of its terms occur. For if M is any term of the series which does not include the whole of H in that dimension, there can be found a fresh term N, consisting of M and a further increment of the D series. And N will be inclusive of M, and will therefore be a term in the inclusion series. This can continue until a term has been reached which includes the whole of the D series. And this term will be H as a whole.

577. What are the relations between the inclusion series and the misperception series? In the first place, the inclusion series will contain at least one term which the misperception series does not include. For we have just seen that the inclusion series will have H as a whole as one of its terms—the term which is inclusive of all others, and which is included by no other. Now we saw earlier (Chap. XLVII, pp. 228–232) that H as a whole must be a correct perception. It cannot, therefore, be a term in the misperception series.

But are all the terms of the inclusion series, with this one exception, terms of the misperception series? Or is it possible that there should be other terms in the inclusion series which are not in the misperception series? Everything in our present experience falls in the misperception series, for it is all perceived as in time, and is therefore misperceived in that respect, even if not in others. But can there be other terms in the inclusion series which are not misperceptions?

578. In the first place, can there be any terms in the inclusion series which are not states of perception at all? This seems clearly impossible. For any term in the inclusion series can be reached from any other by adding or subtracting a certain part of the D series—the terms of which are amounts of perception. Now some terms in the inclusion series are certainly states of perception—$G!H$ as a whole is so, and so are all the terms which fall within our present experience. And it seems impossible to hold that, by adding amounts of perception to or subtracting them from a state of perception, we could reach any other state except another state of perception[1].

579. But could there be, in the inclusion series of $G!H$, various terms, besides $G!H$ as a whole, which is the last term, which are states of perception without being states of misperception? They cannot, indeed, perceive H in just the same way as $G!H$ as a whole perceives H, for then there would be nothing to distinguish them from one another, nor from $G!H$ as a whole. But could the way in which they perceive H differ from the way in which $G!H$ as a whole perceives it by being less complete, without containing any actual error? And could they differ from one another by perceiving H with different degrees of completeness?

It seems to me, however, that this alternative also must be rejected for the following reasons. The terms in the inclusion series are all fragments of the perception $G!H$, and their real relation to each other is inclusive—there are no two of them of which one is not part of the other. The question then arises whether it is

---

[1] In the case in which the whole of the amount of perception in any state of perception was subtracted from it, the result reached would be nonentity, which is not a term of the inclusion series nor a state of perception. For it is not included in any of the terms of the series. It is a boundary of the series, but not a term in it. (Cp. p. 251.)

possible that they should be separate perceptions, if they perceived themselves as having this relation to the other terms in the series. And I think that this is not possible. In order to be a separate perception from the others, it would have to appear to itself as excluding the others—as having no content in common with them[1]. But in fact, as we have just said, it has content in common with each of them. It therefore perceives itself as it is not, in this respect at least. And this probably involves that it perceives itself in other respects as it is not, since that error would bring others in its train. But, at any rate, in this one respect it perceives itself as it is not. And it is therefore a state of misperception.

580. It is to be noticed that our argument is not that it misperceives itself in perceiving itself as a separate perception. This is not misperception, for it *is* a separate perception. The argument is that it could not *be* a separate perception (which it is), unless it *appeared* to itself to have a content unshared by other perceptions (which it has not).

It may be objected that $G!H$ as a whole is a separate perception from the other terms of its inclusion series, and that, being a correct perception, it cannot misperceive those terms as having no common content with itself. But our argument does not require that it should. For, if all the other terms in the inclusion series are separate terms from $G!H$ as a whole, it follows, of course, that $G!H$ as a whole is a separate term from each of them. And their separateness is assured by the fact that each of them misperceives itself as having no content in common with any of the other terms. It is *their* misperception which makes them separate from $G!H$ as a whole, and $G!H$ does not require to misperceive anything in order to make it separate from them.

581. The misperception series, then, is identical with the inclusion series, except that the last term of the latter is not a term in the former. We must now consider the relation of the C series to each of these. G in $G!H$ erroneously perceives H as being in time, as extending, that is, through a B series of terms

---

[1] I am taking here the whole content of the self at any one point in the series as forming one perception.

which are connected by the relations "earlier than" and "later than." The states in H, which are erroneously perceived as forming a B series, do form a C series for G. What is the relation of this C series in H to the inclusion and misperception series in H?

We have seen that all the terms of the C series must be terms of the inclusion series. In fact, it was by considering what the relation between the terms of the C series could be, that we established the existence of the inclusion series. The relations which appear as "earlier than" and "later than" are really the relations "included in" and "inclusive of," though we have not yet determined which relation appears as which. And so when any self, G, at any point in his own misperception series observes H as in time—$i.e.$ as a B series—then it will be the case that some terms, at least, of the inclusion series of H form a C series for G—$i.e.$ are the basis of G's erroneous perception of a B series[1].

582. But will it be the case that *all* the terms of the inclusion series of H will then form a C series for G at that point, or is it possible that some of them should not do so? In other words, if G, at any point in his own misperception series, perceives any of the inclusion series of H as in time, must he perceive, at that point, all that series as in time, or could he perceive some terms of it as not being in time?

I think that we can see that they must all be perceived as in time. For if a self, at any one stage of his misperception series, misperceives, in any one instance, the relations of "included in" and "inclusive of" as the relations of "earlier than" and "later than," it would be necessary that, at that stage, he should misperceive them in the same way in any other instance. What his misperception is, at that particular stage of the misperception series, will be decided by what degree of misperception belongs to that particular stage. And it does not appear possible that the same degree of misperception should produce different misperceptions of the same relation in cases where the terms,

---

[1] It will be seen from this that a C series is always *in* one self *for* another—the other which misperceives it as a B series. The only exception is in the case of perception of self. In that case, of course, the series which is misperceived, and the misperception of it, fall within the same self.

between which the relation holds, are different. If this is so, then at any stage at which any self misperceives any terms in an inclusion series as being in time, he will perceive all the other terms in that inclusion series as being in time.

From this it follows that he will perceive H as a whole as being in time. He will perceive it either as the latest term in the future, or the earliest term in the past, according as the relation "inclusive of" appears as the relation of "later than" or "earlier than." (In point of fact, as we shall see in Chapter LX, he will perceive it as the latest term in the future.)

But H as a whole will never be perceived as present. For, as we saw in Chap. XLVII, p. 227, it could only be perceived as present by something which was at the same stage of the C series as itself. And, when we come to consider what is meant by a common C series (Chap. LI, pp. 274–275), we shall see that the only term in the inclusion series of G which is at the same stage in the C series as H as a whole is the term G as a whole. No other term in the inclusion series of G, then, can perceive H as a whole as present. Neither can that term do it. For G as a whole cannot, as we have seen (Chap. XLVII, pp. 230–232), perceive anything as in time.

583. One point remains in this connection. We have seen that each of the three series is bounded in one direction by the whole of the determining correspondence term within which each series falls, and that that whole is the final term of the inclusion series and the C series, which is the limit of the misperception series. But how about the boundary in the other direction? Since the distinction between the terms in each series lies in the differing amounts of content which are found in each of them, it is evident that the series will be bounded at the other end by nonentity. Is this the final term of each series, or the limit, or is it in some cases the one and in some cases the other?

I think that it is clearly the limit in each case. In the case of the inclusion series the relation between each term and those which are beyond it in the direction of nonentity is that the latter are included in the former. If nonentity were a term of the series, it would be included in all the other terms. And it is clear that nonentity is not included in any of the other terms.

With regard to the misperception series it is obvious that nonentity cannot be a term of the series since it is not a perception at all.

With regard to the C series, we have seen that "included in" is the relation which appears either as "earlier than" or as "later than." If, therefore, nonentity is not included in any of the other series, it cannot appear either as "earlier than" or as "later than," and so will not appear in the B series at all. Besides, we do not hold that anything except the existent can occupy a place in the time-series, and nonentity is not existent. Nonentity will therefore not appear in the B series, and therefore will not be a term in the C series.

584. Thus, to sum up our results so far, the misperception series is identical with the inclusion series, except that the last term in the latter—$G!H$ as a whole—is not a term in the former. And whenever H appears to G in $G!H$ as being in time, then the C series, on which that B series is based, is identical with the inclusion series of H. And thus we have settled the relations of the three series, where there are three series.

585. But are the three series always found together? Whenever there is an inclusion series there must be a misperception series, for we have seen that every member of an inclusion series but one must be a misperception. But is it certain that every member of every misperception series perceives its perceptum as in time? If it does not, there is no C series in its perceptum, so far as that perception is concerned. For a C series was defined as the real series which appeared to someone as a B series. Of course, if the same perceptum was perceived as being in time by one self, and not as being in time by another, or perceived by the same self at different stages as being in time, and not as being in time, then its inclusion series would be a C series with reference to the first perception, but not in reference to the second.

The question is not whether *every* perception of a perceptum perceives it as in time. For we know that the perceptions which form the determining correspondence system, and of which the fragmentary perceptions are fragments, do not perceive their percepta as in time, since to do so would be misperception, and these perceptions are free from error. $G!H$ as a whole, then, cannot perceive H as in time. But the question is whether all the other members of the inclusion series do so, and do so in every case.

586. There are two alternatives, each of which would involve a negative answer to this question. In the first place, it might be possible for the same self at one stage to perceive a perceptum as in time, and at another stage to perceive it as not in time. In the second place, even if it were not possible for the same self at different stages to perceive its percepta, respectively as in time and as not in time, it might be possible for different selves to perceive their percepta respectively as in time and as out of time. G might at every stage perceive H as in time, while J might at every stage perceive L, or perhaps H, as not in time.

Is it possible, then, that a self should misperceive the inclusion series of another self, or its own, otherwise than as in time? We cannot, I think, definitely say that this is impossible. But we cannot say what sort of relation would take the place of the B series as the relation which appeared to connect the terms of the series. And we can say that there are many sorts of relations which could not do so.

587. In the first place, as we saw above (p. 249), the relation which the terms appear to have to one another must be such that the terms related by it shall appear to exclude each other— to have no content in common. For we saw that the terms could not be separate terms unless they perceived themselves as excluding the other terms in their own series. And they could not perceive themselves as having such a relation unless they perceived their other percepta, which are at corresponding points in the common C series, as also having such a relation.

The apparent relation, then, must be such that terms related by it would exclude one another. And it is not to be time. Neither, of course, can it be any relation which is logically dependent on time—which could only hold where time-relations hold. For the question here is to decide what, if anything, can be substituted for the time-relation.

The inclusion series is itself a dimension, the relation connecting the terms of which is transitive and asymmetrical. There is a second dimension in connection with this series, viz. the dimension of occurrence at the same point in the common inclusion series[1]. The relation connecting the terms here is transitive and symmetrical. In the time system there are two dimensions corresponding to these—the dimension whose terms are related by the transitive and asymmetrical relation of earlier and later and the dimension whose terms are related by the transitive and symmetrical relation of simultaneity.

If, therefore, any other relation is to be substituted for the relation of time as the one as which, in some cases, the inclusion series is misperceived, it must be a relation which allows for two such dimensions—one whose relation is transitive and asymmetrical and one whose relation is transitive and symmetrical. And many relations will thus be excluded. For example, it will be impossible that the apparent relation should be spatial, since its dimensions do not differ in this way[2].

588. Neither is it possible that the apparent relation should be one which has any connection with changes in the objects which are apparently known, or with changes in the apparent amount of knowledge, or with the clearness of what is known or of any particular part of what is known, or in the accuracy of the knowledge. Neither can it be a relation of causation. For we saw, in Chapter XLVI, that none of these characteristics show any uniform change in time, but that they all oscillate. Since they oscillate in time, they must, as we also saw in Chapter XLVI, oscillate in the C series. And since the C series is the inclusion series, then, if the inclusion series is misperceived in any other way than in time, they will have to oscillate in that series also, and no change in them can be the transitive and asymmetrical relation which is required[3].

---

[1] Cp. Chap. LI, p. 275.

[2] If the relations of the points in a finite space to something outside them be taken, *e.g.* of the points within a square to one of its sides, then, no doubt, dimensions of the sort required would be constituted. But what we require here are dimensions where the relations connecting the terms are directly between the terms, without bringing in anything else—such as the relations of earlier and later are.

[3] There is another point which might be mentioned. It will be seen in Book VII that there is a certain instability in the inclusion series, and that this instability is reflected in the time-series. And there does not seem to be any other relation, which could in other ways be suitable, which would reflect the instability except the relation of time. But this consideration cannot be conclusive, for it might be possible that the inclusion series should appear in some other form, although it was less adequate to express it than time is.

All that can be said, I think, is that we know of no other sort of relation, except time, as which it would be possible that the relation of inclusion should be misperceived. This does not give, of course, an absolute proof that it can only be misperceived as a time-relation, but it does give a reasonable ground for a belief that, when misperceived, it is always misperceived as a time-relation.

589. Before concluding this chapter, we may notice a consequence which follows from the result reached on p. 248. We saw, in Section 201, that a primary part need not have itself as a member of its differentiating group. And we saw in Chap. XXXVI, pp. 79–81, that it would be possible for a self not to perceive himself, and so not to be self-conscious. Now a self who is not self-conscious can contain no perceptions of his own states, and, consequently, no misperceptions of them. From this it follows by p. 248 that, if a self does not perceive himself, he will have no fragmentary parts which are perceptions—since fragmentary parts can only be perceptions if they misperceive the relations in which they stand to one another. Such a self will have, of course, an infinite series of perceptions, forming his system of determining correspondence parts. But those perceptions will have no fragmentary parts which are perceptions. And as it is only the fragmentary perceptions which are misperceptions, such a self—if there are any such selves—will have no perceptions which are not correct. There is, however, no reason to believe that there are any such selves, though there is no reason to believe that there are not.

CHAPTER L

COMPLIANCE WITH THE CONDITIONS

590. We have now reached a theory as to the terms and relations of the C series, and, at the same time, as to the terms and relations of the inclusion and misperception series. We must proceed to test this theory by enquiring whether it complies with the twelve conditions which we have seen to be indispensable —the eleven conditions mentioned in Chapter XLVI and the twelfth which was added in Chapter XLVIII (p. 240).

The first condition was that the C series must be one which can be found in spiritual substances, and in spiritual substances which, in other dimensions, are divided into parts within parts to infinity by determining correspondence. This condition can certainly be satisfied by our theory.

591. The second condition was that the theory must allow for the existence both of correct and of erroneous cognitions. And, with regard to the latter, it must not only allow for the fact that there is erroneous cognition—that is, erroneous perception, since all cognition is perception. It must, as we have seen, reconcile this with our certainty, the *primâ facie* form of which is that every perception is correct at the time when it is made.

This we are able to do. For "the time at which it is made" is a misperception, the reality misperceived being "the point in the C series at which it occurs." Now all perceptions at any point in the C series are, as we have seen, misperceptions, with the exception of those at the final point of the series—*i.e.* in our previous example, $G!H$ itself as a whole. And the nature and amount of the error in any perception depends on its place in the C series. This does not mean that the nature and amount of the error vary directly with the place in the C series. We saw that we must allow for oscillation in the correctness of our perception as the C series advances. What does vary directly is the amount of perception, in the sense previously explained, and this, as we shall see later in this chapter, is consistent with oscillation in the nature and amount of error. But the amount of the error is determined by the amount of perception, though it does not vary directly with it, and so it depends on the place of the perception in the C series.

Thus, in all cases in which error has to be allowed for, the qualification which appears in the form "at the time at which it is made" turns out to be really "from a standpoint involving a certain error." In other words, a perception will have no more and no less error in it than what is involved in the fact that it is in the place in the C series in which it is.

Every perception, then, in any C series, except the last, will on this view be partly correct and partly erroneous. And this is all that we require in the case of apparent perceptions—those, that is, which, besides being perceptions, appear to be perceptions. For all such will, on our theory of the true nature of reality, be partially erroneous. They none of them, for example, perceive anything as being both timeless and spiritual, while all the objects they perceive, like everything else, are really both. On the other hand, there is no reason to believe that any of these perceptions are entirely false. When we perceive what is really timeless and spiritual as a material thing in time, we are perceiving it, no doubt, as very different from what it really is. But it may well be the case that we are perceiving it, both as to qualities and relations, with some of the characteristics which it really has; and that we distinguish it from other objects, to some extent, by the characteristics which really distinguish it from other objects.

592. So far, then, our second condition has been satisfied. But there is a further question. Our theory tells us that all perceptions in the C series will be partially erroneous and partially correct, and this, as we have just seen, fits in well enough with the nature of those perceptions in the C series which appear to us as perceptions. But besides these, we have states which appear to us as judgments. And, as we have seen, some of these must be accepted as quite correct, and some as quite erroneous. Now all these must, if our theory is correct, be in reality perceptions in the C series, although they appear to be something else. Each of them, then, must be partially correct and partially erroneous. How is this difficulty to be met?

This brings us to the third condition—that our theory must allow, not only for the possibility of erroneous cognition in general, but for the particular forms of erroneous perception which do exist. As it will need a long and detailed enquiry to decide whether this condition has been satisfied, it will be better to postpone it. (It will occupy Chapters LII to LVII.)

593. The fourth condition is that the C series, like the B series, must be a series of one dimension, and that the relation which constitutes it shall be transitive and asymmetrical. Both these requirements are satisfied by our theory. The relation which constitutes the C series is either "inclusive of" or "included in," and either of these relations is transitive and asymmetrical. It is possible, indeed, for those relations to constitute a series in two dimensions. Such is the case, for example, when species are included in genera, genera in orders, and so on. But it is possible to have a series of inclusions which is only in one dimension, and, as we have seen, the series which we have taken is of this nature.

594. The fifth condition is that the C series must have at least as many terms as can be distinguished from each other in the B series. If this were not so, the B series could not be explained as a misperception of the C series. But this condition can certainly be complied with, because there is nothing in our theory to prevent the number of terms of the C series being as large or as small as may be required. They must all, indeed, fall within the limits of the whole—in our example $G!\,H$—but within those limits there can be any number of increments of the D series, which constitute the differences between the terms of the C series. And, consequently, there can be any number of terms in the C series.

We have not determined whether the B series has or has not terms next to each other, and therefore we have no ground for determining whether the D series has terms next to each other. But our theory would allow for either alternative. If there was a minimum increment in the D series which was required to constitute a fresh term of the C series, there would be next terms in the C series. If there is no such minimum increment, there would be no next terms in the C series. And our theory leaves it possible that there should or should not be minimum increments in the D series.

595. The sixth condition is that our theory must allow for the fact that, while in absolute reality my knowledge of substance is differentiated into parts of parts to infinity, in present experience there is no such infinite differentiation of knowledge. Any substance is made up of parts of parts to infinity. And determining correspondence involves that, when I perceive a substance, I perceive all its parts in the determining correspondence system— my perception of the parts being parts of my perception of the whole. In absolute reality, then, all my perceptions will be differentiated into parts of parts to infinity. But this is certainly not the case in present experience. There, when I perceive anything, I may perhaps perceive some sets of its parts, but I certainly do not perceive an infinite number of such sets. (And also, we do not, in present experience, perceive all the past and future parts of those determining correspondence parts which we do perceive. The explanation given in the text will meet this case also.)

How is this difficulty to be met? We cannot meet it by saying that what is not cognized at any one moment of time in present experience can be cognized at other moments of time. For the C series is a third dimension, additional to the two which are involved in the determining correspondence system. Whatever is found at any stage of the C series is itself differentiated by means of determining correspondence, and is therefore differentiated to infinity. The perception, then, which appears as my knowledge at any point in the B series, is differentiated infinitely more than the knowledge itself is. Can this be accepted?

596. I believe that it can. To explain this we shall require a fresh symbol. Let + stand for "and all its determining correspondence parts." Then $H +$ will stand for H and all the determining correspondence parts of H. And $i_x G! (H +)$ will stand for some particular stage x, in the inclusion series of G's perception of H and of all the determining correspondence parts of H. (The inclusion series, we have seen, is identical with the C series.)

Now it seems clear that when $i_x G! (H +)$ is a misperception of $H +$, then, though it will be a perception which is infinitely differentiated, it need not be infinitely differentiated into other perceptions. For, as was said in Chap. XLVII, p. 227 footnote, the fragmentary parts of a determining correspondence perception are states of perception, but not necessarily perceptions. Any number of these parts might form only a single perception. And so it would be possible that a fragmentary part, while divided into states of perception to infinity, was only one undivided perception.

This could not happen in the case of the last term of the inclusion series of $G!(H+)$, which is not a fragment of $G!(H+)$ but is $G!(H+)$ itself. For in the determining correspondence parts, taken as wholes, there can, as we have seen (Chap. XLVII, pp. 228–232), be no error, and there is certainly error when we perceive as undifferentiated what is really differentiated[1]. Nor would the differentiation of $G!H$ to infinity be possible, unless its parts were sufficiently determined by being separate perceptions of parts of H to infinity, since that is the only possible relation of determining correspondence.

But neither of these difficulties will arise in the case of the fragmentary parts. The first will not arise because the fragmentary parts are misperceptions, and therefore can misperceive $H+$ as being an undifferentiated whole, though it is not one. Nor will the second difficulty arise. For when the determining correspondence series of parts is sufficiently determined, the fragmentary parts can be sufficiently determined as fragmentary parts of those determining correspondence parts, in the way shown in Chap. XLVII, p. 227.

597. Now all our present experience is in the misperception series, and therefore this explanation will apply to it. Whenever G perceives H, of which he does not perceive the parts, his perception will be a misperception of $H+$—that is, of H and of all its determining correspondence parts within parts to infinity. This misperception may also be called a confused perception, since the particular form which the error takes is that objects which are distinct and separate are not perceived as distinct and separate. G does not in his perception of $H+$ perceive H as distinguished from its parts, nor the parts of H as distinguished from each other.

---

[1] If it were merely *not perceived as being* differentiated, the perception would be incomplete, but not necessarily erroneous. But there are certainly cases when the perception is *perceived as not being* differentiated into other perceptions to infinity. And in these cases there is positive error.

Further, it is possible that what we misperceive as an un-differentiated whole may not be one unit in the determining correspondence system, but an aggregate of such units. It need not be of the type H, or $H!K$. It may be the aggregate of H and L, or of $H!M$ and $H!Q$, or of $H!M$ and $L!R$.

We are naturally reminded of the illustration used by Leibniz—the waves which fall separately and at different times, but which are heard at a sufficient distance as a uniform murmur. We must remember, however, that this, and any other experience, when interpreted in the ordinary way, does not allow for misperception. The difference of the sound at a distance is not, on the ordinary view, caused by the misperception of a perceptum, but by the difference of the percepta caused by the waves to hearers at different distances. The analogy, then, with the case of the waves is not very close.

598. The seventh condition is that the theory must allow for the persistence and recurrence of certain contents in the time series of our experience. This condition also can be satisfied. What I perceive consists of the selves which form my differenti-ating group, and their parts within parts to infinity. In the C series, at every step of that series, each of these selves and of their parts to infinity will also be perceived. This does not, as we have seen, prevent the content of the different C stages from being dissimilar to one another, because the perceptions are misperceptions, and therefore they can be different, though they are of the same objects. But though they can be dissimilar, they can also be similar. In the first place, the perceptions are only partially erroneous, and so they may resemble one another because they both perceive the object in some particulars as it really is. And, in the second place, although the nature of the error can change, and *must* change in some respects, in order that the different C stages may be dissimilar to one another, it need not change completely at each stage. And thus even an erroneous element may be persistent through various stages.

Persistence, then, is possible. As for recurrence, the positive element in it—the fact that the same content is to be found at several different stages—is accounted for like persistence. The negative element—that these stages are separated from each other by others in which that content is not to be found—is a branch of the more general question of oscillation, to which we must now turn our attention.

599. That there should be change in the content of experience is, as has just been said, easily explicable. But that there should be oscillation in the nature of that change introduces a fresh difficulty. If, in going along the terms of the C series in either direction, we find, first an increase in the amount of a certain characteristic, then a decrease, and then once more an increase, it is obvious that the amount of the characteristic in question cannot be determined simply by the relative positions in the C series of the terms which possess it.

There are three such oscillations which must be considered. The first forms the subject of the eighth condition, which says that our theory must allow for changes and oscillations in the extent of the content of our experience, and in its clearness as a whole. To take first the question of clearness. Is it possible that the amount of this characteristic should oscillate? I see no reason against it. It is true that the characteristic of the amount of perception does not oscillate. Each term, as we go along the series in one direction, is inclusive of all that come before it, and the D series which this involves is one of amounts of perception. Thus, as we take the series from included to inclusive, each term will have a greater amount of perception than the one before it, while, if we take the series in the reverse direction, each term will have a less amount than the one before it. But an increase or decrease in amount of perception need not involve an increase or decrease in clearness of perception—even when the amount of perception is taken, as it is in this case, intensively and not extensively.

For it must be remembered that, although there is a quantitative increase in amount of perception as we proceed from the included terms to the inclusive terms, yet it does not follow that the change is quantitative only. It is quite possible that it is also qualitative—that each amount added may have different characteristics. If this were so, it is quite possible that certain additions, while adding to the amount of the perception, might also, until still further additions were made, make the perception less internally harmonious, and more confused. In that case the perception would be less clear than a perception of smaller amount, which was included in it. And, on the other hand, it is possible that the addition of yet a further amount might make the perception more internally harmonious, less confused, and more clear. It is therefore possible that the clearness of my experience as a whole might oscillate as the series of C terms continues, and would therefore appear to oscillate in time.

600. It is true, no doubt, that oscillations in the clearness of my experience appear to come rather from causes in myself than in the content of my experience. They are due to such causes as fatigue, or illness, or sleepiness, and would not happen to a person perceiving what I am perceiving, unless he was also affected by influences of this nature. But there is no difficulty in the fact that the oscillations depend on the nature of the percipient. For it is only to be expected that the answer to the question, whether a given change in amount of perception would make the perception more or less clear, would depend in part on the nature of the percipient. We have seen that all percipients must have natures differing more or less in the characteristics which constitute them, whether these are qualities or relations, or both. And it may well be that the difference in the characteristics of any two selves are such as to involve the result that a certain change in amount of perception, when they are perceiving the same object, might bring about a clearer perception in the one case, and a more confused perception in the other. And since, when selves are erroneously perceived as in time, certain differences of characteristics will appear as differences in events, we can see why, in the world as it appears to us, the events which appear to happen in us appear to determine changes in the clearness of our perceptions[1].

601. But how about those cases, such as occur in dreamless sleep, in which consciousness appears not only to diminish in clearness, but for a time to vanish altogether? Are we to say that the C series passes, without any intermediate stages, from the stage which corresponds to the last moment of consciousness before sleep to the stage which corresponds to the first moment of consciousness

---

[1] Fatigue, illness, and sleepiness are, of course, states, not of the percipient, but of that which appears as the body of the percipient. But they do not affect the clearness of perceptions directly, but only by affecting the condition of the percipient himself.

after sleep, and that the sleep itself contains no C stages? At first sight this seems the obvious answer, since the stages of any C series are, as we have seen, a series of states of perception, and in the sleep in question there appears to be no perception at all.

602. But there are difficulties in the way of this view. If the series of C states does go on unbrokenly from the consciousness before sleep to the consciousness after sleep, it seems difficult to account for the very strong suggestion, which our experience undoubtedly presents, that these two states are not continuous. And again, when I wake up after some hours of what would usually be called dreamless sleep, I find the content of my experience altered, in many respects, in just the same way as it would have been altered by some hours of conscious life. The position of the sun in the sky, the position of the hands of my watch, are altered in the same way as they alter when I am awake. And this, also, suggests very strongly that there are stages of the C series which correspond to every period of sleep.

Besides this, we shall see in the next chapter that we must admit that there is a certain phenomenal validity in the assertion that two events in two different selves may be simultaneous. And the only way of allowing any such phenomenal validity is by showing that two selves have in a certain sense a common C series. But this solution would be impossible if the C series did not run through dreamless sleep. For then, if Smith and Jones had experiences simultaneously, and subsequently had again experiences simultaneously, and if Smith kept awake all the time between the experiences, while Jones slept dreamlessly for some hours, there would be, between the pairs of simultaneous points, terms in Smith's C series which had no terms corresponding to them in Jones' C series. And this would destroy the possibility of explaining a common time series as the appearance of a common C series.

For all these reasons it seems that we must conclude that, when apparently dreamless sleep occurs, what really happens is that the clearness of my experience sinks to so low a level that it cannot be remembered after waking, but that the consciousness does continue; and that there is therefore a continuous C series between the last moment before the sleep and the first moment after it.

There is no empirical consideration, as far as I know, which renders this view either impossible or improbable. We do not remember the experiences in the sleep, which we therefore suppose to be dreamless, but we do not remember everything, and the less clear our experience is, the less likely we are to remember it. Our experiences in falling asleep, or in waking gradually, have very little clearness, and if the experience between should be still less clear, it is natural enough that we should fail to remember it, and should suppose that there had been no consciousness between sleeping and waking. Even on empirical grounds, it has sometimes been maintained that sleep is never completely without consciousness, and at any rate there is no empirical difficulty to prevent our holding this, if we are led to that conclusion on other grounds.

603. But, besides oscillations in the clearness of knowledge, we had also to allow for oscillations in its extent. We should, however, rather say, in its apparent extent. *Primâ facie*, no doubt, I appear to observe much more at one time than at another. My consciousness may at one moment be diffused over a large field, and next moment concentrated on a small part of it, so that much seems to have dropped out, while nothing has come in. And this presents a difficulty, since the whole of my differentiating group, and all their determining correspondence parts, is represented at each point in the C series, and therefore, it would seem, the extent of my field of consciousness ought to be the same at each of those points.

We saw, in discussing the sixth condition, that it was possible for an object which was differentiated to infinity to be misperceived as not being infinitely differentiated. If we apply this result here, our difficulty can be removed. For then it will be possible to hold that the field of objects which I perceive has always the same extension at every stage in the C series, and that the apparent changes in the extent arise from parts of that field being sometimes perceived more clearly, so that they are perceived as separate objects, and sometimes less clearly so that they are not perceived as separate objects. Thus what seemed to be an oscillation in the extent of the field of experience would turn out to be only an oscillation in the proportion of that field which was above a certain standard of clearness.

604. How then is that part of the field of objects perceived at all, which, at any point in the C series, is not perceived as separate objects? It seems to me that there is to be found in our consciousness a perception of a vague background to our more definite perception, which is itself not a perception of any definite object or objects as being a definite object or objects. It would, I think, often be described as one's consciousness of being alive, and although the phrase is loose, it is perhaps suggestive—though of course such a consciousness must be clearly distinguished from the percipient's perception of himself as a definite object, which has been discussed in Chapter XXXVI. I believe that the background, of which we are thus conscious, is the whole of that part of the field of experience which at that point is not perceived as separate objects. The existence of such a background does not, certainly, prove this, since its existence might perhaps be accounted for otherwise. But it does give us something which is perception without being perception of definite and differentiated objects, and which could enable us to explain the oscillations in the extent of that part of our experience which is experience of definite objects[1].

The same explanation would account for the fact that our experience of objects changes and oscillates in apparent extent in the C series. Not only do we sometimes perceive objects which at other times we appear not to perceive, but we sometimes perceive points in the C series of any one object (appearing as its states in time) which at other times we appear not to perceive[2]. The explanation here also would be that what appeared not to be perceived was really perceived as part of the undifferentiated background[3]. And thus we should always perceive all the C states of any self which we perceive at all.

605. The ninth condition is that our theory must allow for change and oscillation in the clearness of our knowledge of particular objects, including the extreme case when something

---

[1] Cp. Dr Ward's *Principles of Psychology*, Chap. IV, Section 6.

[2] I may, *e.g.*, at one time think of Napoleon at Austerlitz, and at another time of Napoleon at Waterloo.

[3] Of course only those perceptions, which are the same point in the C series as their percepta, *appear* as perceptions. The others will appear as judgments, assumptions, or images, though they are really perceptions.

which did not appear to be in consciousness at all, appears to come into consciousness, or *vice versâ*. This latter case we can account for, on the same principle which we have just employed, by holding that the consciousness which appears to arise or cease has really only risen above or sunk below the degree of clearness which is required in order that separate objects should be perceived as being separate objects, and that when it appears not to have arisen or to have ceased it is really a part of the consciousness of background, of which we have spoken.

This case, then, can be reduced to a case of change and oscillation in the clearness of our knowledge of particular objects. Can we explain this? The apparent causes of such oscillations seem to be sometimes in the object perceived—as when, in ordinary language, I cease to see a thing because it moves away from me. Sometimes they appear to be in the observer—as when I cease to notice a thing because my interest is called off to another matter. Sometimes they appear to be in something which is neither the observed object nor the observer—as when I cease to see a thing because my body moves away from it. The cause cannot be assumed to be really what it appears to be, but these facts create, no doubt, a presumption that the real cause of the oscillations in the clearness of G's perception of H are sometimes in H himself, sometimes in G, and sometimes in some other substance.

606. We are able to satisfy this ninth condition, and to satisfy it in such a way that we can accept the presumption just mentioned that the cause of the oscillation may be found either in G, the observer, or in H, which is observed, or somewhere else. We have seen (p. 262) that the change, as we pass from included terms to inclusive terms, need not be only quantitative, but might also be qualitative, and that in this case an addition to the amount might increase or diminish the internal harmony of the perception, and might render it less or more confused. Now in the case of the perception of two objects by the same subject, say $G!H$ and $G!K$, it is clear that the characteristics of $G!H$ and $G!K$ will depend in part upon the characteristics of H and K, and will be different if those characteristics are different. And, since H and K are distinct substances, they must be dissimilar, and have characteristics which are in some degree different. $G!H$ and $G!K$ then will be different, and consequently an increase in the amount of perception, as both of them pass from one stage to another in the C series of G, might change the amount of internal harmony in one, while leaving it unchanged, or changing in the inverse direction, in the other. And this will account for changes and oscillations in the clearness of G's perceptions of particular objects, and will accord with the presumption that the cause of such changes and oscillations is sometimes to be found in the objects perceived.

But, again, G has a nature of his own, and a perception $G!H$ will differ from a perception $L!H$ because G and L have different natures. And thus it might happen that a change in the amount of perception in the C series of $G!H$ would make a change in the amount of its clearness, when, at a corresponding point in the C series of $L!H$, a corresponding change of amount would make no change in the amount of the clearness, or would make it in the inverse direction. And thus we should have changes and oscillations in the clearness of the objects perceived by G, which would accord with the presumption that the cause of some of these changes and oscillations is to be found in the percipient.

It must be considered, in relation to this last possibility, that the other members of the C series of $G!H$ and $L!H$ may vary more than the wholes $G!H$ and $L!H$ vary. The latter are both correct perceptions of the same object, H, and, though this does not prevent their being different from one another, since they are perceptions in different percipients, it limits the possible variations. But the other C stages in each of these are made up of misperceptions, and it is quite possible that the stages may vary more than the wholes which they make up vary, and that the variation may be due to the difference between G and L.

Finally, the natures of G and L may be different, by reason of their different relations to other substances. If, for example, G has M as one of its differentiating group, and L has not, this will make a difference between them[1]. And, again, H and K may be different from one another in the same way.

---

[1] We saw, in Chap. XLII, p. 171, that such differences as these could not be the

It is therefore possible that, when a change of the amount of perception takes place in passing along the C series of $G!H$, and the perception alters by becoming more or less clear, this may be due to the relations which G, or H, or both of them, bear to some other substance or substances, M, N, etc. And this would accord with the presumption that the real cause of the changes and oscillations in the clearness of my perceptions of any substance may sometimes lie partly in a third substance.

It is true, of course, that the presumption immediately suggested by the facts is that the clearness of the perception is causally determined by *events* in the percipient, in the object, or in some third substance, and that we have spoken here of their determination by timeless characteristics of those substances. But, as we saw when we were speaking of the eighth condition (p. 263), certain differences of characteristics will appear as differences in events when the substances to which they belong are themselves erroneously perceived as in time.

607. The tenth condition is that our theory must allow for changes and oscillations in the accuracy of knowledge. It is plain that if we have accounted for changes and oscillations in the clearness of G's perception of H, a similar explanation can be employed to explain the possibility that the perceptions should change and oscillate in respect of the degree in which they were erroneous. If an increase in the amount of perception could either increase or diminish the internal harmony of the misperception, it could increase or diminish the extent to which it misrepresented the facts.

But the only case in which we were absolutely certain that, in our present experience, the accuracy of knowledge changes and oscillates was the case of the knowledge which appears as judgments. If our theory is correct, all this knowledge is really perception. Whether it is possible that what is really perception can appear as judgments, and sometimes as true judgments, and sometimes as false judgments, is part of the question whether our theory can satisfy the third condition. This will be discussed

only differences between selves, as that would involve either a vicious infinite series, or a vicious circle. But we saw in the same chapter that selves could be differentiated in other ways. And, when once G, L, and M are otherwise differentiated, then an additional difference between G and L will arise from their different relations to M.

in Chapters LII–LVII. But, if this general question should be settled in the affirmative, there would be no difficulty in the change and oscillation of the truth of our judgments on a given subject, since there can be change and oscillation in the correctness of our perceptions, and since the reality which appears as judgments is really perceptions.

608. The eleventh condition was that our theory must allow for some relation of the content of perception to its place in the time series. And this has been satisfied. For that which appears as a place in the time series is really a place in the C series, and we have seen that the place of a term in that series will modify its content, though the effect of the modification will not always be in the same direction in respect of extent, clearness, or accuracy.

609. Finally, we have complied with the twelfth condition— that the terms of the misperception series, while all parts of the correct perception $G! H$, do not form a set of its parts.

CHAPTER LI

FURTHER CONSIDERATIONS ON TIME

610. The theory of error which we have now reached is, as will have been seen, very closely connected with the illusion of time. All error is misperception. Every term in the misperception series is a term in the C series, and so appears as a term in the B series, and appears to be in time. Conversely, every term in the C series, except the last, is a term in the misperception series, and, as this last term never appears as being present, it follows that every term which appears as being present is a term in the misperception series.

The nature of the time illusion is that, from the standpoint of each stage in the misperception series, all that is on one side of that stage in the C series appears as future, and all that is on the other side of it as past, while the stage itself appears as present. Moreover, the future appears to be continually becoming present, and the present appears to be continually becoming past. This series of past, present and future is what we have called the A series, on which the B series of earlier and later is dependent. The term P is earlier than the term Q, if it is ever past while Q is present, or present while Q is future[1]. But both

---

[1] Two terms may both be present together, although one is earlier than the other. This is due to the fact that the present is a duration, and not an indivisible point. But the statement in the text remains an adequate definition of "earlier than," for although P and Q may at one time be in the same present, yet, before that, P is present while Q is future, and, after that, P is past while Q is present.

Since the present comprises different terms, of which any one will be earlier or later than any other, it might be thought that the fact that P was earlier than Q would be perceived when they were both present, and that "earlier than" need not be defined in terms of the A series. After this, it might be thought, the future might be defined as what is later than the present, and the past as what is earlier than the present. Thus the A series would be defined in terms of the B series, instead of the B series in terms of the A series.

But this would be a mistake. For the series of earlier and later is a time series. We cannot have time without change, and the only possible change is from future to present, and from present to past. Thus until the terms are taken as passing from future to present, and from present to past, they cannot be taken as in time, or as earlier and later; and not only the conception of presentness, but those of pastness and futurity, must be reached before the conceptions of earlier and later, and not *vice versâ*.

these series are only appearance. The C series is real, but no terms are really past, present, or future, and there is no real change. And, since the A series is not real, the B series is also not real.

611. The present, however, of which we speak is not an indivisible point. It has a certain duration, which comprehends more than one term. The C series, as we have seen, consists of simple terms, so that the duration of such a present cannot be divided into parts of parts infinitely. But the number of terms in such a present may be infinite, if no simple term is next to any other simple term. If this is not the case, the number of terms in a present will be finite.

Thus presents overlap one another. We have a present, for example, containing the terms WX. But a later present will contain XY, where W has ceased to be present, and become past, and Y has become present instead of future. And a still later present will have lost both W and X, and will contain Z as well as Y.

There is no fixed magnitude for such presents. Their magnitude may vary from self to self, and from one part to another of the time-series of the same self.

Presents like this are frequently called Specious Presents. It will be convenient, I think, to employ the same name occasionally, in order to remind ourselves that the present of which we are speaking is a duration, and a duration which may vary in length. The special need, however, for the use of such a term, has been removed by the adoption of our theory. If time were real, it would be difficult to avoid accepting the existence of an absolute present—whether a point or a fixed duration—as distinguished from the observed present, which is a duration varying from individual to individual, and from time to time. (Cp. Chap. xxxiii, pp. 28–29.) And thus it would be necessary to have a special name for the observed present, to distinguish it from the absolute present. But, if there is really no time, but only an illusion of time, we have no present to consider except the observed present, and to distinguish it by the name of specious present is no longer necessary, though it may be convenient.

612. In this chapter we shall discuss four questions relating to time. The first is whether, on our theory, we can in any sense assert that two or more selves have a common time-series. The second is whether, on our theory, we can in any sense maintain the ordinary distinction between real and apparent length of time. The third is whether time is to be taken as infinite in length. And the fourth is whether it is to be taken as infinitely divisible.

613. There is, on our theory, no time-series, for nothing is in time. There is no series of events, but a timeless series of misperceptions which perceive a series of timeless existents as being in time. Now every misperception, since it is a perception, must fall within a self, and no misperception can be common to two selves. Two selves can no more have the same misperception than the same fit of toothache. They may have misperceptions which are similar in various respects. In particular they may have misperceptions which are similar in being each of them a misperception of L, and misperceiving it as being X. But they cannot have the same misperception.

It would seem therefore as if there could not be a common time-series. But, on the hypothesis that time was real, there certainly would be a common time-series. My remorse and yours might be really simultaneous, just as my remorse would be really later than my crime. Now we have seen that, although time is not real, the appearance of time is a *phenomenon bene fundatum*, since the order of the apparent events in time is the same as the order in the inclusion series of the realities which appear as events in time. If, for example, the apparent event of my crime appears, in the time-series, between the apparent events of my temptation and my remorse, then the stage in the inclusion series which appears as the event of my crime will be really, in the inclusion series, between the stages which appear as the events of my temptation and my remorse. And therefore a common time-series between different selves, which would be absolutely real if time were absolutely real, ought to be a *phenomenon bene fundatum* if time is a *phenomenon bene fundatum*. In other words, we must be able to say that it is as true that my remorse is simultaneous with yours as it is to say that my remorse is later than my crime.

If all percipient selves perceived nothing but one self, and the parts of that self, the question would be simpler. If we take H and G as percipient selves, and L as the perceived self, then the perceptions which H and G have of any given state in the C series of L as present, may be taken as simultaneous with one another, since, *sub specie temporis*, they stand in the relation of simultaneity to the same thing. And, when points in two different time-series can be taken as simultaneous, the two series form a common time-series.

614. But the hypothesis in the last paragraph does not represent the facts. For we have seen that, in order to establish determining correspondence, there must be primary parts which have differentiating groups consisting of more than one primary part—in other words, that there must be selves, each of whom perceives more than one self. And we are not certain that *any* two selves do perceive exactly the same group of selves. We must allow for the possibility, for example, that H perceives only himself, L, and M, while G perceives only himself, N, and O. They have thus no percepta in common. Can they have a common time-series?

I think that we can legitimately speak of a common time-series in this case. The time-series of H and the time-series of G have not, as their respective C series, the *same* inclusion series. But they have, as their respective C series, *correspondent* inclusion series, for all inclusion series correspond to one another. And this will give us a common time-series which will be, like the time-series in each self, a *phenomenon bene fundatum*.

The correspondence of different inclusion series is not only essential in order to give a common time-series, but even for the explanation of the time-series in a single self. For we said in Chap. XLVII, p. 227, that a perceptum which was at a correspondent stage in the C series—*i.e.*, the inclusion series—to the stage of its perception would be perceived as present, while others would be perceived as future and past. It was mentioned there that the question of correspondent series would be discussed in the present chapter.

615. In what manner can inclusion series correspond? Each of them has, as its last term in one direction, a term which contains all the content which falls in any of the terms. Each of them is bounded at the other end by nonentity. The terms which fall in between differ from each other in the amount of content which falls in each of them, or, in other words, in the amount of the D series of increments which falls in each of them.

This gives us a ground for establishing correspondence. The last terms of each series—terms which have the common quality that each of them contains all the content which falls anywhere in its series—will correspond to each other. And, of the rest, any two terms in different series will be correspondent if each of them contains the same proportion of the content of its series as the other does of the content of its series.

It is, as was maintained in Chapter XLVI, correspondence of this sort which causes any term to be perceived as present—that is, as simultaneous with its perception. And, since this correspondence exists, not only between perceptions and their percepta, but between terms in every inclusion series in the universe, it gives us a common time-series. When terms which are at correspondent stages of different inclusion series are taken as events in time, they are taken as simultaneous events in time. They are not really events in time, but their simultaneity is a *phenomenon bene fundatum*. They are as really simultaneous as two things in a single self can be—for example my perceptions of taste and smell when eating an orange.

Since a point can thus be found in any individual time-series which is simultaneous with any given point in any other individual time-series, and since terms which are earlier or later than any term are earlier or later than any term simultaneous to it, we should thus have reached a common time-series, which would include everything which ever entered into any individual time-series as an apparent event in time.

616. It has been seen that the last term, in one direction of the time-series in every self, will be a term which contains all the content which falls in any term in that series. And, in the common time-series, all these terms will be simultaneous. At the other end, each series will have terms, each of which is simultaneous with a term in each other series, until the series stops at the boundary, which is nonentity. It follows that, when they are looked at as terms in a common time-series, each self will appear to have the same duration—a duration which will stretch from the beginning to the end of time. This is, no doubt, only an appearance, since selves are not in time at all. But it is a *phenomenon bene fundatum*, and it is as true as any other statement as to temporal duration. It is as true that I endure through all time as it is that my repentance came after my crime[1].

617. The second question which we had to discuss was whether, on our theory, we can in any way maintain the ordinary distinction between real and apparent length of time. On the theory that time is real, such a distinction must be made, and presents no difficulty. Let A and B stand equally near a clock which goes perfectly accurately, from the time it strikes one till the time it strikes two, and let the duration of their experiences from one striking to the other be called m for A and n for B. Let them do the same next day, but now let A be more bored with whatever occupies his thoughts than he was the first day, while B is more interested in whatever occupies his thoughts than he was the first day. Let us call the durations of their experiences for the second day, o for A and p for B. Then o will *appear* longer than m, and m will *be* equal to n, while n will *appear* longer than p, and p will *be* equal to o. The oscillation of "appear" and "be" in this statement prevents it from being contradictory. If the word had been "be" in each case there would have been a contradiction.

If time is not real, it follows that the word in each case must be "appear." This does not in itself involve a contradiction. But, if in all four cases the appearance is taken as on the same level, it is impossible that time should be a *phenomenon bene fundatum*. For if o appears longer than m, and m appears equal to n, and n appears longer than p, and p appears equal to o, then it is clear that the appearances of equality or inequality in time cannot have a uniform one to one relation to any characteristics of that timeless reality which appears as in time, and in that case time cannot be a *phenomenon bene fundatum*.

We can, however, escape from this difficulty if we give a different meaning to "appear" the first and third times that it occurs, from that which we give it the second and fourth times

---

[1] The consequences which arise from this conclusion will be considered in Chapter LXII. Cp. also Chapter XLIII.

that it occurs. And, when we look into the matter, we see that the meaning is different.

618. What are the causes which produce the result, which would be expressed, on the theory that time is real, as the appearance of equal periods of time as unequal? Periods of boredom and periods of intense expectation seem longer than the normal. Periods in which we are deeply absorbed in what we are doing seem shorter than the normal. And, again, periods into which many exciting events have been crowded seem on retrospect to be longer than periods of quiet tranquillity—at any rate, provided that these latter have not been wearisome.

The reason for this result in the first case seems to be that in periods of boredom or expectation we pay more attention to the passage of time than usual, because we are more than usually anxious for it to pass, and because we have little else—or little else on which we can fix our minds—to which to attend. And, since we pay as much attention to time in a short period as we should usually pay in a longer period, we judge the period to be longer than it is. It is not, therefore, the apparent greater length which leads to tediousness. It is the tediousness which makes the length seem greater than it is. Again, in periods in which we are deeply interested in what we are experiencing, we have little attention to spare for the lapse of time, and so we tend to judge that little time has elapsed.

In the second case—when a period which has been full of exciting events seems longer than an equal period of tranquillity—the reason seems to be that we remember more of the content of the exciting time than of the tranquil time, and so tend to judge that the greater number of events must have been spread over more time than the smaller number of events.

In the second case, the effect is clearly only produced in retrospect. It is only when we look back on a crowded period that it seems longer than a tranquil period. It is not so clear in the first case that the effect is only produced in retrospect, but I think that we must conclude that it is so. It is true that, in a period of boredom or expectation, we say that the time seems to be passing slowly, but the real fact is, I think, that the part of the period which has already passed seems to us now to have passed slowly. When we take a period approximating to a specious present—for example, the period between two ticks of a clock— I do not think that it seems longer when we are bored or expectant, though the period between the striking of two hours does seem longer under those circumstances.

619. We can now see that, according to our theory that time is unreal, there will be a different meaning for "appears," when we say that m appears equal to n, and p to o, from the meaning which it has when we say that o appears longer than m, and n appears longer than p. And the first appearance will be real relatively to the second, and will be a *phenomenon bene fundatum*, while the second is not a *phenomenon bene fundatum*, and is not real even relatively to the first.

So far as the first sort of appearance goes, equal stretches in the C series will appear as equal stretches in the B series—that is, as equal periods of time. And stretches which have any given proportion to each other in the C series will appear as periods of time having the same proportion to each other. Two stretches in the C series are equal if they are both the same proportion of the whole series from zero to the term which contains all the content. Thus two stretches in different C series, each of which stretched from a term which contains $\dfrac{x}{z}$ of the content to a term which contains $\dfrac{x+y}{z}$ of the content, would be equal.

This sort of appearance, therefore, is erroneous in respect of its representing two equal stretches of the C series as periods of time, which they are not, but not erroneous in representing them as equal, which they are. But when the period between the two strikings of the clock appears long to A, who is bored, and short to B, who is interested, the appearance is doubly erroneous—as representing stretches in the C series as periods of time, and as representing them as unequal, when they are really equal. Thus, while they are both appearances from the standpoint of absolute reality, the second is appearance from the standpoint of the first, while the first is reality from the standpoint of the second.

And the first sort of appearance will be a *phenomenon bene fundatum*, since all its assertions about the magnitude of durations in time will have a uniform one to one relation to the real magnitudes of the stretches which are misperceived as durations in time. But this is not the case with the second sort of appearance, and so it is not a *phenomenon bene fundatum*.

It is thus possible to retain the common-sense distinction between real and apparent time, even if all time is appearance, in the same way that, even if all space is appearance, we can find a meaning for the statement that the sun appears larger at sunset than at noon, but is really the same size.

620. Our third question was whether time is to be taken as infinite in length. What is meant, in ordinary language, by time being infinite in length? What, for example, does a man mean who believes in a heaven persisting in time, when he says that heaven persists for an infinite length of time?

It is clear that we do not mean by an infinite time a time which has an infinite number of parts. For it is generally held that an hour has an infinite number of parts, and no one would say that an hour was an infinite length of time. Nor does it mean a time which is not divided into a finite number of parts. For every time is so divided. Every time, for example, is divided into two parts, one of which consists of any given moment together with all earlier moments, while the other consists of all later moments.

If we put what is ordinarily meant by the distinction between finite and infinite time into precise language, I think that it comes to this. Take any finite length of time, and make a series of periods of this length, each beginning where the one before stops. If in either direction, any finite number of such periods reaches to a point beyond which there is no more time, then time is, in that direction, finite. If, however, in either direction no finite number of such periods reaches such a point, then time is infinite in that direction[1].

The answer to this will of course depend upon whether time has, in that direction, either a last term or a limit. If it has, the number of such periods will be finite. For a finite number of

---

[1] The series of such periods could have next terms even if the time series was a series of indivisible parts which had no next terms. If one period included all the indivisible parts up to and including the indivisible part M, and another included all parts beyond M up to and including P, the two periods would be members of a series of next terms.

stretches of the same magnitude must eventually reach any boundary. But if there is no last term or limit, the number of such periods will be infinite.

There are, of course, no real periods of time. What appear as periods of time are really stretches of the inclusion series. Now every inclusion series is bounded in both directions. In one direction it has, as its last term, the whole of the substance in which the inclusion series falls—the term, in other words, which contains all the contents which fall anywhere in that series. In the other direction it is limited by nonentity. From any point in the series, therefore, a finite number of such stretches will reach the end of the series. And, therefore, while the series of such stretches appears as a series of periods of time, time will appear as being finite in each direction[1].

621. There remains the fourth question—whether time is to be taken as infinitely divisible. It could be infinitely divisible on either of two hypotheses—that it had no simple parts, or that it had simple parts which formed a series without next terms.

The first of these alternatives is untenable. For we have seen that it is impossible that a whole should be divided into parts of parts to infinity unless those parts are determined by determining correspondence. And we saw in Chap. XLVII, p. 226, that the terms of the C series cannot be determined by determining correspondence. They cannot, therefore, be divided into parts of parts to infinity and therefore the time-series cannot be divided in such a way.

There remains, however, the other possibility. The C series may consist of indivisible terms, none of which is next to the other, and which are therefore infinite in number. Whether this is the case or not is a question which, as far as I can see, we must leave undetermined.

There is no logical contradiction in the series having no next terms. It is true that each of its infinite number of terms must have a sufficient description. But then each of the terms in, for example, $G! H$, could be sufficiently described as that term in the inclusion series of $G! H$ which had precisely a certain pro-

---

[1] The practical result which would seem to follow from this conclusion will be affected by certain considerations which will be brought forward in Chapter LXII.

CH. LI] FURTHER CONSIDERATIONS ON TIME 281

portion of the content of $G!H$. On the other hand, there is no contradiction in the view that the number of indivisible parts is finite.

Nor do we get any further guidance when we consider that the C series appears as the B series. It must, of course, have at least as many parts as we can distinguish in the B series. But that number is always finite. And thus, so far as this goes, the C series and the B series may be only finitely divisible. On the other hand, it is quite possible that the C series may have parts infinitely smaller than those which we perceive as separate. And so the finite number of such separate perceptions of time does not prove that time is not infinitely divisible.

The ordinary view of time would, I conceive, agree with ours in holding that time had simple parts, but would hold that they were infinite in number, and that time was therefore infinitely divisible. But I can, as I have said, see no reason why the number of simple parts should not be finite.

CHAPTER LII

APPARENT MATTER AND APPARENT SENSA

622. We have now to discuss whether our theory complies with the third condition laid down in Chapter XLVI, and this discussion will occupy us for the next six chapters. That condition is that our theory shall allow, not only for the existence of some erroneous cognitions, but for the different sorts of erroneous cognition which actually exist. We perceive certain objects of our perception as being matter, sense data, judgments, assumptions, and so on. And these perceptions are erroneous, since nothing exists but spirit, and it has no contents except perceptions. Our theory must be consistent with this.

In this chapter we shall discuss apparent matter and apparent sensa. In Chapter LIII we shall enquire why certain perceptions are apparent perceptions—that is, are themselves perceived as what they are—while others are not. Chapter LIV will deal with apparent judgments, and Chapter LV with the fact that some apparent judgments appear to be reached by inference. In Chapter LVI we shall consider other apparent forms of cognition, and in Chapter LVII volition and emotion.

The objects of this discussion do not all bear the same relation to reality. For, as we saw in Book V, we really have perceptions, and perceptions really have the qualities of being volitions and emotions. We do not perceive perceptions, volitions, and emotions as being in all respects exactly what they are, but we are correct in perceiving them as being perceptions, volitions, and emotions. But when we cognize anything as being matter, sensa, judgments, assumptions, or imagings, we are cognizing them erroneously, since nothing really has any of these qualities.

623. With regard to matter there is no difficulty. For, as we pointed out in Chapter XXXIV, we do not, *primâ facie*, perceive anything as being matter. Our belief in matter is *primâ facie* a judgment, and a judgment which can only be justified by an inference. Now there is no difficulty in the supposition that a judgment which has been inferred has been inferred wrongly. Everyone admits that judgments can be erroneous, and that many are.

It is true that we have come to the conclusion that there are really no judgments, but only perceptions. And thus what appears to us as an erroneous judgment will be really an erroneous perception. But any difficulties which arise from this do not follow from the denial of the reality of matter, but from the denial of the reality of judgments, and will form part of the question to be discussed in Chapter LIV.

624. We pass now to sensa. The view which we adopted in Chapter XXXV was that, when we appear to perceive a sensum, we do really perceive something, but that we misperceive it. The object which we perceive has not the nature which it appears to have. And as a "sensum" is generally taken to mean something which has this nature, it seems better to say, not that we misperceive sensa, but that sensa do not exist, though some percepta are misperceived as having the nature of sensa. In the later chapters of Book V we came to the conclusion that whatever was perceived by any self was a self or part of a self. Whenever, therefore, we perceive anything as being a sensum, this is a misperception of a self or selves, or part of a self or selves.

The theory that there are really no sensa would clearly be untenable, unless we accepted the possibility of erroneous perception, since undoubtedly we do perceive certain percepta as being sensa. But when once the possibility of erroneous perception is accepted, there is no impossibility in the view that no sensa exist.

It must be noticed, however, that the error involved in perceiving these realities as sensa does not prevent our perception of them as sensa from being a *phenomenon bene fundatum.* We saw that the *primâ facie* form of the guarantee of the accuracy of perception was "this that I perceive is as I perceive it, while I am perceiving it." And we saw that the true form of the guarantee must be "this that I perceive is as I perceive it, subject to the degree of error which is involved by the place of my perception in the inclusion series." And, in so far as this degree of error is governed by laws, the result, which depends jointly upon it and upon the real nature of the thing perceived, will also be governed by laws.

625. But, it may be asked, have we really got rid of the sensa? Do they not come back again? If I perceive a self, or a part of a self, and my perception is so erroneous that I perceive the self or the part of a self as a sensum, does not this, it may be asked, involve that a sensum is real? If I have a perception of anything as being *XYZ*—taking that to represent some description which would only be applicable to a sensum—does not this show that there is something which is *XYZ*? How can I think of such a thing unless there is such a thing?

Similar questions could be proposed about matter, and about judgments, assumptions, and imagings. If I perceive anything as being a piece of matter, or as being a judgment, or an assumption, or an imaging, does not this show that there is something which is a piece of matter, a judgment, an assumption, or an imaging? How can I think of them unless they are there to be thought about?

The question now raised is, of course, quite distinct from the question which we have previously discussed, whether perception involves the *existence* of what is perceived such as it is perceived. The question now is whether any form of cogitation[1] does not involve the *reality*, as distinct from the existence, of what is cogitated, and as it is cogitated. Everyone would admit that we can have cogitations of things which do not exist, and that we can have cogitations of things as existent which do not exist. A man may believe that I was hanged yesterday, and I myself may make the assumption that I had been hanged yesterday, and yet my hanging yesterday does not exist. But how can there be a cogitation without an object, and how can anything be the object of a cogitation without being real?

The consideration of this question, then, is not strictly necessary for the purpose of this chapter, or of the chapters which follow it.

---

[1] On our theory, of course, there are no cogitations except perception. But as the argument in the text depends on the quality of being a cogitation, and not on the quality of being a perception, it would apply to other cogitations, if there were any.

Their purpose is to consider whether it is possible that matter, sensa, judgments, assumptions, and imagings do not exist— since their non-existence is required by our general theory. And this might be the case even if, without existing, they were real.

But if they were real it would invalidate the conclusion which we reached in Chapter II—that nothing was real except existent substances, and characteristics standing in certain definite relations to existent substances. It is necessary, therefore, to consider the question somewhere, and this seems a convenient place in which to consider it.

626. I cannot see any reason for supposing that every cogitation requires a real object which is cogitated, and there does seem to me to be a very good reason for holding that this is not the case. To begin with, a cogitation in general does not require such a real object in order that it may have an independent reality to act as its standard. A true cogitation does require this, for, as we saw in Chapter II, the truth of a cogitation consists in its correspondence to a standard which is independent of it. A correct perception, a true judgment, or a true assumption, does require that what is cogitated shall be, and shall be as it is cogitated. But this is not required by an incorrect perception, or by a false judgment or assumption.

627. There remains, however, another line of argument. It may be said that the cogitation of H as XYZ has, as one of its constituents, H as XYZ, and that therefore, if the cogitation of H as XYZ is real, H as XYZ must be real also.

It is to be noticed, in the first place, that if this argument were accepted, we could scarcely avoid results which are obviously false. For, if the constituents of what is real must be real, surely it is equally obvious that the constituents of what is existent must be existent. And as the cogitation of H as XYZ is existent, then H must exist, and exist as XYZ. Everything, therefore, would exist which anyone has ever believed to exist. The belief that a Lord Chancellor wrote *Hamlet* exists in some people's minds, and therefore it would follow that a Lord Chancellor who wrote *Hamlet* existed. And this is not the case.

The solution of the difficulty, I submit, is that the cogitation of H as XYZ has not H as XYZ as one of its constituents. And, therefore, however valid may be the argument from the reality of a complex to the reality of its constituents, we are not entitled to argue from the reality of the cogitation to the reality of H as XYZ.

If H as XYZ is a constituent of the cogitation, it must fall within it, for a constituent is, after all, nothing but a part, though it may be a part of a special nature. And H as XYZ would be a sensum, for XYZ must be taken as forming a description which would only be applicable to a sensum—and, if it were otherwise, the reality of H as XYZ would not prove the reality of sensa, which is what it is supposed to prove. Thus, if H as XYZ is a constituent of the cogitation, a sensum would be a part of a cogitation, and part of a cogitation of itself.

Now it seems clear that a sensum cannot be part of a cogitation. A cogitation, and all the parts of a cogitation, are parts of a self, and therefore spiritual. And we saw in Chapter xxxv that visual sensa have the qualities of colour and shape, auditory sensa the quality of sound, and so on. It follows that, if they are constituents of cogitations, there will be spiritual substances, parts of selves, of a green colour and of an approximately circular shape. And this is surely an absurdity. We have already rejected it, in another connection, when we spoke of sensa.

Again, if a cogitation has its object as a constituent, then a cogitation of the universe will have the universe for a constituent. And a constituent is a part. Therefore the universe will be part of a cogitation, which itself is part of the universe.

But if the object of a cogitation is not a constituent of it, the argument which we are considering breaks down, and we cannot argue from the reality of the perception of H as a sensum to its reality as a sensum.

628. Why has it been supposed that H as XYZ is a constituent of the cogitation of H as XYZ? The reason is, I think, that the cogitation and the sensum have been confused with the descriptions of the cogitation and the sensum. It is no doubt the case that the description of a perception as "a perception of H as XYZ" has as a constituent of it, "H as XYZ," which is a description of the sensum which is perceived.

But the description of a thing is not the same as the thing

itself. The perception of which we speak is a substance. And the sensum of which we speak would be a substance[1]. But the description of each of them is not a substance, but a complex quality. And the fact that one description is a constituent of the other does not prove that one thing described is a constituent of the other.

629. It follows from what we have said that the existence, and the consequent reality, of a description do not imply that it describes anything real. I do not see that there is any objection to taking this view. If, indeed, we ask whether a thing can be described without being real, it might seem as if the answer must be in the negative. But that is because we have assumed the reality of the thing when we said that *it* was described. In order to put the question without this illegitimate assumption, we must ask whether there can be a description which does not apply to any reality.

If we put the question in such a way, I do not see any reason for answering in the negative. And there are certainly weighty reasons for answering in the affirmative. For we can include in a description characteristics which are known *à priori* to be incompatible. When, for example, we say that a round square is an impossibility, or that there is no mention of a round square in Wordsworth's sonnet on Venice, we are joining round and square in the same description. If there is a real thing (though not an existent thing) corresponding to this description, the result would be that the *à priori* incompatibility of roundness and squareness applied only to the existent, and not to those spatial realities which do not exist. Nor is this all. We can include in the same description characteristics which are logical contradictories, *e.g.*, round and not round. If there is a real thing corresponding to this description, there must be reality which is exempt from the law of contradiction. Or, again, we can include in our description the characteristic of unreality, and then there will be a real unreal thing. Such results as these are clearly

---

[1] I say *would be* a substance and not *is* a substance, because if it were a substance it would at any rate be real, if not existent, and, as I have been trying to show, *H* as *XYZ* is neither existent nor real, when *XYZ* is a description of a sensum.

absurd, and they follow inevitably from the position that every description is a description of something real. The only way to escape them is to accept the position that there are descriptions which are not descriptions of anything real[1].

[1] Since I first wrote this section I have had the pleasure of finding that I am in agreement on this question with Mr Russell: "Such a proposition as 'x is unreal' only has meaning when 'x' is a description which describes nothing," *Introduction to Mathematical Philosophy*, p. 170.

CHAPTER LIII

APPARENT PERCEPTIONS

630. In this chapter we shall deal with apparent perceptions—that is, with those substances which not only *are* perceptions, but which, when they are themselves perceived by introspection, appear as perceptions.

Here, then, the appearance is not so different from the reality as in the cases of apparent matter, apparent sensa, apparent judgments, and apparent assumptions and imagings. For these are not really matter, sensa, judgments, assumptions and imagings. But apparent perceptions are really perceptions. They are not, of course, the complete perceptions which are determined by determining correspondence, but they are among those fragmentary perceptions which are parts of the complete perceptions, and form their inclusion series. We have not, therefore, to undertake the task of showing how what is really not a perception can appear to be one.

These apparent perceptions, however, appear differently from what they are in reality. They do this in three respects. Firstly, they appear to be in time, while they are really not in time, though they are in the C series. Secondly, none of them appear as erroneous, while really they are all so, more or less. Thirdly, they do not appear as an inclusion series—each term including all those on one side of it, and being included by all those on the other side of it—although they really are such a series.

631. There seems no difficulty in accepting the view that the appearance and reality of apparent perceptions differ to this extent. Whether the reality is as it is here stated to be has been discussed previously. Granted that the reality has such a nature, and granted that perceptions, like other things, can be misperceived in introspection, there seems no reason why we should regard any of these three differences as impossible, or even as suspicious.

As to the first—that the terms of the inclusion series should appear to form an A series, and consequently a B series, is, no doubt, an ultimate fact. We cannot explain why it should be so. But there is no reason why it should not be so. And therefore there is no reason why we should suspect the conclusion, to which our argument has led us, that the reality which appears as a time-series is really a timeless inclusion series.

Nor is there any difficulty about the second difference—that the apparent perceptions do not appear to be erroneous, while in fact they are so. The difficulty, which arose from the apparent self-evidence of the proposition that every perception is correct at the time that it occurs, has been seen to disappear when we considered what the reality is which appears as simultaneity of perception and perceptum. And, apart from this, there is nothing more remarkable in a perception which is really erroneous not appearing as such to the percipient than there is in the admitted fact that every judgment appears true to the person making it, while many of them are really false.

There remains the third difference. My perception at any point of the inclusion series is a part of my perception at any further point in the inclusion series. And, when they appear as states in time, they certainly do not seem to overlap in this manner. But there is nothing in the way in which they do appear to us which is incompatible with the conclusion which we have reached that they *do* overlap in this manner.

It is true, no doubt, that the content of the terms of the time-series, as we observe it, does not at all suggest that the terms do in reality stand in the relation of including and included. For the content does not appear to show any uniform change in either direction. Everywhere we find persistence of content, recurrence of content, and oscillation of content. But we saw in Chapter L that all these features—persistence, recurrence, and oscillation—could be accounted for on the theory that the terms of the inclusion series were related as we have held them to be related.

There is thus nothing in the appearance which is inconsistent with our theory of the nature of reality, while the fact that the perception does not appear as it really exists is not to be wondered at. For it only appears to us when, in introspection, it becomes itself an object of perception, and this secondary perception, like all other perceptions in our present experience, is more or less erroneous.

632. But another problem now arises. We have seen that there is no reason to distrust our theory because those perceptions, which appear as perceptions, appear in some respects otherwise than as they really are. But can anything be said as to the causes which determine some perceptions to appear as perceptions, and others to appear as judgments, assumptions, and imagings?

The question, indeed, why there should be misperceptions of perceptions at all, cannot be answered. It must be accepted as an ultimate fact that all the stages of the inclusion series, except the last all-inclusive stage, are misperceptions. That the fact is so is proved by the arguments which show us what the nature of the reality is, together with the fact that it appears to us as something different.

But there is still the question why some perceptions should be misperceived more than others. Those which are apparent perceptions are perceived as being perceptions, which they are, while those which are apparent judgments, assumptions, or imagings are perceived as being judgments, assumptions, or imagings, which they are not. Can anything be said about this?

The most striking difference, other than the intrinsic nature of the appearances, which we find between those perceptions which appear as perceptions, and those which do not, is that the scope of the latter is so much wider. We can have apparent judgments about everything as to which we can have apparent perceptions. But there are many things about which we can have apparent judgments, but not apparent perceptions. In the first place, apparent perceptions can only be of something as existent, while apparent judgments may deal with the non-existent. In the second place, I can have apparent perceptions only of what exists within the same specious present as my perception (that is, in reality, that which exists at a corresponding point in the inclusion series). Thus I cannot have an apparent perception of the square root of 81 or of the death of Caesar, but I can have apparent judgments about both of them. And there are other restrictions on apparent perceptions. I cannot just now have an apparent perception of Mount Everest, though it is existing simultaneously with me. But I can have an apparent judgment about it. I can have an apparent assumption about anything as to which I can have an apparent judgment. And, although the limits of apparent imaging are narrower than those of apparent judgment and assumption, yet they are wider than those of apparent perception. I can, for example, image the death of Caesar.

The question of apparent judgments which are non-existential had better be postponed to the next chapter, but about the other two limitations of apparent perceptions we must say something here.

633. That a perception should be an apparent perception seems to depend on conditions both in the percipient and in the object. For while, as has just been said, there are many objects of which we cannot have apparent perceptions, there are none of those objects of which it is not possible to have an apparent judgment without having an apparent perception. And this suggests that conditions both in the object and in the percipient must concur in order that there should actually be an apparent perception.

And it would seem that a perception which is an apparent perception must indicate a closer relation between the object and the percipient than occurs in the case of a perception which is not an apparent perception. For, in the first place, it can occur in only a few of the cases in which the other can occur. In the second place, the perception is itself perceived more correctly, since what is really a perception appears as such, and not as something else. And this suggests that such a perception is in some way more forcible, and less easy to mistake.

Now with regard to the limitation of apparent perceptions to cases where the perception and the perceptum are temporally simultaneous, it is not only possible that there should be a closer relation, but we can see that there is one, and can see what it is. For it is obvious that two substances which are at corresponding points in the inclusion series (which is, of course, the real relation which appears as simultaneity) are in one respect more closely related than two substances whose places in their inclusion series do not correspond. Thus we may infer that this particular closeness of relation is essential for apparent perception—though, of course, not sufficient for it, since our cognition of objects simultaneous to the cognition is often not an apparent perception, but an apparent judgment.

634. And there is another fact which enables us to see that it is impossible that an apparent perception should be of any object which, *sub specie temporis*, is earlier or later than the apparent perception. When the object and the apparent perception are simultaneous, the certainty takes the form "this that I perceive is as I perceive it while I now perceive it," in which there is no absurdity. But if the object was earlier or later, we should get "this that I perceive was as I perceive it while I now perceive it," or else "this that I perceive will be as I perceive it while I now perceive it." And both of these are absurd. The period, for which the guaranteed correctness is asserted, is declared in each of them to be simultaneous with the perception, and also not to be simultaneous with it.

635. There are also the cases in which we do not have an apparent perception, although the perceptum and the perception are simultaneous, as in my present cognition of Mount Everest, or of the books behind my chair. Here we have the same reason as before to suppose that the cases where an apparent perception does occur indicate a closer relation than the cases where the perception is not an apparent perception. But in these cases we are not always able to see what relation is closer in the one case than it is in the other. And therefore we cannot see why our perceptions of some simultaneous objects are not apparent perceptions, while we could see why perceptions of objects not simultaneous to them are not apparent perceptions. But there is nothing in the facts to exclude the possibility that there may be a closer relation in every case of apparent perception, especially when we remember that the increased closeness of the relation may depend on some internal change in the percipient (it might, for example, depend on a change in the direction of the interest or volition).

There remains the case in which I have simultaneously an apparent perception and an apparent judgment of the same object. It will be more convenient to consider this in the next chapter (p. 299).

636. One more difficulty remains. Whenever, in our present experience, a perception is an apparent perception, it is always partially erroneous. For it always perceives its object as in time, and often misperceives it in other respects. But an apparent judgment can be absolutely true. For one thing, it need not judge its object to be in time—it may even judge it not to be in time. And it need not judge it to have any of the other qualities which we know cannot be true of its object. Moreover, the apparent judgments, "to be a self implies immortality" and "to be a self does not imply immortality," have both been made, and one of these must be quite true. Does not this fact—that an apparent judgment can be quite true while an apparent perception must be partially erroneous—cast a doubt on our view that there is a closer relation between the object and the percipient in the case of the apparent perception?

This difficulty will be answered in the next chapter. For we shall see there that what appears as a perfectly true judgment is in reality part of a partially erroneous perception. There is, therefore, no reason to suppose that it is in a closer relation to the percipient than apparent perceptions. The possibility that the apparent judgment should be completely true (or completely false), while the apparent perception cannot be, is due, as we shall see, to the fact that the perception which appears as the judgment is itself more misperceived than the perception which appears as the perception.

CHAPTER LIV

APPARENT JUDGMENTS

637. We must now pass on to consider that portion of the contents of the self which appears to consist of judgments. In reality, if our theory is correct, they are not judgments. They are, like everything else in the self, perceptions. We must now enquire whether it is possible that anything which is really a perception should appear as a judgment.

What ground is there for the belief that judgments do exist? The ground for this is obviously introspection. When a man turns his attention to what is passing in his own mind, part of what passes appears to be judgments. Now introspection is perception, and we have seen that perception can be erroneous. It is therefore possible that what we perceive as being a judgment is not really a judgment. The perception of anything as a judgment, in our present experience, is certainly erroneous in some respects, since it perceives it as being in time, which it really is not. But of course it does not follow from its being erroneous in some respects that it is erroneous in presenting its perceptum as a judgment, and not as a perception. Still the possibility has been opened.

638. It might be objected to this that any theory that judgments did not exist would imply their existence, and must therefore be rejected. For such a theory would itself appear as a judgment, or as a series of judgments. And therefore, it might be said, in asserting that judgments do not exist, we should be making what was *primâ facie* a judgment, and, to avoid inconsistency, we should have to follow it up by the assertion that it also was not a judgment, though it appeared as one. This second assertion would then be in the same position, and so, it might be maintained, an infinite series would be generated, and that series would be vicious.

But this would be a mistake. There would be no vicious infinite series. We have seen (Chap. XLIV, p. 198) that a vicious infinite series would be generated by the attempt to deny all error. For an apparent error can only be got rid of by asserting that the appearance of error is itself erroneous. The denial of one error thus affirms the existence of another. If this is got rid of in its turn, the denial of it involves the existence of a third, and so on. Thus the denial of all error is impossible until the infinite series has been completed. And, as it cannot be completed, the infinite series of denials is vicious.

Here the matter is different. The knowledge of a certain mental state as not being a judgment may itself appear as a judgment, but it does not involve that it, or anything else, *is* a judgment. It is not necessary, therefore, to go on to a further denial that it is a judgment, in order to save the original denial. And so no vicious infinite series has been generated[1].

639. What appears as a judgment professes to give information about facts. And the information which is given is sometimes completely true, and sometimes completely false. "The square of three is greater than twice four" is completely true. "Redness involves sweetness" is completely false.

Now there is no difficulty about an apparent judgment giving us information. For an apparent judgment is really a perception. And a perception can give the same information which is given in a judgment, though it gives it in another form. We can see this in the case of those perceptions which, in our ordinary experience, appear as perceptions, and not as judgments. For we constantly make apparent judgments which have no other ground than apparent perceptions, and so the information expressed in the apparent judgment must also have been given in the apparent perception. This is obviously the case in such judgments as "this is a sensum of red," or "I am angry." If this information were not given in perception, we could not get it at all. And, again, in such judgments as "this table is red," though other elements besides apparent perceptions are introduced, we could neither judge that the object was a table, nor judge that it was red, except on the ground of information given us in apparent perception. Apparent perception, then, can give the same information which is given in judgments, and since, on any theory, these apparent percep-

---

[1] In this respect the denial of the existence of judgments resembles the denial of the existence of time, which, as we saw in Chap. XLIV, p. 198, does not involve a vicious infinite.

tions are real perceptions, it follows that such information can be given in perceptions.

I am not asserting, so far, that *all* the information which is given in judgments could be given in perceptions—that will be considered later. My contention at present is only that *some* information which is given in judgments can be given in perceptions, and that therefore the possibility of all information being really given in perceptions remains so far open.

640. But the question remains as to the truth or falsity of the information. It is not surprising that the information given to us in apparent judgments should be partly correct and partly erroneous. For all our apparent judgments appear successively as being in the present. They will therefore, by the results obtained in Chapter XLIX, all be terms of the inclusion series, while none of them will be the final term of the series—the term which includes and is not included. It follows that they will all be members of the misperception series, and that they will all perceive their objects, partly as being what they are and partly as being what they are not.

But a difficulty remains. Every perception in the misperception series is partially correct and partially erroneous. But some apparent judgments are completely true, and some apparent judgments are completely false. A true judgment is correct, a false judgment is erroneous. If what appear as judgments are really partially erroneous perceptions, how can they give information which is completely correct, or completely erroneous?

It is impossible to doubt that some of our apparent judgments are completely true. In the first place, there are some which are so clearly self-evident, that, if we doubted their truth, we should have to doubt everything. And, in the second place, the assertion that it is, or could be, the case, that no apparent judgments were completely true, refutes itself. For in making that, or any other assertion, the person who makes it is asserting it as completely true, and so denies that no judgments are completely true.

Of course no judgment expresses the whole of what is true, or even the whole of what is true about any one subject. And it has sometimes been asserted that, for this reason, no judgment is ever completely true. But there seems no reason whatever for holding this. It is far from being the whole truth about Skakespeare that he knew Ben Jonson, but in what way does that prevent the proposition that he did know him from being absolutely true? And the theory is inconsistent with its own assertion, for it asserts that no partial truth is quite true, and this is itself a partial truth[1].

There must, therefore, be some apparent judgments which give us absolutely correct information. And there must also be others which give us information which is not correct. To deny this would not be suicidal, as the denial of correct judgments has been shown to be. But it would be clearly false. For if anywhere within the universe there have ever been two apparent judgments, the content of which are contrary or contradictory to one another, then there must be at least one judgment in the world which gives information which is not correct. And there certainly do exist apparent judgments whose contents are contrary and contradictory to those of other existent apparent judgments.

641. Judgments can be divided into those which do, and those which do not, assert existence, and it will be convenient for us to deal with the two classes separately. Let us take first those which assert existence, which we may call for shortness "existential judgments." (This phrase must of course be distinguished from the phrase "existent judgments," which means, not judgments which assert existence, but judgments which exist.) Whatever appears as an existential judgment is really a perception. We have to enquire, firstly, whether a perception could give us that information which is given in an apparent existential judgment, and, secondly, whether we can explain why certain apparent existential judgments are completely true or completely false, while the perceptions which appear as those judgments are partially correct and partially incorrect.

---

[1] It might be said that it is superfluous to talk of complete truth, since, if a proposition is not completely true, it is not true at all. In the strict sense of "true," this is correct. But, in ordinary language, a false statement which approximates fairly closely to the truth is often called an incomplete truth. And, in the theory mentioned in the text, partial truth is maintained to be the only truth which our judgments can have. It seems desirable therefore, to avoid mistakes, that we should emphasize the fact that some of the judgments in our ordinary experience possess a truth which is not incomplete.

An existential judgment, like all other judgments, asserts something. And what it asserts, as with all other judgments, is one or more characteristics. It asserts that something has some quality or that something stands in some relation to something.

It is true that, in the case of an existential judgment, that about which the assertion is made is always substance. It may be that a quality is asserted of a single substance, as "Socrates is good," or "Socrates was the inspirer of Plato." It may be that a quality is asserted of each of a group of substances, as "every Cambridge college in 1922 has a dining-hall." Or it may be that a relation is asserted between two substances, as "Brutus killed Caesar." But the assertion is always about one or more substances. (This is not equivalent to saying that every such judgment has a subject and predicate, as will be seen by the last example.)

But what is asserted is always characteristics. And the substance or substances about which the assertion is made are usually described by characteristics. This is clear with such descriptions as "the largest ship in the world," or "Cambridge colleges in 1922." But it is also the case when the assertion is made about Brutus and Caesar. For when a person is described by a name, what is meant is that he is the person on whom that name was conferred in a particular manner, or who is usually called by that name. And these, of course, are characteristics. The only case where the substance is not described by characteristics is that in which the subject is something perceived at the moment when the judgment is made, as "I am happy," "this is red." And even here, what is asserted of the subject is a characteristic.

642. The nature of a perception is very different from this. It is a cognition, like a judgment. But it is in all cases a direct cognition of a substance. It knows the substance directly, and does not identify it by describing it by means of characteristics. And, although it gives information about the characteristics of the substance which is perceived, it does not do so by making an explicit assertion about them. It does not assert that B is X, but, in perceiving B, it perceives it as having the quality X.

These are, no doubt, very important differences, but they are not such as to be incompatible with our theory. For, as we have just pointed out, a perception does give knowledge of the characteristics of the substance perceived. This, as we saw, is evident in the case of those perceptions which are apparent perceptions, and, if it were not so, we should never gain information such as "I am happy," "this is red." The same information, then, about existent substances, which appears to be given to us in a judgment, can really be given to us in a perception.

And this is a point of vital importance. For we certainly do have the information which is given us in the apparent judgment, and if this could not be given us in a perception, then the apparent judgment could not really be a perception, as our theory requires it to be. But, for the reasons given above, it seems clear that such information can be given in a perception. The perceptions of ordinary life—those which we have called apparent perceptions—are universally admitted to be really perceptions. And, if they did not give us information about the characteristics of the objects perceived, we should have no knowledge about these characteristics. (Indeed we should have no knowledge about the individual characteristics of any other existent substances. For, with the exception of a few general characteristics, known à priori to belong to everything, all our knowledge of the characteristics of unperceived existent substances depends on inferences from our knowledge of the characteristics of substances which are perceived.) We may therefore affirm with safety that perception does give us information as to the characteristics of the objects perceived.

The information is, no doubt, given in a very different form in a judgment and in a perception. But it is given in both. And thus there seems no difficulty in the view that what is really a perception may appear as a judgment. When we say that a perception appears as a judgment, we mean that another perception perceives it as a judgment. And we have seen that perceptions can be erroneous, and that all perceptions in our present experience are erroneous. There is therefore no reason why we should not accept the conclusion, to which our argument has led us, that what is perceived as a judgment is really a perception.

643. When a perception is perceived as a judgment, what

determines, we may ask, which of the characteristics which the substance is perceived as possessing are singled out to appear as asserted of the substance, and which are singled out as describing the subject of the assertion[1]? There are three ways in which this may happen. It seems probable that all three occur on different occasions, and again that any two of them, or all three, may co-operate on other occasions. In the first place, it may be due to some grouping of the characteristics in the perception, which is due to various relations between the characteristics in the object perceived. Or, in the second place, it may be due to some grouping of the characteristics in the perception, which is due, not to anything in the object, but to something in the percipient self. Or, finally, it may be due, not to any grouping of the characteristics in the perception, but to some feature in the way in which the perception is itself perceived, when it is misperceived as an apparent judgment.

644. There remains the question how the partially correct perceptions can appear as apparent existential judgments which are completely true or completely false. That some apparent existential judgments are completely true, and others completely false, cannot be doubted. There do not seem, indeed, to be any existential judgments which are self-evident. But some must be true, and some must be false, since some are contradictory to others. The judgments "Francis wrote the *Letters of Junius*" and "Francis did not write the *Letters of Junius*" have both been made, and one of these must be true and the other false.

The explanation which I should give is as follows. When we perceive a perception as a judgment, the second perception—the perception of the first perception as being a judgment—is, of course, partially erroneous. If it were not so, it would not perceive as a judgment what is, in fact, a perception. Now it might be the case that its misperception of the original perception might consist, not only in perceiving the information in the form of a judgment, but in not perceiving part of the information in the original perception at all.

---

[1] It is not necessary, as we shall see later (p. 312), that every characteristic which the substance is perceived as possessing should fall into one or the other of these two classes. It may be the case that some of them do not enter into the apparent judgment at all.

And if it only perceived part of the information, then that part of the information might be completely correct or completely incorrect, although the information given in the whole of the original perception was partially correct and partially incorrect. And thus the original perception would appear as a judgment which was completely true or completely false.

Thus, in the case of a true judgment such as "the author of *Hamlet* is the author of *Othello*," we should hold that the perception which was misperceived as this judgment would be one in which the substance was perceived, correctly, as the author of *Hamlet*, and as the author of *Othello*, but was also perceived, incorrectly, as something which it really was not since it must have *some* error. But, we should say, when, in introspection, we erroneously perceive this perception as a judgment, we only perceive it as giving part of the information which it really does give. And as, in this case, this part of the information is all correct, it is perceived as a judgment which is completely true. In other cases, when the information which it is perceived to give is partially correct and partially incorrect, we get a false judgment of the type "the author of *Hamlet* is a king of Utopia," where it is correct that there is an author of *Hamlet*, but not that he is king of Utopia. If there are judgments such that none of the characteristics given are true of an object, this would be due to the fact that, while the perception gave information about its object which was partially correct and partially incorrect, we only perceived it, when we misperceived it as a judgment, as giving information which is incorrect.

645. This theory must not be interpreted as meaning that the truth of the apparent judgment arises from misperceiving a misperception. It would be possible, no doubt, that H, which had really the nature X, should be misperceived as having, not the nature X, but the nature Y. And it would be possible that this misperception should itself be misperceived as being a cognition of H as X. But although chance would, in that case, lead us to the belief that H was X, which it really was, the belief would be unjustified, and would have no value as knowledge. The case here is quite different. H is perceived in a way which is partially erroneous—as having some characteristics which it really has, and as having some which it really has not. When this perception is again perceived, it may be misperceived as being a cognition of H as what it really is, and not a cognition of H as what it really is not. In this case it will be perceived as a true apparent judgment. If, on the other hand, it is misperceived as a cognition of H as more or less what H really is not, it will be perceived as a false apparent judgment. But the misperception, which is necessary in order that the apparent judgment should be one which is completely true, is not necessary to account for the fact that truth appears in the apparent judgment, but to account for the fact that falsehood does not appear in it.

646. The theory, as has been seen, involves that we can never recognize in introspection the full amount of cognition in a perception which is an apparent judgment, since there are always elements of cognition—correct or incorrect—in the cognition, which do not appear when the perception appears as a judgment. Thus in every such case we have knowledge which we never know.

It might perhaps be objected to this conclusion that it would be impossible for me to have knowledge which I did not know. But this is due to a confusion. If the words "which I did not know" are taken as meaning "the object of which I did not know," then anything of which they were true could not be my state of knowledge. But if they are taken in their correct meaning of "which is not itself an object of my knowledge," then they might be true of something which is my state of knowledge. For I can have a state of knowledge which is not itself an object of my knowledge.

And our theory does not go so far as to assert even this. It remains quite possible that every state of my knowledge is an object of my knowledge[1]. All that the theory involves is that states of my knowledge which are objects of my knowledge are

---

[1] Indeed, if I know myself, or any of my parts, it is not only possible, but necessary, that every state of my knowledge should be an object of my knowledge. For I cannot know myself at all, unless I am a member of my own differentiating group. And if I am a member of my own differentiating group I shall perceive all my perceptions through an infinite series.

not perceived as possessing all the qualities which they do, in fact, possess.

647. It must also be noticed that, if H is perceived as being X, Y, and Z, it is not by any means necessary that its perception as each of the three is equally prominent. The qualitative difference made in the perception by the fact that it is a perception of H as X may be intrinsically much more important than the difference made by the fact that it is a perception of H as Y. Or, secondly, the perception of H as Y, even if not less intrinsically important, may be much less important for the general interests of the percipient. In either of these cases, my failure to see that the perception perceives H as Y may be much less important than my success in seeing that it perceives it as X. And the fact that I do succeed in seeing it as the one and not the other gives a certain presumption that what was observed was, one way or another, more important than what escaped observation.

648. It might be suggested, however, that a perception which appeared as a true apparent judgment perceived its object as having only those characteristics which appear in it, when it, in its turn, is perceived as an apparent judgment. Since the object really had all those characteristics, the perception would be a completely correct perception, which would appear as a perfectly true judgment. In the case, for example, of the judgment "the author of *Hamlet* is the author of *Othello*," the perception which appears as such a judgment would be the perception of a certain substance as possessing two characteristics, the authorship of *Hamlet* and the authorship of *Othello*, and not as possessing any other characteristics[1]. Since there is a substance which does possess these two characteristics, though it also possesses many others, the perception would be, so far as it went, completely correct.

In this way, it might be said, we should get a theory which was simpler, and therefore more probable. When we perceived such a perception as a judgment, we should misperceive it in

---

[1] This does not mean that it is perceived as *not* possessing any other characteristics. This would be an incorrect perception, since it possesses many others. But it is only these two characteristics which it is perceived as possessing.

that respect, since it is not a judgment. But we should not misperceive it as being completely correct, since it really is so. Nor should we misperceive it as only giving us knowledge of two of the characteristics of its object, since it really only gives us knowledge of two of them.

On this theory it would no longer be the case that all perceptions in the inclusion series, except those at the final stage, would be partially erroneous. Their common characteristic would be inadequacy, which would take in some of them the form of being more or less erroneous, and in others the form of being partial, though correct.

649. I do not think, however, that there would be any advantage in this theory even if we could accept it. The gain which would be involved is only that the amount of misperception involved, when a perception was itself perceived as an apparent judgment, would be diminished. But even then there would be a great deal of misperception still involved. For what is really a perception is perceived as a judgment, and this is so much misperception that we should not make things much simpler by a theory which got rid of the other misperception.

And, when we come to non-existential judgments, we shall see that it is impossible to explain them except on the view that the perceptions which appear in such judgments are misperceived in such a manner that not all the information given in the perception appears when the perception is perceived as a judgment. And we shall also see that the explanation would involve the possibility that a perception which was partially erroneous might be perceived as a perfectly true judgment.

Thus the additional elements of misperception which the present theory would eliminate in the case of existential judgments could not, on any theory, be eliminated in the case of non-existential judgments. And thus their removal in the case of existential judgments would not make our theory of reality in general simpler, but, on the contrary, more complicated.

And the presumption is against this theory on the ground of analogy. For we know that those perceptions which are apparent perceptions are partially erroneous, and we have seen that the only explanation of this is to take the error as dependent on the place of each perception in the inclusion series. And this certainly gives a presumption against the view that there are other perceptions, at the same places in the inclusion series, in whose case their position in those places does not involve error. It would appear, then, that we must keep to our earlier theory, that all perceptions, at every stage in the inclusion series except the last, are partially erroneous.

650. The question remains of how we are to deal on our theory with the cases where we have simultaneously an apparent perception and an apparent judgment of the same object. There are three classes of such cases. In the first class are those in which one or more of the characteristics are the same in the judgment and the perception. Thus I may perceive something as red and square, and simultaneously judge that it is red and square. In the second class the characteristics are neither the same nor incompatible. Thus Washington might simultaneously have perceived himself as being in pain, and also have made the judgment "the first President of the United States will be remembered in history." As to these two classes, the natural explanation seems to be that there is only one perception, which, by an error of introspection, is itself perceived both as an apparent perception and as an apparent judgment.

651. But there is a third class of cases, in which the apparent perception and the apparent judgment cognize the object as having incompatible characteristics. Thus I may perceive something as in time, and simultaneously judge that it, like all other things, is really timeless. Here also the apparent perception and the apparent judgment must each, in reality, be a perception. Can they be the same perception?

I think that they can. If, while I perceive anything to be X, I judge it not to be X, this can only be because I perceive it to have some quality incompatible with its being X. For example, if, while I am perceiving something as in time, I judge it to be timeless, it must be the case that I perceive it to have some quality incompatible with its being in time. (For any person who should accept the conclusion which we reached in Chapter XXXIII that nothing could be in time, it would be sufficient that he should perceive the object as existent, since existence would be incompatible with being in time.)

What happens, then, I hold, is this. I perceive A as being X, and also as being W. And, in perceiving this, I perceive W as being incompatible with X, either directly, or as involving Y which is incompatible with X[1]. There are thus two elements in the perception—my perception of A as X, and my perception of A as W and therefore not X. (It is not impossible to perceive a thing as having two incompatible qualities, but of course the perception must be a misperception, since the thing cannot have two incompatible qualities.) Thus, in the cases we are discussing, in which the apparent judgment and the apparent perception are incompatible, what has happened is that the perception of A as X and as W and therefore not X appears as an apparent perception of it as X and as the judgment that it is not X.

652. We come now to those apparent judgments which do not assert existence, and which we have therefore called non-existential. And about these, as about existential judgments, we shall have two enquiries to make. Firstly, we must enquire whether a perception could give us that information which is given in an apparent non-existential judgment. And, secondly, we must enquire whether we can explain why certain apparent non-existential judgments are completely true or completely false, while the perceptions which appear as those judgments are partially correct and partially incorrect.

653. In the first enquiry, we shall meet a difficulty which did not exist in the case of apparent existential judgments. All perceptions are cognitions of existent substances. And this is what apparent existential judgments profess to be. So that with them the only problem lay in the fact that what was really a perception appeared as a judgment. But apparent non-existential judgments profess to be cognitions of something else, and not of existent substances. Here, therefore, the difference between the appearance and the reality is greater than in the other case. The difficulty, however, will be found not to be insuperable.

654. It was pointed out in Section 26 that all non-existential judgments are assertions that the presence of one characteristic, positive or negative, implies or does not imply the presence of another characteristic. And this, as we saw later (Section 108), is

---

[1] The possibility of this will be discussed later in the Chapter, in connection with non-existential judgments (pp. 310–313).

equivalent to an assertion that one characteristic does or does not intrinsically determine another characteristic.

We have seen earlier in the chapter that perception of a substance can give knowledge about some of its characteristics. But we may go further, and assert that perception of a substance can give us information about the characteristics of its characteristics. We can see that this is so, in the case of various perceptions which are apparent perceptions. For we continually make judgments such as "My present happiness is more intense than my happiness yesterday," or "the shade of red in A resembles the shade of red in B more closely than it resembles the shade of red in C." And it would be admitted that these judgments are in some cases well founded.

Now such judgments as these are assertions about the characteristics of characteristics. It is not my happiness yesterday which is asserted to be greater than my happiness to-day, but the intensity which one feeling has which is asserted to be greater than the intensity which the other feeling has. In other qualities, for example in duration, the greater magnitude might perhaps be attached to to-day's happiness and not to yesterday's. And, in the same way, it is not A which is asserted to resemble B more closely than C, but the quality of redness in A which is asserted to resemble that in B more than that in C.

The materials for such judgments can only be given us in perception. It is only from materials given us in perception that I can know that my state has the quality of happiness, and it is only from materials given us in perception that I can know what intensity it has. Similarly, it is only from materials given us in perception that I can know the three shades of red, and it is only from materials given us in perception that I can know that the shade in the first datum resembles that in the second more closely than that in the third.

Thus we can see in the case of apparent perceptions that information can be given in perception about the characteristics of characteristics. And there is nothing to suggest that this is confined to those perceptions which are apparent perceptions. Now if the characteristic X implies the characteristic Y, it is a characteristic of X that it implies Y (just as it is a characteristic

CH. LIV] APPARENT JUDGMENTS 309

of Y that it is implied by X). Thus it is possible that, in perceiving a substance B, I may perceive it as having the characteristic
X having the characteristic of implying Y. And thus the information that X does imply Y will be given in perception. If, further,
this perception should be misperceived as a judgment, and if, in
the process of misperception, we should fail to perceive any of
the information in the perception except the information of the
implication of Y by X, then the perception would appear as an
apparent judgment that X implied Y. And this apparent judgment is non-existential.

In this way, then, non-existential apparent judgments can be
explained. The explanation involves that a perception, which
appears as such a judgment, is misperceived to a greater extent
than one which appears as an existential judgment. But there
seems no difficulty in this. All that is essential is that the whole
of the information given in the apparent judgment should be such
as could be given in the real perception. And this, as I have just
shown, is the case.

And in this way we can account for false non-existential judgments as well as for true ones. For, since our perceptions are
always more or less illusory, we shall naturally in some cases
perceive characteristics as having characteristics which they
really have not. And when these erroneous perceptions are themselves misperceived as separate apparent judgments, they will
naturally be perceived as non-existential judgments whose assertions are, in point of fact, false.

655. But, it might be said, this will only account for one sort
of non-existential judgment—that in which the characteristic of
which a characteristic is asserted does actually exist. It will account for such judgments as "whatever is conscious has value,"
because the characteristic of consciousness is a characteristic of
various existing substances, and, when we perceive one of these,
we may perceive it as being conscious, and so gain information
about consciousness. And among the information we gain may
be the fact that consciousness possesses the characteristic that
whatever has it has value. But many non-existential judgments
assert the implication of characteristics which do not exist. "No
phoenix could be a hippogriff" would be an example of this. And

if we have been right in concluding that nothing but spirit exists, another example would be "whatever is material is extended."

In discussing this objection we must notice, to begin with, that there will be no difficulty with regard to those characteristics which substances appear to have, though they do not really have them. For in such cases the substance is perceived as having them, though the perception is in this respect erroneous. And thus the perception will give information about the characteristic, and can also give it, in the way explained above, about the characteristics of the characteristic. This will account for such judgments as "whatever is material is extended," since various substances do appear to us as material. But how about "no phoenix could be a hippogriff"? Shall we say that every person who appears to make this judgment is really misperceiving some particular substance as a phoenix, and that his information about the impossibility of a phoenix being a hippogriff is a part of what is given in this misperception? I do not see that it is possible to refute such a hypothesis, but it would be, at any rate, very wild and improbable.

Nor could we avoid the difficulty by saying that the belief could be deduced from the fact that no bird could be a hippogriff, and that the characteristic of being a bird is one which various things are perceived as having. The question is not how the apparent judgment can be justified, but what it really *is*. And the apparent judgment is about the characteristic of being a phoenix, not about the characteristic of being a bird.

656. But there is a further possibility. Every non-existential judgment, as we have seen, asserts a characteristic of a characteristic (the characteristic which is asserted being that the characteristic of which it is asserted does or does not imply something). Now with regard to the characteristics which are asserted we have fallen back on our perception of the characteristics of characteristics. And it is possible to do the same with regard to the characteristics of which they are asserted.

The characteristic of which the assertion is made, X, may not be a characteristic of the substance perceived. But the substance perceived may have a characteristic, Y, which has a certain relation to X. And, in perceiving the substance as having the characteristic Y, we may perceive it as having the characteristic Y having the relation W to the characteristic X. And in that case the perception will contain information as to the characteristic X.

It is clear that every characteristic must have some relation, of some sort or the other, to every other characteristic. But it does not follow that, in gaining information about any one characteristic, we shall gain information about every other characteristic. If I perceive B as having the characteristic Y, which has in fact a relation W to the characteristic X, I *may* perceive B as having-the-characteristic-Y-which-has-the-relation-W-to-X. But it is also possible that the relation of Y to X should not enter into the perception at all. For, in order to perceive B as having the characteristic Y, it is not necessary to perceive it as having a characteristic which has all the nature of Y, as distinct from the *meaning* of Y. It is possible that B is only perceived as having a characteristic which has part of the nature of Y, and this part of its nature may not include its relation to X.

We should perhaps mention specially the case of negative characteristics. No doubt, if B has not the characteristic X, it must have the characteristic not-X, and, if it is perceived as having the characteristic not-X, it must be perceived as having a characteristic which has a definite relation to X. We cannot have information about the characteristic not-X, which does not also give information about the characteristic X.

It seems clear that we *can* perceive substances as having negative characteristics. For we can make apparent judgments that they have not certain characteristics, and these apparent judgments must be, in reality, perceptions of them *as* not having those characteristics, which is the same thing as perceiving them as having the corresponding negative characteristics. But it is not at all necessary that we should perceive B as having all the negative characteristics which, in fact, it has. And therefore we cannot be certain that we shall gain information in that perception as to all characteristics.

657. We cannot, then, be certain that our perceptions give us information about every characteristic. But this is not required.

All that is required is that each man's perceptions should give him information about every characteristic which enters into one of his apparent judgments. There will be many characteristics which do not enter in this way, and it is not necessary that his perceptions should give him information about them. There may also be many characteristics which do not enter into any apparent judgment of any man. And it is not necessary that information as to these should be given in any man's perceptions.

658. With this limitation our task becomes practicable. For when we consider the apparent judgments which we do make about characteristics which we have reason to believe non-existent, we find that in every case the apparent judgment is suggested by some characteristic which we have found in apparent perception. I do not, of course, mean that the validity of the judgment depends on any observation of the existent, but that the occurrence of the judgment depends on such an observation. The validity of Euclid's 10th axiom is not dependent on any apparent perception, but we may safely say that it would not have occurred to anyone who had not seen two objects whose outlines approximated to straight lines, and who had not considered whether these two things could be placed so as to enclose a space. The suggestion need not always be direct—it may suggest one apparent judgment from which another is deduced. But it seems clear that, in one form or another, it is always present.

This connection of non-existent characteristics in apparent judgments with existent characteristics in apparent perceptions gives us a clue as to the nature of the real perceptions which appear as such judgments. For we have seen that each of the non-existent characteristics in question has such a special and definite relation to an existent characteristic that, under certain circumstances, the thought of that existent characteristic suggests the thought of that non-existent characteristic, rather than of any other. And since non-existent characteristics have their special and definite relations to existent characteristics, we can see how the information in the apparent judgment can be given in a perception. For Y may have a special and definite relation of this kind to the non-existent characteristic X, which it has not to other non-existent characteristics. And thus when we perceive the substance B as having the characteristic Y, our perception of B as having Y, though it would not give us information about all the characteristics to which Y stands in some relation or other might, in virtue of this special relation, give us information about X

Let us take the case of the apparent judgment that a phoenix cannot be a hippogriff. It would be possible for this to be asserted by a person who did not perceive any substance as being a phoenix. For he might perceive a substance as being a bird, or, more in detail, as being an eagle. This gives information about the characteristic of being a bird, or of being an eagle. And both of them stand in closer and more intimate relations with the characteristic of being a phoenix than they do with many others. It will not, therefore, be surprising if the information given about the characteristic of being a bird, though it does not include its relation to all other characteristics, should include its relation to the characteristic of being a phoenix, and so give information as to the characteristic of being a phoenix. And, when this is given, it may very well include the fact that that characteristic is incompatible with the characteristic of being a hippogriff.

659. Thus we have seen that we can gain in perception information both as to existent and non-existent characteristics, and as to their mutual implication. And this is all the information which is given in non-existential judgments. Nothing, therefore, appears in the form of a non-existential judgment which cannot appear in the form of a perception, and the apparent existence of non-existential judgments, as of existential judgments, can be due to the misperception of a perception as being a judgment.

It should be noticed that the object, the perception of which gives us this information, need not be any object which, *sub specie temporis*, is contemporary with the perception. Apparent perceptions, as we have seen, can only be of something which is so contemporary. But the perceptions which appear as judgments are not, of course, apparent perceptions. It will be sufficient, therefore, if such an object occurs at any stage in the C series.

660. We have thus explained how a perception can give us that information which is given in an apparent non-existential judgment. Our second question was whether we could explain why certain non-existential judgments are completely true or completely false, while the perceptions which appear as those judgments are partially correct and partially incorrect. The solution of this question is exactly the same as that which was given for existential judgments (pp. 301–302), and need not, therefore, be repeated.

661. One more point remains. Whatever appears as a judgment is really a perception, and has therefore parts within parts to infinity, all of which are perceptions. A judgment is divided, but not into parts of parts to infinity. The judgment contains as its ultimate constituents simple characteristics which cannot be divided into parts. Can anything which is really divided in the first way appear as something which is divided in the second way?

This seems to be quite possible. On the one hand, the division into parts which does exist in the perception is not perceived when the perception is misperceived as a judgment. And in this there is nothing surprising. Anything can be perceived without perceiving its parts. Even when a perception is also an apparent perception—that is, when it is correctly perceived as being a perception—not all its parts are perceived. If they were, we should perceive it as infinitely divided, which we do not do.

When a perception is misperceived as a judgment, then, it is only the perception as a whole which is misperceived as a judgment. The parts of the perception are not perceived at all. And this accounts for the fact that the apparent judgment does not appear as divided into parts of parts to infinity.

But it does appear as divided into some parts, as was said above, and these apparent parts are not any of the real parts. For the real parts are all perceptions, while the ultimate apparent parts, as has been said, are simple characteristics. And, since this plurality is not part of the real plurality, how is it that the apparent judgment appears to have it? Its appearance is due to misperception—that is, there are really no such parts in the perception as appear to be there. But it is due to a plurality of a sort in perception, though not to a plurality of parts. In the perception, the object is perceived as possessing a plurality of characteristics. And it is these characteristics which form parts of the judgment. The erroneous element, as was explained above, consists in taking the plurality of characteristics as if it were a plurality of parts, and so getting an apparent judgment.

CHAPTER LV

APPARENT INFERENCE

662. Judgments, as they appear in our ordinary experience, can be divided into those which are self-evident, and those which are only to be justified by inference from other judgments. Some, therefore, appear to have direct certainty, and others only indirect certainty. But all apparent judgments are in reality perceptions, and whatever certainty they have is direct. Yet it is scarcely possible to suppose that the *primâ facie* distinction in the apparent judgments is merely erroneous—that there is no reality behind the appearance of inference. Can we find an explanation which will give a reasonable account of the appearance of inference, while preserving the direct certainty of perception?

Not all perception, doubtless, is direct perception in one sense of the word. We saw (Chapter XXXIX, p. 126) that, if B perceives C's perception of D directly, he perceives D indirectly, since, in getting information about C's perception of D, he is also getting information about D. And if he perceives C's perception of D's perception of E, he will perceive E indirectly, and so on.

But this will not help us here. For what we are now considering is the *primâ facie* dependence of a judgment asserting one thing on a judgment asserting something else—as when our judgment that a proposition of Euclid is true depends, *inter alia*, on our judgment that one of the axioms is true, or as when our judgment that all lead sinks in water depends, *inter alia*, on our judgment that some particular piece of lead has sunk in water. And the indirectness here is obviously quite different from the indirectness which we get when B perceives D reflected in C's perception of D.

663. Again, when B perceives C as a bird, the information that C is a bird, and the information of what is meant by that characteristic, may be said to be given more directly than the information—which the perception can also give—of the relation

of the characteristic of being a bird to the characteristic of being
a phoenix. And the further information—which the perception
can also give—of the incompatibility of the characteristic of
being a phoenix with the characteristic of being a hippogriff may
be said to be still more indirect.

But, once more, this sort of indirectness is not what is
needed here. The information that B is a bird, the information
as to the nature of the characteristic "bird," as to its relation to
the characteristic "phoenix," and as to the incompatibility of the
latter with the characteristic "hippogriff," are all given in the
same perception. This will not help us with regard to the case we
are considering, which is, at any rate *primâ facie*, a case of the
dependence of one cognition on one or more other cognitions.

664. It seems to me that the required explanation can be
found without difficulty. In the first place, let us note that the
content of a judgment—the fact asserted—is quite independent
of the question whether the judgment is self-evident or inferred,
and whether, if it is inferred, it is inferred *à priori* or empirically.
These are questions of how the fact is known, not of what fact is
known.

This becomes evident when we consider that the same fact can
be known in different ways. That which is known self-evidently
can also be known by inference, both *à priori* and empirical. Thus
the validity of a particular syllogism in Barbara is a self-evident
truth, and can be known as such, without reference to anything
else. But the general principle of the validity of all syllogisms in
Barbara is also a self-evident truth, and the validity of this
particular syllogism can be inferred from it. Again, the axiom of
Euclid that two straight lines cannot enclose a space is a self-
evident truth. But a person who did not see that it was self-
evident might infer it empirically. For if he took many pairs of
things, each thing in each pair having an outline which was
indiscernible from a straight line, and if in each case he endeavoured to make these outlines enclose a space, and failed, he
would have good grounds for believing, as a result gained by
induction, that two straight lines cannot enclose a space. And he
would be believing in just the same fact as the person who saw
the self-evidence of the truth.

Again, take a proposition which is not seen to be self-evident, but which can be inferred à *priori*—for example, that two sides of a triangle are always longer than a third. A person who should measure the sides of different figures which were approximately triangular, varying as much as possible the nature of the figures in other respects, might arrive inductively at the conclusion that two sides of a triangle were always longer than the third. And the fact which he was believing, as a result of induction, would be just the same fact as that which a student of geometry believes, as an inference à *priori* from self-evident axioms.

665. I do not wish to suggest that the distinction between what is self-evident and what is inferred, or between what is inferred à *priori* and what is inferred à *posteriori*, has no relation to the nature of the fact asserted, and is only a question of the way in which the particular thinker has reached his judgment about the fact. For it might be the case, as it appears to be *primâ facie*, that it is only certain facts, which, when asserted in apparent judgments, can appear as self-evident, while others can only appear as inferred. And, again, it might be the case that only some facts can appear as inferred à *priori*, and that others can only appear as inferred à *posteriori*. And thus self-evident and à *priori*, in the sense of *"capable* of being known as self-evident," and *"capable* of being known à *priori*," might be characteristics of the facts, independently of the way in which they *are* known. But although, in that case, they would be characteristics of the fact asserted in the judgment, they would not be asserted of it in that judgment. If I know, as a self-evident truth, that X is Y, what I know self-evidently, in that cognition, is that X is Y. The facts that I know the proposition self-evidently, and that it is capable of being known self-evidently, are other judgments about other facts. And this is made clear, as was said above, by the consideration that another person may know, as the result of an induction, this same fact which is capable of being known self-evidently, and which I know in that manner.

666. Let us apply this result to the problem before us. The information which appears to be given in a judgment reached by inference can, as we have seen, be given in a perception. In the apparent judgment the information appears to be given indirectly, while it is really given, in the perception, directly. But this difference, as we have just seen, does not affect the information which is given, and so the information which appears to be given in one way can really be given in the other. What remains is to explain what corresponds in reality to the appearance of the dependence of the judgment on those others from which it is inferred.

When we assert that the judgment P has been inferred from the judgments Q and R, what do we mean? We mean that the making of the judgments Q and R by the thinker was part of a group of facts which determined him to make subsequently the judgment P. We must not say that they are a necessary antecedent of P, since the same conclusion might be reached in other ways. (For example, it might be inferred from other premises.) Nor must we say that they are by themselves an antecedent which necessitates the making of the judgment P, since it is possible to hold the premises without drawing the conclusion. But we can say that there is a group of circumstances, of which the occurrence of Q and R forms part, which does render necessary the subsequent occurrence of P. (The other circumstances would include sufficient interest and insight to draw the conclusion, the continued existence of the thinker for long enough to draw it, and others of a similar nature.)

But, secondly, more than this is required. Even if a set of circumstances, of which the occurrence of Q and R forms part, should render necessary the subsequent occurrence of P, this would not constitute an inference of P from Q and R, unless it were also true that the thinker held a belief that the truth of Q and R implied the truth of P.

This, then, is what is meant when we say that P has, as a matter of fact, been inferred from Q and R. But when we say in addition that the inference is a valid one, we mean in addition that the truth of Q and R really does imply the truth of P.

667. Let us consider first the third of these points. It is quite clear that the implication of the truth of P by the truth of Q and R is not in the least inconsistent with the fact that the content of P, as well as that of Q and R, is really known to us by perception. If I know the content of P by perception, I shall know it independently of P's being implied by Q and R, and may know it without knowing that P is implied by Q and R, but this will not be incompatible with the fact that P *is* implied by Q and R. If I contemplate a particular syllogism in Barbara, I may recognize its validity directly, and so independently of its being implied by the general principle of the validity of all syllogisms in Barbara, but this will be quite compatible with the fact that it *is* implied by that general principle.

668. Let us now return to the first point. The *primâ facie* position is that the judgment P has been inferred from the judgments Q and R. In reality these apparent judgments are perceptions, and their contents are known directly. But this will not prevent them from having to one another in reality the relation which *primâ facie* they do have. For that relation is that the cognition P would not occur where it does unless certain groups of occurrences had been so related to it in the C series, as to appear to precede the cognition P in time; and that if one of those groups did so occur, the cognition P must occur where it does; and that, of one of these groups, the occurrence of the cognitions Q and R is an essential part. This is what the relation turns out to be, when we have substituted for "judgments," which the cognitions P, Q, and R really are not, the more general term "cognitions," which does really apply to them, and when we have substituted for the apparent temporal relations the real relations which appear as temporal.

Now in all this there is nothing whatever incompatible with the fact that the content of P is perceived, and therefore known directly. For a direct cognition may, like anything else, be so connected with other things by causal laws, that its occurrence is determined, or partially determined, by the occurrence of those other things. And so the direct cognition P may be determined by the direct cognitions Q and R.

669. There remains the second condition. We saw that P would not be said to be inferred from Q and R unless the thinker, who was determined to the belief P by his beliefs Q and R, believed also that the truth of Q and R implied the truth of P. But here again there is no difficulty. For it is possible to know a fact directly, and also to know that it is implied by another fact which is known directly, as when I recognize immediately the validity of a particular syllogism in Barbara, and also know that its validity is implied in the validity of the general principle, which I likewise recognize immediately. And so, notwithstanding that the contents of the cognitions P, Q, and R are really known directly, we may know that the truth of the cognitions Q and R implies the truth of the cognition P. (Of course our cognition of this implication will not be in reality a belief, but a perception.)

Thus P, Q, and R may really have all those relations to one another which were asserted when it was said that P was inferred, and truly inferred, from Q and R, and inference, accordingly, offers no difficulty to our theory.

CHAPTER LVI

OTHER APPARENT FORMS OF COGITATION

670. The apparent forms of cogitation, other than perception and judgment, which we enumerated in Chap. XXXVII, p. 87, were awareness of characteristics, assumptions, and imagings. If our theory is correct, whatever appears as any one of these is, in reality, perception.

It will be convenient to begin with assumptions. Now what we may call the internal nature of an assumption is exactly the same as that of a judgment. When we say "Smith is bald," we have the same connection of the same qualities with the same subject, whether we make the assertion that Smith *is* bald—which is a judgment—or whether we say "it is said that Smith is bald," "it will be surprising if Smith is bald," "is it true that Smith is bald?"—in all of which "Smith is bald" is an assumption. All that differentiates the judgment from the assumption is that the judgment is an assertion, and the assumption is not.

A perception cannot be said to be an assertion, but, like an assertion, it is a cognition, since it professes to give information. An assumption is not a cognition, and differs, therefore, from a perception more than a judgment does.

We saw in Chapter LIV that it was possible that a perception, when itself perceived, should be so far misperceived as to appear as a judgment. If, when this misperception took place, it should be carried still further, so that the element of cognition, which exists in the perception, and which is still evident in the apparent judgment, should no longer be evident, then the perception, instead of appearing as a judgment, would appear as an assumption.

There seems nothing improbable in the view that such further misperception could take place. If, for any of the many reasons which might produce such a result, the relation between the original perception and the perception of that perception was less close in some cases than in others, it might well happen that in the former cases the element of cognition would cease to be apparent, and the perception would appear, not as a judgment, but as an assumption.

The fact that the assumption appears as divided into indivisible parts, while the perception is divided into parts of parts to infinity, can be explained in the same way as in the case of apparent judgments (Chap. LIV, p. 314).

671. We can thus account for the fact that we have apparent assumptions. But can we account in this way for all the apparent assumptions which we do have? Anyone who has read *Through the Looking Glass* has made the assumptions "the sea is boiling hot" and "pigs have wings." Must we suppose that every such person has so misperceived some substance as to perceive it as a boiling hot sea, and other substances so as to perceive them as winged pigs? This is wild and improbable. Among other consequences it would involve that each man committed all the errors which he contemplated. For he can only contemplate an error by making the corresponding assumption, and if that assumption is really the perception, he will have committed the error.

Nor could the difficulty be avoided by the suggestion that he perceives the sea, not as boiling hot, but as having the negative quality of not being boiling hot. For such a perception, if it was misperceived as an assumption, would appear, not as the assumption that the sea is boiling hot, but as the assumption that the sea is not boiling hot—an assumption which is also sometimes made.

The occurrence of such apparent assumptions can, however, be satisfactorily explained. However wild an assumption may be, yet, if it actually does occur, we can always find a judgment which contains that assumption as an element, and which, whether true or false, is one which it is not improbable would be made. The assumption that pigs have wings is fantastic enough, but such a judgment as "if an example of fantastic suppositions is wanted, we cannot do better than take the supposition that pigs have wings" may very probably be true, and has certainly been made. Or again we might take the judgment "if a walrus could talk to oysters, he would be likely to discuss whether pigs have wings." These judgments are non-existential, but we might also have such an existential judgment as "there are many readers in the world whose nature is such that they would be amused by contemplating the assumption that pigs have wings."

Since there is nothing wild or improbable about the occurrence of such apparent judgments as these, it follows that there may be perceptions which are such that they may be perceived as such judgments. And then, if such a perception is misperceived to a still greater extent, we shall get the assumption. The assumption is part of the content which appears in the judgment. That part may be perceived without the other part of the content which forms its framework—for example, "there are many readers in the world whose nature is such that they would be amused by contemplating...." And, if it is so perceived, we shall get the assumption.

This, I think, will give us a sufficient explanation of how assumptions can arise. For, with assumptions, as with judgments (cp. Chap. LIV, p. 312), we have not to account for every cogitation which could be produced by every possible combination of characteristics. All that we have to do is to account for those cases where an assumption does appear to occur. There is no necessity to explain how such an appearance can arise, except in those cases where it actually does arise. And whenever it does arise there must be some judgment of this kind which is true. Whenever an assumption comes to my mind, there must be some reason why an assumption does come, and why this one comes rather than another. And this reason can always be expressed in some judgment, such as the three given above, of which the assumption is a constituent. Thus the conditions necessary for the solution are present in every case in which they can be wanted—the cases in which assumptions do actually appear to occur.

672. In a similar way we can explain the occurrence of apparent awareness of characteristics. There are occasions when we appear not to be judging or assuming any proposition, but only to be contemplating a characteristic. We appear, for example, to be conscious of what is meant by "red," without either asserting or assuming any proposition in which that characteristic enters as a constituent. It has been already shown how it is possible for what is really a perception to be misperceived as a judgment or an assumption. Judgments and assumptions contain characteristics as constituents. And the perception may be so misperceived as to be perceived with still less of the information which it gives than would be the case if it appeared as a judgment or as an assumption. In that case the only information which appears may be the meaning of one of the characteristics; and thus we shall get an apparent awareness of the characteristic, as distinct from any proposition into which the characteristic enters.

673. We now pass to those things which appear to us as imagings. Here, as elsewhere, the reality behind this appearance must be perceptions. Imagings stand in a relation to perceptions which is similar to that in which assumptions stand to judgments. The internal nature of an imaging is similar to that of a perception, as the internal nature of an assumption is similar to that of a judgment. But while a perception, like a judgment, is a cognition, and gives information, an imaging, like an assumption, gives no information and is not a cognition.

It follows from this that, for what is really a perception to appear as an imaging, it is only necessary that it should be so misperceived that we fail to perceive the element of cognition. A perception, which is not perceived as being a cognition, will appear as an imaging.

Thus one sort of misperception will make a perception appear as a judgment, while another sort of misperception will make it appear as an imaging. If its internal structure is incorrectly perceived, while it is correctly perceived as being a cognition, it will appear as a judgment. If, on the other hand, it is not perceived as being, as it really is, a cognition, while its internal structure is correctly perceived, it will appear as an imaging. The two sorts of misperception are so different and each of them so fundamental, that it is impossible to say that one of them involves a greater amount of error than the other. But there is a sense in which it can be said that an apparent assumption is more erroneous than either an apparent judgment or an apparent imaging, because, when a perception appears as an assumption, both sorts of error are involved. For it is not perceived as being a cognition, and it is perceived as having a different internal structure from that which it really has.

674. The perception which appears as an imaging may be a perception of something which is at a different stage of the C series from the perception itself—that is, of something which appears as past or future. In this point apparent imagings, like apparent judgments and assumptions, differ from apparent perceptions, which can only be of something which is in the same stage of the C series as the perception, and which therefore appears as present. This difference between apparent imagings and apparent perceptions presents no difficulty. For, as we saw in Chap. LIII, pp. 292–293, the reason why an apparent perception must always be of what appears as present is to be found in circumstances connected with the guarantee of the correctness of the perception. And as an imaging is not a cognition, and gives no information, there can be no question of the guarantee of its correctness. There is therefore no reason why an apparent imaging should not be of something which is not at the same stage of the C series.

The imagings which are involved in memory are of this type. We saw (Chap. XXXVII, p. 112) that, *primâ facie*, memory consists of a judgment about an imaginatum. It follows from the results which we have since reached that there is only an apparent judgment and an apparent imaging, and that the reality of both is perception. We have also seen that an apparent perception and an apparent judgment can both be really the same perception. And, in the same way, it will be possible for a single perception to appear both as the apparent imaging and the apparent judgment which are involved in memory.

The imagings which occur in memory are always, I suppose, less full and exact than were the original perceptions, on which the memory is based. And in some cases they are positively erroneous. We remember a thing inaccurately, as when Fouché "remembered" Robespierre addressing him as Duke of Otranto. There is no difficulty here. All our perceptions in our present experience are more or less erroneous, and it may well be that my present perception of a past object (which appears as memory) may be more erroneous than my past perception of it, which was simultaneous with the object, and which appeared as an apparent perception.

675. At first sight it might seem as if we could explain all imagings in the same manner. Any imaging representing anything existent, past, present, or future, might be really a perception of that thing, and any imaging which represents something which never exists might be really a misperception of something which does exist.

It is possible that this explanation may be the true one in some cases. I may image an imaginatum as to which I make no judgment that it corresponds to anything which I have experienced in the past. It is therefore not a memory imaginatum. And yet it may correspond to something which I have experienced in the past, and my imaging may, in reality, be a perception of that past something.

But this explanation will not serve for most of the imagings, other than those involved in memory, which we have in our present experience. I may image the execution of George III in Berkeley Square. This never happened, and now never can happen. Is it probable that the true nature of this imaging is that it is a misperception of some existent event in the present, past, or future?

And, in addition to this, there seems positive evidence that such imagings do arise in another way. It seems impossible to deny that my imagings of what I have never experienced are— at least in very many cases—imagings of imaginata composed of elements furnished out of what I have experienced, arranged in accordance with some judgment or assumption which I have had.

If, for example, I image the Capitol in the days of Caesar, it seems clear that I first make a judgment, or an assumption, that it had certain qualities, and then image an imaginatum formed out of my own past experiences. I may have seen the Capitol last year, I may have seen other buildings which I judge that the Capitol resembled, and I may have seen pictures which were conjectural restorations of the Capitol. It is out of these that the elements of my imaginatum are taken.

So, if I am thinking about the French Revolution, I may make the judgment, "if Charles I had not been executed at Whitehall,

George III might not improbably have been executed in Berkeley Square." This contains the assumption "that George III was executed in Berkeley Square." This may lead to my imaging the execution. Is it not certain that the imaginatum can be analyzed into elements supplied by memory—my memory of Berkeley Square with no one being executed, my memory of the statue of George III on horseback in Cockspur Street, and my memory of a guillotine at Madame Tussaud's waxworks?

676. If imaginata are built up in this way, can an imaging be a single perception, which must have a single object? And yet it is essential to our theory that an apparent imaging should be really perception.

To this difficulty, however, there is a solution. We have seen (Chap. L, p. 261) that a plurality of perceptions can, when they are themselves perceived, be misperceived as a single perception. The same principle will apply in this case. The previous judgment or assumption—perhaps accompanied by a volition to form an image—will make those perceptions more prominent which have qualities correspondent to those given in the judgment or assumption. In the case of the execution, these perceptions will be those of the Square, the statue, and the guillotine. My interest in each of them, excited by the judgment or the assumption, will be in the qualities which are compatible, and not in those which are incompatible. My interest, in the first case, will be in the quality of being Berkeley Square, not in the quality of being devoid of guillotines. My interest, in the third case, will be in the quality of being a guillotine, and not in the quality of being outside Berkeley Square. If, under these circumstances, the three perceptions should be misperceived as having an imaginal not a perceptual nature, and as being a single imaging of a single imaginatum, with the qualities which are more prominent for me in each of the three percepta, I shall get an image of George III's execution in Berkeley Square. If, on the other hand, this does not happen, the three perceptions will appear to me as imagings of three separate imaginata. This also often occurs. It often happens that a man fails to image something which he wishes to image, and has to content himself with contemplating various imaginata, each conforming in some respects, though not in all, to the judgment or assumption with which he starts.

677. Is it, then, the case that no perceptions can appear as imagings unless they, or their constituents, are perceptions of objects of which the same self has previously had apparent perceptions? It is not certain that this is the case, even with the imagings which we have to-day. It is the case, no doubt, with such imagings as those which we have just been discussing—the imaging of the Capitol in the past, or the imaging of the execution of George III. But I can see no reason for absolutely denying the possibility that among the imagings which we have to-day, some may really be perceptions of objects of which the percipient never has an apparent perception, or of which his apparent perceptions are still in the future[1].

And supposing that this limitation does apply to all imagings which we have to-day, it does not follow that it must necessarily apply to all imagings throughout the misperception series of all selves. Other selves than we, or ourselves at another position in our C series, may have apparent imagings which are really perceptions of objects of which they never have apparent perceptions, or only have them later than the imagings.

678. But at any rate the limitation extends to most, if not all, of our experience to-day, and it may be absolutely universal. Can we find anything which may possibly be the reason of this circumstance? We saw in Chap. LIII, pp. 292–293, that there is reason to believe that a percipient is more closely related to a perceptum of which his perception is an apparent perception than to one of which his perception is not an apparent perception. And this may be the reason why it is only such percepta of which any perceptions can appear as imagings. With regard to the limitation to percepta of which the apparent perceptions are in the past, it will be convenient to postpone any consideration till we again deal with the time-series in Book VII.

---

[1] It might be said that there was a strong presumption against this from the fact that we never find that any imaging gives us any simple quality which has not previously been given to us as a quality of an object of an apparent perception. But it must be considered that the number of simple qualities which we know is comparatively small, and that much the greater part of the differentiation of our experience is produced by the variety of the ways in which they are combined. It would not, therefore, be very remarkable if a range of perceptions much wider than the range of apparent perceptions showed no additional simple qualities.

CHAPTER LVII

EMOTION AND VOLITION

679. We have now to consider how far the facts which we observe in our present experience as to volition and emotion are compatible with those conclusions as to the true nature of the existent which we reached in Book V. It will, I think, be convenient to reverse the order in which they were taken in that Book, and to deal first with emotion.

When we considered the nature of the existent in Book V we found that the universe consists of selves, which form a set of parts of the universe, and that each self contains a set of parts which are perceptions, and so on without end. We found that each of those perceptions has the quality of being an emotion. The direct perception of another self is an emotion of love. The perception of my own self is, in consequence of this, an emotion of self-reverence. The indirect perception of other selves is an emotion of affection. The perception of parts of other selves, or of myself, is an emotion of complacency.

These were the only emotions which we were able to assert must exist. But we saw that it is also possible that such perceptions should have the qualities of sympathy, approval, disapproval, pride, humility, gladness, and sadness. On the other hand, it was found impossible, for various reasons, that they should have the qualities of hatred, repugnance, malignancy, anger, courage, cowardice, jealousy, envy, regret, remorse, hope, fear, surprise, or curiosity (Chap. XLI, pp. 166–167).

When we look at our present experience, we find that besides perceptions, we have, or appear to have, awareness of characteristics, judgments, assumptions, and imagings. Some of these states do not appear, on introspection, to have any emotional qualities, and although it is not impossible that they might possess them in so slight a degree as to escape observation, there is no reason to believe that this is the case, and it is probable that they actually are without emotional qualities. And of those

which do have emotional qualities, some appear to have none of
the qualities of which it was previously determined that each of
the perceptions must have one—love, self-reverence, affection,
or complacency. And, again, many of these states appear to
have emotional qualities—hatred, malignancy, hope, fear, and so
forth—which we had previously decided could be possessed by
none of the perceptions.

680. As we have seen, all these states, both apparent perceptions, and also apparent judgments, assumptions, imagings, and
awareness of characteristics, are really perceptions. They are not
the perceptions, discussed in Book V, which form the system of
determining correspondence, but fragmentary perceptions, which
fall within the others, and are members of their inclusion series.
And, while the perceptions of the determining correspondence
system are correct perceptions, all these fragmentary perceptions
are more or less erroneous.

This shows us how it is possible that these states may differ, in
respect of their emotional qualities, from the perceptions of the
determining correspondence system. These latter, as we saw, had
their emotional qualities, not because they were perceptions, but
because they had the content which they do have. Now the
content of the fragmentary perceptions is different. It is true
that they perceive the same objects as the others. But they all
perceive the objects more or less incorrectly, and therefore their
content will be different. And this is the reason why their
emotional qualities will, in many cases, be different. If, for
example, an object, which is correctly perceived as another self,
is incorrectly perceived as a material body, it is clear that, while
the correct perception can be a state of love, this particular incorrect perception cannot be such a state, since love can only be
felt towards what is recognized as other selves.

It may therefore sometimes be the case that a fragmentary
perception not only differs from the complete perception of which
it is a part in its emotional qualities, but may have emotional
qualities which are incompatible with those of the complete
perception. G, in his present experience, may hate H. This hatred
is really a fragmentary perception of H, which forms a part of
a complete perception of H. And this complete perception of H

will be a state either of love or of affection, according as the perception is direct or indirect. Thus a state of hatred of H may be a part, in this dimension, of a state of love of H.

681. There does not seem any difficulty in this. It is, of course, a matter of common observation that a whole and its part can have respectively qualities which would be incompatible in a single subject. A nation, for example, is impersonal, but it has as its parts citizens who are personal. And, with regard to the particular qualities with which we have to deal when we are considering emotions, there does not seem any special difficulty. The case which we have just taken—a state of hatred as part of a state of love—is the case which is perhaps most doubtful. But I do not see any difficulty even here. It must, of course, be remembered that when we say that the hatred is part of the love, we do not mean that the quality of hatred is part of the quality of love, but that a state which has the quality of hatred is part of a state which has the quality of love. And it must also be remembered that the terms of the inclusion series do not form a set of parts of the complete perception, and that therefore our theory does not involve that two states of hatred could be added together, so as to produce, or to co-operate in producing, a state of love.

682. Passing to volitions, we find a similar problem to that with which we have dealt as to emotions. The whole content of the selves, which themselves form the whole content of the universe, does in reality consist of perceptions forming systems of determining correspondence, each of which perceptions is a volition—a state of acquiescence in what is perceived. And we saw that none of these volitions could be ungratified. When, on the other hand, we regard our present experience, we find that we appear to have, not only perceptions, but judgments, assumptions, imagings, and awareness of characteristics. Some of the members of each of the first four classes appear to be states of volition, and some do not, while acts of awareness of characteristics never, so far as I know, appear as states of volition. All these states are really, as we have seen, fragmentary perceptions, falling within the inclusion series of perceptions of the determining correspondence system. And all of these fragmentary perceptions are more or less erroneous.

The solution here is analogous to the solution as to emotion. The reason that we held that all perceptions in the determining system are states of acquiescence is not because they were perceptions, but because of their content. They were all, we found, perceptions of other selves, or of the parts of other selves, or of the percipient, or of parts of him, and they perceived their objects as having the qualities by which we have just described them. And they were consequently all states of love, or of self-reverence, or of affection, or of complacency. And, as a consequence of their having these emotional qualities, and of their not having certain other emotional qualities, we held that they must all be states of acquiescence.

The fragmentary and erroneous perceptions have the same objects as the perceptions in the determining correspondence system of which they form part. But they none of them, since they are erroneous, perceive their objects in quite the same way as the complete perceptions which form the determining correspondence system. For these latter are correct. And consequently it is not necessary that the former should all be states of acquiescence. They may be states of acquiescence for the same reasons that their wholes are—as when, in my present experience, my cognition of someone whom I love has the quality of acquiescence. Or they may be states of acquiescence because of some of the qualities which they are erroneously perceived as having. Thus I may misperceive something as a mountain, and as a mountain which is beautiful, and I may acquiesce in it because I perceive it as beautiful. But it is also possible they should not be states of acquiescence at all.

683. In our present experience some of our volitions are ungratified. It makes, of course, no difference to the internal content of a volition whether it is gratified or not. It only means that the cogitation, which is the desire, is not in accordance with the facts. An assumption or imaging could be known by the self who has it not to be in accordance with the facts, and so a desire, if it were cogitatively an assumption or an imaging, could be known to be ungratified by the self who has it, while he has it. But this could not be the case with a judgment, since it is impossible for a man to make a judgment which he knows, while he makes it, not to be in accordance with the facts.

We know that judgments, assumptions, and imagings need not be in accordance with the facts; and thus, if the fragmentary perceptions in the inclusion series were really judgments, assumptions, and imagings, there would be no difficulty about the fact that ungratified desires occurred among the fragmentary perceptions, though not among the complete perceptions of which they are parts. But then they are not really judgments, assumptions, and imagings, but perceptions. This, however, introduces no difficulty, for there is nothing in the fact that a cogitation is a perception to prevent it from being an ungratified volition, provided that it is more or less an erroneous perception. If G perceives H as having the characteristic X, and if this perception is also a desire, G will acquiesce in the existence of H as, *inter alia*, having the characteristic X. And, if H has really not got that characteristic, then the desire will be ungratified. It is true that the acquiescence is in H, which is perceived, and not in the proposition that H has the characteristic X. But the acquiescence will be in H as G perceives it to be, and so, if H does not exist as he perceives it to be, the desire will be ungratified[1].

And such an ungratified volition might be known to be ungratified, even by G while he entertained it. For G might have at the same time an apparent perception of H as X, and a perception which appeared as a judgment that H could not be X. And then the apparent perception, if it were a desire, would be known by G to be ungratified. He might, *e.g.*, perceive H as in time, and he might—if he had a passion for time as against eternity—acquiesce in his being in time. But he might also have a perception which appeared as a judgment that nothing was in time, and then he would know that the desire was ungratified. And the same would be the case if, instead of an apparent perception, he had an apparent assumption that H was in time, or an imaging of H as being in time.

684. We see, then, that the perceptions which appear as apparent perceptions, judgments, assumptions, and imagings, can

---

[1] It might, of course, be the case that, if G perceived H correctly, that perception would also be an acquiescence. And this acquiescence would, of course, be a gratified volition. But this does not alter the fact that the acquiescence in the perception of H, which he does have, would be an ungratified volition.

have those emotional and volitional qualities which they appear to have, in spite of the fact that many of these are qualities which cannot be possessed by the complete perceptions of which they are parts[1]. But another question arises. Have we any reason to believe that, in any particular case, they do have the particular emotional and volitional qualities which they appear to have?

With regard to apparent perceptions, there does seem to be such a reason. For everything must be supposed to be what it appears to be, unless there is a definite reason for doubting it. And in this case there appears no reason for doubting it.

But with regard to the other apparent forms of cogitation, the matter is different. What really exist here are not judgments, assumptions, or imagings, but perceptions. Perceptions are cognitions of substances. We have seen, in Chapters LIV and LVI, that part of the information given in perception may be misperceived—still as cognition—in existential or non-existential judgments, or—no longer as cognition—in assumptions or imagings. But the emotional and volitional qualities of the perception will not be preserved in the appearances.

Let us take the existential judgment "Smith is a swindler." If I am a normal person, this judgment will not be an acquiescence, since the normal person does not desire that other people should be swindlers. But my perception of Smith may be an acquiescence, though I perceive him as being, among other things, a swindler. For I may love him, and then my perception of him will be an acquiescence, whatever qualities I perceive him as having.

The case is still stronger with assumptions. I may make the assumption "that Smith is a swindler," and this may have—probably will have—the quality of being a disapproval, and the quality of not being an acquiescence. But the perception which is misperceived as being such an assumption may be a perception of Smith as having some quality which is perceived as having the quality of incompatibility with the quality of being a swindler[2]. Such a perception, whether I love Smith or not, may very probably

[1] There is strictly only one volitional quality—acquiescence. But it will be convenient to speak of the negative quality of not being an acquiescence as a volitional quality.

[2] It will be remembered that part of the misperception of the perception has been held to be the failure to perceive part of the content of the perception.

have the qualities of being an approval and an acquiescence. The same case may arise with regard to imagings.

Again, with non-existential judgments we find the same thing. The non-existential judgment "no poet can be an aeroplane" would probably, in the normal man, have no positive emotional or volitional quality. If it occurred to him, he would admit its truth, but he would not be in the least interested in it. But the perception which was misperceived as being such a judgment might be the perception of Smith as having the quality of being a poet which is perceived as having the quality of incompatibility with the quality of being an aeroplane. And such a perception of Smith might be volitionally an acquiescence, and might have various positive emotional qualities.

Thus we see that, when a perception appears as a judgment, assumption, or imaging, with certain emotional and volitional qualities, we have no reason to suppose that the perception really has those qualities. They stand or fall with the state in question being, as it appears to be, a judgment, assumption, or imaging. And as the state is not really a judgment, assumption, or imaging, but is only misperceived as such, it follows that it is only misperceived as having those particular qualities.

Of course the perception may itself have emotional and volitional qualities. But we do not know what they are, since we know nothing of the perception except our misperception of it. They may in some cases be the same as the qualities which it is misperceived as having. A perception of Smith which is an acquiescence may be misperceived as the judgment "Smith is a hero," and as being, as that judgment, an acquiescence. But this may be, or may not be, and we do not know in any particular case whether it is so or not.

685. This result increases, no doubt, the difference we have already found to exist between appearance and reality. But we shall have an exaggerated idea of the amount of error in ordinary experience which it involves, unless we attend to an important fact. It is only an error to say that a particular judgment or assumption is a state of hope or of acquiescence, in the way in which it is an error to say that a particular building—Westminster Abbey, for example—is Gothic. Since nothing exists but spirit, no buildings exist, and to say that Westminster Abbey exists—since the name implies that it is a building—is erroneous. But it is not an error to say that Westminster Abbey is Gothic in the same sense that it would be an error to say that it was Palladian. If what appears as Westminster Abbey really was, as it appears to be, a building, it would be really Gothic. And it is not the case that the appearance of the reality as a building appears to be Gothic. What is the case is that the reality appears to be a building and Gothic.

In the same way, it is not the case that the appearance of a particular perception as a judgment appears to be a state of hope and acquiescence. It is the perception which appears to be a judgment and a state of hope and acquiescence. When, in ordinary language, I judge that the world is getting better, and acquiesce in it, it is just as true that I acquiesce as that I judge. It is not, indeed, as certain that I acquiesce as that I cognize, for perception is really cognition, and I really have the perception, and I am not sure that the perception is a state of acquiescence. But the assertion of the acquiescence has the same phenomenal truth as the assertion of the judgment.

686. It may be said that this result will destroy the whole value of our emotional and moral life. All acts of love in our present experience are judgments, since we have no apparent perceptions of persons[1]. How, then, can we know that we really do love any of the people whom we appear to love? And does not a great part of our moral life consist in our emotions and volitions which are cogitatively judgments or assumptions? If I cannot be sure that I really regard the conduct of Nero with disapproval, and the conduct of Spinoza with approval and acquiescence, if I cannot be sure that I really disapprove of any suppositious crime or approve of any suppositious act of virtue, is not the whole of my moral life reduced to a chaos?

However calamitous such consequences of our doctrine might be, they would not, of course, justify us in rejecting that doctrine as false. But I do not think that these consequences do follow.

687. With regard to love, G's perception of H will be really a state of love if the connection of G and H is particularly close

---

[1] No apparent assumptions or imagings can be states of love, since love requires a cognition of the beloved as existent.

and intimate. Now if G's perception of H is misperceived as an apparent judgment with H as its subject, there is nothing in this misperception as an apparent judgment which could make the connection of G and H appear more intimate than it really is. If, therefore, the perception, when it appears as a judgment, appears as a state of love, the connection of G and H really has the closeness and intimacy which is necessary for love, and the perception is really a state of love.

688. With regard to our moral natures, the matter is rather different. It is not certain that there is really any disapproval of crimes, or any approval of, or acquiescence in, acts of virtue, but the fact that there are apparent approvals, disapprovals, and acquiescences is sufficient for morality. It is certain that, from the *primâ facie* standpoint, the assumption "that I shall become rich by swindling," will be regarded with approval and acquiescence by some men, and with disapproval and the absence of acquiescence by others. If our theory is true, it follows that these cogitations, besides not really being assumptions, are not really approval and acquiescence in the one case, and disapproval in the other. But the fact that one appears as approval and acquiescence, and the other does not, must be due to differences in the characters of the two men, since the assumption contemplated by each as to his possible action is the same. And those differences are real differences of moral character. It is not the fact that there is approval or disapproval. But it is the fact that, when something appears as an assumption of a crime, it appears for one man with the quality of approval and for the other man with the quality of disapproval. This is sufficient to justify the assertion that the first man is, in this respect and at this time, of a vicious character, and that the second man is, in this respect and at this time, of a virtuous character.

CHAPTER LVIII

APPEARANCE AND REALITY

689. We have now considered the various differences which, if our theory is true, are found between the apparent nature of the existent and its real nature. It remains to sum up our conclusions.

We have found the explanation of the difference by considering the inclusion series. The inclusion series really exists. But the perceptions which fall in it, except those falling in the last term, are all more or less erroneous, and form a misperception series; and therefore when I perceive any object by a perception falling within any term of the series, except the last, that object will to some extent appear to me as having a different nature from that which it really has.

Since the inclusion series really exists, there really exists, within each part of the determining correspondence system, a series of parts which have natures different in many respects from the natures of the parts in the determining correspondence system. For all the members of each inclusion series—except the last, which is a part in the determining correspondence system—are more or less erroneous, while the perceptions in the determining correspondence system are correct. Some of the erroneous terms are states of acquiescence, and some are not. Some of them have the same emotional qualities which the parts of the determining correspondence system have. Others have emotional qualities which the parts of the determining correspondence system cannot have. Others, again, have no emotional qualities at all.

These terms really exist, but, since they are all misperceptions, the objects which they perceive appear as having a very different nature from that which they really have. There appears to be time in the universe, but in reality all that exists is timeless. There appear to be matter and sensa in the universe, but there is really only spirit, and the selves of which this spirit consists perceive themselves and one another. There appear to be, within various selves, judgments, assumptions, imagings, and simple acts of awareness of characteristics. But in reality the whole content of each self consists of perceptions. There are no judgments, assumptions, imagings, or simple awarenesses of characteristics. And, finally, those perceptions which appear as being judgments, assumptions, or imagings, appear as having volitional and emotional qualities which it is not certain that they do have.

Most of the objects, which are thus erroneously perceived, are themselves members of misperception series. It is therefore important to keep clear the distinction between those qualities—given in the last paragraph but one—which they really possess, and those qualities—given in the last paragraph—which they are erroneously perceived as possessing.

690. What justification have we for believing the theory of the relation of appearance to reality which we have reached in this Book? The justification is that we can find no other explanation of the facts, consistent with the conclusions arrived at in the five previous Books. If those conclusions are right, the true nature of reality is in many respects very different from what it appears to be. And therefore there is error. There is error in apparent judgments, since we do sometimes believe that things are as, if our theory is true, they cannot be. And there is error in apparent perception, since, as we have seen, it is impracticable to explain all the error involved merely as an error in apparent judgment. Place, then, must be found for error. But the perceptions of the determining correspondence system give no place for it. And such perceptions comprise the whole of the content of every self. The erroneous cognition, then, must have content which falls within the content of the determining correspondence system. And, while it cannot consist of the perceptions which form that system, it must consist of perceptions, and of perceptions only, because otherwise we could not without contradiction ascribe to it that divisibility into parts of parts without end which belongs to every substance. And, once more, our account of its nature must comply with the twelve conditions which we have found to be necessary. (Chap. XLVIII, p. 239.)

Our reason for accepting the theory of appearance which we

340 APPEARANCE AND REALITY [BK VI

have put forward is that it complies with all these requirements, and that no other theory is to be found which does so. Such a proof is, of course, of a negative nature. It rests on the assumption that there is not some other alternative theory which has failed to occur to us. If there were such another theory which equally fulfilled all the requirements, it would be doubtful which of the two theories was correct.

691. As we have seen, the negative character of the proof need not lead us to any distrust of our conclusion. It is shared by every argument, in philosophy, in science, or in everyday life, in which we arrive at a conclusion because it complies with certain conditions which have been established as valid, and because we can find no other that does so. Such arguments are frequent, and the degree of certainty to which they reach is often very great. And in the present case it must be remembered that the number of requirements to be satisfied is considerable, which diminishes the probability of any other theory being able to satisfy them all.

But the whole force of our argument rests, as has been said, on the conclusions established in the earlier Books, as to the real nature of the existent. Now the arguments in the first four Books professed to be absolute demonstrations, but it is, of course, possible that there may be some undetected error in them. And the arguments of the fifth Book, in so far as they are positive and not negative, do not claim to be absolute demonstrations, although they do claim to give sufficient reasons for accepting their conclusions. If, therefore, the results which we have reached were such as to be in themselves highly improbable, the question might be raised whether it was not more probable that there was some undetected error in the earlier argument than that such results should be true.

Our results certainly differ considerably from that view of the universe which presents itself to us *primâ facie*, and which is sometimes called the position of common sense. But I cannot see that they differ in such a way as could reasonably cause us to feel distrust of them.

The rejection of the existence of matter, as was pointed out when we discussed it, is less far-reaching than our other conclusions, because it does not by itself involve the existence of erroneous perception. Matter, *primâ facie*, is not perceived, but inferred. (It is only as a result of our further conclusion that all cogitation is perception that we reach the conclusion that our cognitions of matter are really perceptions.) But all the other differences between appearance and reality which were enumerated above (pp. 338–339) involve the existence of erroneous perception.

692. The possibility of erroneous perception, then, is essential to our theory. But the explanation of its possibility which we have given seems to have nothing improbable about it, when once the unreality of time has been accepted. No one has ever asserted that the self-evident correctness of a perception meant more than that it was self-evident that the perceptum was as it was perceived *while* it was perceived. If time is unreal, this must be re-stated in order to get a proposition which is not merely phenomenally true, but really true. And in that case there seems nothing improbable about the theory that the reality which appears as a perfectly correct perception occurring at a certain time is really a timeless perception which is only partially correct.

This explanation of the possibility of erroneous perception depends, indeed, as has just been said, on the unreality of time. But the assertion of the unreality of time can scarcely be said to be so improbable as to throw doubt on any theory which includes it, especially when we consider how many philosophers, from Descartes to the present day, have agreed, while differing on so many other points, to deny the reality of time.

693. Of all the results which we have reached, the most paradoxical, no doubt, are those which deal with apparent non-existential judgments, with apparent assumptions and imagings, and with the apparent emotional and volitional qualities of judgments, assumptions, and imagings. A view which holds that the contents of my own mind are so very different, in many respects, though not in all, from what they appear to me as being, must be pronounced to be so far paradoxical. But, after all, the paradoxical is only what is surprising and unexpected. That a result should be surprising and unexpected ought to lead the thinker—as it almost certainly will lead him—to re-examine the steps by which he has reached it. But if, after this, he can still find no flaws in his argument, he ought not to reject his results because they are surprising and unexpected. No philosophy has ever been able to avoid paradox. For no philosophy—with whatever intentions it may have set out—has been able to treat the universe as being what it appears to be.

BOOK VII

PRACTICAL CONSEQUENCES

CHAPTER LIX

THE FUNDAMENTAL SENSE OF THE *B* SERIES

694. In this Book we shall consider various questions of practical interest, as to which it is possible to draw conclusions by means of the results reached in the previous Books. We shall consider what can be determined as to the relative amounts of good and evil in the universe as a whole, and also as to the relative amount of good and evil in the apparent future as compared with the amount in the apparent present and apparent past. As a preliminary step to the solution of this question, it will be necessary to consider whether, when P appears to stand to Q in the relation of being earlier than Q, the real relation in which it stands is the relation of being included in Q, or the relation of being inclusive of Q. We have seen in Chapter XLVIII that it must be one of the two, but which of the two has not yet been determined. This will form the subject of this chapter, and of the next.

695. In mathematics, the letters of the alphabet, taken from A to Z, would be said to form one series, and the same letters, taken from Z to A, would be said to form another series. But it is clear that these two series are closely related. It is not merely that the terms of the two series are the same. This would also be the case with such a series as D, R, W, K, and so on irregularly till all the twenty-six letters had been used up. But the series from A to Z and the series from Z to A are connected, not only by the similarity of their terms, but by a certain similarity in their arrangement. If any letter is between two others in the one series, it will be between those two others in the other series. And if a letter comes before another in one series, it will come after that other in the other series.

It would be possible to express this relation by saying that the two series had the same order of terms. But I think it will be more convenient, and more in agreement with ordinary language, to depart so far from mathematical usage as to say that there is only one series with two opposite senses.

Every series has two such opposite senses, for where there is one generating relation there must always be a converse generating relation. If there is one relation which P has to Q, Q to R, and R to S, then there is another relation which S has to R, R to Q, and Q to P. And while the first relation generates the series in one sense, the second generates it in the other sense.

We shall say, then, that there is one B series, and that it has two senses, according as we take the generating relation to be "earlier than" or "later than." In the same way we shall say that there is one C series, and that it has two senses, according as we take the generating relation to be "included in" or "inclusive of." It makes no difference which relation out of each pair we take as the generating relation of the series. But it is a question of great theoretical importance, and, as we shall see, of great practical importance, which of the relations in the B series corresponds to which of the relations in the C series.

696. How, if at all, shall we be able to determine this? The simplest way, if it were practicable, would be to discover some similarity between "earlier than" on the one hand, and either "included in" or "inclusive of" on the other hand, which should lead us directly to the conclusion that "earlier than" corresponded, in the B series, to the similar relation in the C series—which would, of course, imply that "later than" corresponded to the remaining relation in the C series. I cannot, however, detect any such similarity.

697. But there is another way which is possible. The two converse generating relations are inseparable. But they need not be equally important. In one sense of the word, indeed, they must be equally important, since they are both indispensable. The one cannot be there without the other. But it is possible, as we shall see in this chapter about the B series, and in the next chapter about the C series, that one may be more important than the other in another sense—that the process from P to Q, R, and S successively, may express the nature of the series more adequately than the process from S to R, Q, and P. And, if in each series one sense is more important than the other, we may find it possible to argue that the more important sense in the B series corresponds to the more important sense in the C series.

698. Is, then, one sense of the *B* series more important than the other? There is one fact which is clearly of great significance. The series, like other series, can be taken in either direction. We can go from earlier to later or from later to earlier. But this series is a series of changes. And these changes go in one direction and not in the other.

When we say that the *B* series is a series of changes, we do not, of course, mean that the terms change their places in the series. If one term is ever earlier than another, it is always earlier than that other. But the *B* series is a time-series, and time involves change. And the change in the terms of the *B* series is that they are successively present (passing from futurity to presentness, and from presentness to pastness). It is first an earlier term which is present, and then a later one[1].

Now this fact—that there is a change in the *B* series which goes from earlier to later—is sufficient to show that, of the two generating relations of the series, the relation of the earlier term to the later is more important than the relation of the later term to the earlier. For the relation of the earlier to the later takes us in the direction from earlier to later. Starting from any point *P* in the series, we shall find it to be earlier than *Q*, and shall then find *Q* to be earlier than *R* and so go on continually from earlier terms to later terms. On the other hand, if we start from *P* and take the relation "later than," we shall find that *P* is later than *O*, which is later than *N*, which is later than *M*, and so go on from later to earlier terms. And thus the relation of "earlier than" gives us a sense of the series which agrees with the direction of the change, while the relation "later than" gives us a sense opposed to the direction of the change. The first of these relations, then, expresses the nature of the series more adequately than the second, and we may call the sense from earlier to later the Fundamental Sense of the series[2].

---

[1] When we say that there is change in the time-series, of course we mean that there is apparent change in an apparent time-series. Time is unreal, and it does not, therefore, give us real change. Nor is there real change anywhere. For, while there can be no time without change, there can be no change without time.

[2] It is possible, as was said in Section 218, that this fact may be one of the causes producing the common but erroneous belief that the earlier determines the later in a way in which the later does not determine the earlier.

699. We have further to enquire whether the C series also has a fundamental sense, and, if so, which of its two senses is fundamental. And, if one sense is fundamental, we shall have further to enquire whether we can infer that it is this which corresponds to, and appears as, the fundamental sense of the B series. But before doing this it will be well to consider certain peculiarities of the time-process on which the practical importance of these questions depends.

700. The earlier and later stages in the B series are found in our experience, as far as that goes, to differ qualitatively. And they may be expected to do so throughout the whole series. For the earlier stages differ from the later in being the appearance of stages in the C series which are more or less inclusive. Whether it is more or less has not yet been determined, but it must be one or the other, according as the one or the other relation in the B series corresponds to each relation in the C series. In the C series a more inclusive term will differ from a less inclusive term, not only by its position relatively to it, but by containing more content—that is, it will differ qualitatively. A similar qualitative variation may be expected to accompany different positions in the B series.

701. Should these different positions vary, among other things, in the amount of good and evil which is to be found in them, then their position in the B series will be a matter of great practical importance. I do not mean that a good or evil state becomes better or worse in any particular position in the B series than it would be in any other position in the B series. But it is an undisputed fact that anticipation of future good or evil affects our happiness or unhappiness in the present far more than the memory of past good or evil.

Let us take, for simplicity, a case which is more clearly cut than those of actual life. Let us suppose that, on a certain day, G and H can each remember perfectly all that has happened to them in the ten years preceding that day, and can anticipate with perfect certainty all that will happen to them in the ten years succeeding that day, but have no more distant memories or anticipations. Let us suppose that G looks back on ten years of intense misery, and anticipates ten years of intense happiness, while *H* anticipates misery as great as *G* remembers, and remembers happiness as great as *G* anticipates. What will be their position during the neutral day which intervenes between the two periods of intense experience? Each will be contemplating equal amounts of good and evil in his own life. But it is obvious that the intermediate day will be a period of happiness for *G* whose good is in the future, and whose evil is in the past, while it will be a period of unhappiness for *H*.

Thus future good and evil are more important to us in the present, supposing we know about them, than past good and evil of equal amount. No doubt even future good and evil are less important than present good and evil, when the amounts are equal. But this can be accounted for by the fact that experience of anything is more vivid than anticipation, and more certain than any anticipation which is possible to us in our present state. But anticipation has not the same advantages over memory which experience has over anticipation. Anticipation is not more certain than memory, but the reverse. I am almost always more certain of what has happened to me at a certain time in the past, than I am of what will happen to me at an equal distance in the future. And anticipation tends to be less vivid than memory. It is generally much easier to form a vivid picture of what I have experienced than of what I shall experience. Yet, in spite of this, anticipation affects present happiness much more than memory does.

702. I am not asserting that past good or evil does not tend to produce *some* happiness or unhappiness in the present. It may do so, to begin with, for incidental reasons. We may be saddened by the results which past evil has left behind in the present, or which may be expected to appear in the future—if those results are themselves evil, which of course is not always the case with results of past evils. Or the remembrance of evil may remind us that the universe is not wholly good, and may make us fear evil in the future. Again, if past evil has been caused by the wickedness of any person, the fact that the evil has passed away will not affect the fact that the responsible person is still wicked—unless, indeed, he has improved. And, on the other hand, remembered good often derives happiness from reasons corresponding to these.

And, apart from these incidental consequences, I do not think it can be denied that past good and evil, when known in the present, do intrinsically tend to produce some happiness or unhappiness in the present. The past is, after all, part of the universe, and every piece of goodness which we find in the universe tends to make us happy by knowing of it, and every piece of evil unhappy. And I do not think that even those persons who regard time as real, regard the past as so absolutely non-existent, that its character is of no present importance to us. We should, I think, always desire of any past evil, just because it was evil, that it should not have existed—unless of course it has produced something whose good outweighs the evil of its cause.

703. But, while thus allowing that past good or evil does tend to produce present happiness or unhappiness, it remains the case that future good or evil tends to produce them to a much greater extent. I do not know that any reason can be assigned for this greater present importance of the future. It may have a reason which is not yet discovered. Or it may be an ultimate fact. But it cannot be denied to be a fact.

Whether it has a reason or not, there is no ground for regarding it as unreasonable—that is, as contrary to reason. It would no doubt be contrary to reason if we judged a certain good to be greater, because it was in the future, than we should have judged it, had it been in the past. For the goodness of a state is not affected by any difference in my temporal relations to it. But this is not what happens. What happens is that the anticipation of a good in the future produces greater happiness in the present than the memory of an equal good in the past. There may be no more reason for this than there is for a man's preference of burgundy to claret, or of claret to burgundy. But absence of reason is not here contrariety to reason. It would, no doubt, be possible for a man to hold that his nature would be more admirable if good and evil affected him to the same extent when they were past as when they were future. But I do not know any reason why anyone should hold this. And, even if he did hold it, the fact would remain that for him, as for the rest of us, future good and evil *are* more important than past good and evil.

704. And this greater importance of the future is not destroyed by the belief that time is non-existent. It will still be the case, for the person who holds such a belief, that the states which appear to him as future, though he does not believe them to be really future, will have a greater effect on the present than those which appear to him as past. It is, indeed, probable, that the difference of importance between the past and the future will be *diminished* by the belief that time is unreal. The fact that past good and evil do belong to the universe, and are therefore real, as present and future good and evil are, will be more prominently before a person who believes that time distinctions are only apparent, and that all things are really timelessly coexistent. And the result of this, no doubt, will be to increase the effect on our present happiness of the good and evil of the past, since their pastness will be regarded as only apparent, and not real[1].

But, when every allowance has been made for this, it will still remain the case that good or evil in what is recognized as being only an apparent future will have more influence on happiness than good or evil in the apparent past. It is not only that a person who believes time to be unreal recognizes that there is such a greater influence in the lives of people who do not believe time to be unreal, and that such an increase of their happiness or unhappiness is itself a real good or evil. In the life also of the thinker who believes time to be unreal, the greater importance of the apparent future will show itself.

705. This greater importance is always of a later event as against an earlier, since, if one event is future while another is past, the future event is the later of the two. But the greater importance belongs to it because it is future and the other is past, not because it is the later of the two. This can be seen by considering the case of good and evil states, one of which is later than the other, but both of which are in the past, or both in the future. In these states we are not more affected in the

---

[1] The truth of this appears, I think, more clearly in regard to the increase of the positive effect on our happiness from the memory of past good, than it does by the increase of the negative effect on our happiness from the memory of past evil. This seems attributable to the fact that the belief in the unreality of time is often, though not always, accompanied by a belief in a close and intimate unity of the parts of the universe, such as may be called mystical. And all mystical belief, it may be maintained, tends to hold that the good is better than, and the bad not so bad as, they respectively appear to be from a non-mystical standpoint.

present by that state which is the later of the two than by that state which is the earlier of the two. Of course if between two states, one good and one evil, there was a third state in which the one was remembered and the other anticipated, that state— now itself, like the others, in the past or the future—would be more affected by the state which was future to it than by the state which was past to it. And for this reason we may say that the whole past or the whole future would be better if the good in it followed the evil than it would be if the evil followed the good. But in doing this we are not taking the later state as the more important to us now, but only recognizing that it is more important to a state to which it was future while the other was past.

706. Thus there seems no reason to consider that the later, as such, is more important than the earlier. But the later an event is, the more points there are at which it is future, and the fewer points are there at which it is past. If, therefore, an event is anticipated to end after a certain point, then, the later it is, the longer will it have the greater importance of the future before it passes on to the lesser importance of the past. And thus the later it is, the more important will it be. If a certain quality, of a good or evil nature, is true of all states later than a certain state, then some states which possess this quality will be always future or present, and they will never all be in the past. Thus the nature of this quality, if its occurrence can be anticipated, will be especially important.

707. And thus, though the later as such is not the more important, yet the later, if it can be anticipated, is the more important, by reason of its effect on the present in which it is anticipated. We have, therefore, justified the assertion made earlier (p. 348) that, if the different positions in the B series should vary, among other things, in the amount of good and evil which is to be found in them, then their position in the B series will be of great practical importance. And we now see that it is desirable that the states which are better should come later in the series. We shall see further on (Chapters LXV and LXVII) that there is reason to think that the term in the C series which is inclusive of all the others and included by none of them—the final term of the series at the inclusive end—has a value which greatly, indeed infinitely, exceeds the value of any other term. And we shall see that there is reason to believe that this value is good. It would therefore be very desirable that it should be the relation "inclusive of" which appears as "later than" and the relation "included in" which appears as "earlier than."

I believe, for reasons which will be given in the next chapter, that we have grounds for holding that the desirable result is the true result, and that it is the relation "inclusive of" which appears as the relation "later than." It will, however, be necessary to scrutinize the argument with special care, just because it does lead to a result which we should accept as desirable. For the fact that it is desirable does not make it more likely to be true. And, on the other hand, the fact that we desire it does give rise to a real danger that we should be unconsciously misled by our desires, and so accept too easily arguments which lead to the desired result, and reject too easily objections to that result.

708. But, it might be said, although the fact that a proposition is desirable does not, as such, make it more likely to be true, is it not possible in this particular case to argue that the desirability of the proposition is a ground for inferring conclusions which would involve that the proposition is true? The proposition which is desirable is that the more inclusive terms should be those which appear as later. And the reason why this is more desirable is that the good or evil of the future is more important for present happiness or unhappiness than the good or evil of the past. Now, it might be argued, the fact that the future is more important than the past, for which we have not yet found an explanation, could be explained on the hypothesis that what appeared as later was in reality more inclusive. For then a state which was future while another was past, being later than that other, would have the other as a part included in it. And since the future state would contain the past state and more also, our greater interest in the future would thus be accounted for. We are justified, therefore, it might be concluded, in adopting a hypothesis which gives the only explanation of an undoubted fact.

709. But this argument would be invalid. In the first place, although the fact is undoubted, it is possible that it is ultimate, and that the greater importance to the present of the future is something which neither admits nor requires an explanation. And, in the second place, the proposed explanation would prove too much. For it would prove that of two states, both of which were future, or both past, the later would be the more important to the present state. And we have seen that this is not the case.

We must fall back, then, on the method indicated on p. 346. We have ascertained that the *B* series has a fundamental sense, and that it is the sense from earlier to later. If we find that the *C* series has also a fundamental sense, we may find ground for holding that it is this sense which corresponds to the sense from earlier to later.

CHAPTER LX

THE FUNDAMENTAL SENSE OF THE C SERIES

710. Is there a fundamental sense of the C series, and, if so, which sense is it? Can the first question be answered by means of the result which we have already reached—that the B series, which is the appearance of the C series, has a fundamental sense? This fact does give a certain presumption that the C series also has a fundamental sense. For, if it has not, then, when the C series is misperceived as the B series, the fundamental nature of one sense is part of the misperception. Now it may fairly be said that the presumption is that no element in a perception is erroneous, except those which can be shown to be so. And in that case the presumption is that the sense, whichever it may be, which corresponds in the C series to the fundamental sense in the B series, is itself fundamental.

711. To this it might be objected that what makes one sense in the B series fundamental is the fact that this is the direction in which change goes, and that, since the perception of change is certainly a misperception, we have no reason to believe that the possession of a fundamental sense by the series is in any better position than the change. I do not, however, think that this objection is valid. For there are certain characteristics which we know that the C series has, and which are only manifested in the B series in the form of change. If, in the B series, M is succeeded by O, and O by P, we know that this is a manifestation of the fact that, in the C series, O comes between M and P. And, since a characteristic which certainly belongs to the C series is manifested in the form of change, it is clear that the fact that something else is manifested in that form does not prove that it is not a characteristic of the C series.

There is, therefore, a presumption that the sense in the C series which corresponds to the sense from earlier to later is the fundamental sense of C. But this would give us no help in determining which of the two senses, "included in" and "inclusive of," was the fundamental sense in question. We must, therefore, enquire more directly into the nature of the C series.

712. It is clear, to begin with, that neither sense of the C series can be taken as fundamental for a reason similar to that which led us to hold that the sense from earlier to later was the more fundamental sense of the B series. Our reason there was that the B series is a series of change, and that the direction of change in the B series is from earlier to later. But it is only in a time-series that there can be change, and the C series is not a time-series.

713. But it cannot, I think, be denied that there are some series, which are not time-series, in which one sense is more fundamental than the other. Take, for example, a series of inferences. If, from the self-evident premises A and B, we infer C, and then, from C and the self-evident premise D, infer E, and so on to Z, the sense of the series from A and B to Z expresses the nature of the series more adequately than the series from Z to A. It is not that the series cannot be taken in either direction. When we have once got the terms we can go through them in whatever order we like. Nor is it that the terms must be discovered in this order. For it is possible to start with a conclusion, as to which we desire to know if it could be proved, and to proceed by enquiring from what premises it could be inferred. The reason why the sense from A and B to Z is more fundamental is that the series is essentially one of inference—the generating relation in the case of each third term is that it can be inferred from the two previous terms. And when terms are connected as premises and conclusion, the sense from premises to conclusion is more fundamental than the sense from conclusion to premises. The series has no temporal change in it, and, as has just been said, when it is contemplated in a series of time-states, the conclusion may be contemplated before the premises. But, although it is not a temporal series, it agrees with a temporal series in having the characteristic of an intrinsic direction of its own—the direction from premises to conclusion. And therefore the sense of the series from premises to conclusion will be the more fundamental, because of its correspondence with the intrinsic direction.

If we take the Hegelian dialectic, we get a specially striking example of such an intrinsic direction from premises to conclusion, because, if Hegel is right, the chain of propositions is independent of any premise except the one proposition which asserts the validity of the category of Pure Being. Once that is admitted, the validity of every other category follows from it by inference.

714. But it is not merely the case that the dialectic, on its author's theory, is a specially striking example of the fundamental sense from premises to conclusion. There is an additional reason in the case of the dialectic why that sense should be fundamental. And this is that, if Hegel is right, every category in the dialectic, except the Absolute Idea, has an element of instability in it, and of instability in the direction of the Absolute Idea—that is to say, in the direction from lower to higher categories, and in the direction from premises to conclusion.

By saying that a term is unstable in a certain direction, I mean that the fact that that direction is the intrinsic direction of the series can be seen from the nature of that term alone, without taking into account its place in the series relatively to other terms. There is no such instability in the non-dialectical chain of argument of which we spoke first. For it is only when one proposition is taken in connection with others—as forming part of a chain of argument, and as filling a particular place in it—that we can know that the intrinsic direction is from D to E, and not from E to D. But, if Hegel is right, there is an inherent contradiction in each category of the dialectic, except the Absolute Idea, which compels us on pain of contradiction to pass from that category in a certain direction—towards the next highest category. And thus the intrinsic direction of the series can be seen from the nature of any single term in it, except the final term, the Absolute Idea[1].

It is clear that when there is instability in a series, there is an additional reason to say that the series has an intrinsic direction, and the sense of that direction is the more fundamental.

715. Our examples so far have been taken from series of inferences. The *C* series, however, is not a series of inferences. The terms in it are not propositions at all, and cannot therefore be related by the relation of inference. If, indeed, a proposition which asserted the nature of one term in the *C* series could be inferred

---

[1] I am not asserting that Hegel is right in the claims he makes for the dialectic. The example is as good for our purpose whether the claims are or are not justified.

from a proposition which asserted the nature of another term, we might find that the C series, although not itself a series of inferences, had a one-to-one relation with such a series. Now we have no reason to believe that the nature of any one term in the C series could be inferred merely from the nature of any other term or terms in the C series. For we cannot infer the nature of any term in the B series merely from the nature of some term or terms which are earlier or later than it, and we have no reason to suppose that, in this respect, the real series differs from the apparent series. But, on the other hand, the analogy of the B series suggests that it is possible to infer the nature of a term in the C series from the nature of another term standing in a definite relation to it in the series, with the help of general propositions, gained by induction, as to the laws by which certain sequences take place among the terms of the series.

716. Thus one term in the C series can be related to another by the fact that a proposition asserting the nature of the one can be one of the premises for an inference to the nature of the other. But this will not enable us to take either sense of the C series as more fundamental than the other. For, if we can make such inferences, it is because the nature of the one term implies that another term, in a certain positive relation to it, will have a certain nature. And this is a causal relation between the two terms. Now we saw in Section 210 that in the B series such implications do not run exclusively in one direction. If, of two events in the B series, the earlier is described as "drinking of alcohol by a man," and the later as "the same man becoming drunk," the later will imply the earlier, and the earlier will not imply the later. But if the earlier event is described as "drinking of the amount X of alcohol by a man when his body is in the state Y," and the later, as before, as "the same man becoming drunk," the earlier will imply the later, and the later will not imply the earlier.

In causal relations, then, the implication can go in either direction in the B series—from earlier to later, or from later to earlier. And, as we have no reason to hold that the real series differs in this respect from the apparent series, we have no reason to hold that causal implication goes in one direction in the C series more than in the other. We cannot therefore hope to discover, by means of such implication, whether one sense in the *C* series is more fundamental than the other.

717. Another set of considerations, however, will, I believe, enable us to determine one sense of the *C* series as fundamental. Each *C* series, as we saw in Chapter XLVIII, falls within some determining correspondence part of the universe. And each *C* series is an inclusion series, each member of the series, except the last, containing more or less of the content, in one particular dimension, of the determining correspondence part within which the *C* series falls. The last term of the inclusion series is the determining correspondence part itself, which we spoke of as the whole of the series.

The whole in question is the final term of the inclusion series in one particular direction—the direction from less inclusive to more inclusive. And we saw in Chap. XLIX, p. 251, that it is also the final term of the *C* series, in one direction. For, although it is never perceived as present, it is either perceived from the standpoint of every other term as future, or else perceived from the standpoint of every other term as past. It is therefore part of what is misperceived as the *B* series. And that which is misperceived as the *B* series is the *C* series.

718. Now what is it which assigns to each term in the *C* series, other than the whole, its place in the series? The characteristic of each term, to which it owes its place in the *C* series, is the amount of the content of the whole which is included in it—if the amount is greater it will be nearer to the whole, if it is less it will be nearer to the other end of the series. Thus the characteristic which determines the place in the series of each term other than the whole is a characteristic which has reference to the nature of the whole, and must be stated in terms of that nature[1].

But, on the other hand, the characteristic to which the whole owes its place in the series need not be stated with any reference

[1] At the other end the series is bounded by nonentity, which is the boundary of the series without being a term in it. The characteristic cannot be stated by reference to this boundary. For it is the boundary of *all* *C* series, in whatever wholes they are. If, therefore, we should endeavour to assign a place to a particular term in a particular *C* series—the *C* series, for example, within *G ! H*—by saying by how much its content exceeded the content of nonentity, we should not in this way determine in which series the term in question was to be found. For this we should have to refer to the whole—*G ! H*—in which the series falls.

to the nature of any other term in the series. The place of the whole in the series is that it is one of the extreme terms of the series, and that it is the extreme term in the direction from included to inclusive. The characteristic to which it owes this place is that it is a term in the determining correspondence system—that it is, for example, $G! H$. This characteristic can be stated without any reference to the other terms of the C series. The fact that a certain state is $G! H$—the perception of H by G in the determining correspondence system—can be understood without any knowledge of the fact that it contains a C series of other perceptions. And thus, while the characteristic which determines the place of the other terms refers to the whole, the characteristic which determines the place of the whole does not refer to the other terms.

719. Of course any *description* of the place of the whole in the series *does* refer to the other terms. It is impossible to describe the place of any term in any series without reference to the other terms. When we say that the whole is an extreme term of the series, and the extreme term in a certain direction, we are making a reference to the other terms. But the question here is not about the place itself which the term occupies in the series, but about that characteristic of the term which determines it to hold that place. And that characteristic in the case of the other terms refers to the whole, while in the case of the whole it does not refer to the other terms.

720. I think that it follows from this that a relation which the other terms bear to the whole, and which the whole does not bear to the other terms, will express the nature of the series more adequately than a relation which the whole bears to the other terms, and which the other terms do not bear to the whole. For such a relation relates the other terms to the whole, while it leaves the whole unrelated, so far as that relation is concerned, to the other terms. Of course this involves that the whole *is* related to the parts, but this is by the converse relation. And thus it emphasizes the fact that the dependence of the other terms on the whole is greater than the dependence of the whole on the other terms—a fact which we have noticed in considering the respective characteristics which assign them their places in the series.

721. There is another respect, also, in which the dependence of the other parts on the whole is greater than that of the whole on the other parts. *G! H* itself, as we have seen, is a correct perception of *H*. All the other terms in its series are misperceptions of *H*. Now that anything is a misperception of *H*, or a particular misperception of *H*, can only be stated by a comparison of it with a correct perception of *H*. But that anything is a correct perception of *H* can be expressed without any reference to any particular misperception of *H*. And, on this ground also, the nature of the series will be more adequately expressed by a relation which the other terms bear to the whole, and which the whole does not bear to the other terms.

Now the relation "included in" fulfils these conditions, since all the other terms in the series stand in this relation to the whole, while the whole does not stand in this relation to any of the other terms. And the relation "inclusive of" does not fulfil this condition, for the whole stands in this relation to every other term, while none of the other terms stand in this relation to the whole. And therefore the sense of the series from included to inclusive, which is generated by the relation "included in," is the fundamental sense of the series. When the series is taken in this sense, it is limited at the beginning by nonentity and has the whole as its final term.

722. Can we go further, and say, as we did in the case of Hegel's dialectic, that the terms, other than the final term, are unstable in the direction of the final term? I think we can. We said (p. 357) that a term was to be called unstable in a certain direction, if the fact that that direction is the intrinsic direction of the series can be seen from the nature of that term alone, without taking into account its place in the series relatively to other terms. Now in the case of the *C* series this is what does happen. In a non-dialectical argument, where there is no instability, we cannot tell how the fundamental sense runs by inspecting a single term in the argument. We have to know from what it is proved, and what is proved by it, before we can answer the question. But in a dialectical argument it can be answered by the inspection of a single term. And so it can with the *C* series. For to know what the term is at all, we shall have to know it as that member of the series within $G!H$ which has a given amount of the content of $G!H$. And this, as we have seen, is sufficient to determine the fundamental sense of the series.

723. There is, however, an important difference between the instability which we have here, and that which would be found in the Hegelian dialectic. In the case of the dialectic, each term is unstable towards all the terms beyond it in one direction— primarily to the next term, through this to the next, and so on till the Absolute Idea is reached. But in the C series the instability of each term is only towards the whole, and not to any of the terms intermediate between it and the whole. The characteristic which determines the place of a term in the series cannot be stated by reference to a more inclusive term, any more than it can be stated by reference to a less inclusive term, unless that more inclusive term is the whole—the last and most inclusive term. But since the instability does lead from each term towards what is more inclusive than itself, it will remain the fact that the instability determines the sense which goes towards the more inclusive as the fundamental sense.

724. The C series, then, has a fundamental sense—the sense which goes in the direction from less inclusive to more inclusive. The B series, as we have seen, has also a fundamental sense— the sense which goes in the direction from earlier to later. Now this, I maintain, gives us, not an absolute demonstration, but good reason for believing that the relation "included in" in the C series appears as the "earlier than" in the B series, and that any state of $G!H$ which appears as later in time than a second state of it, does, in the timeless reality, include that second state.

We saw on p. 355, that the fact that the B series had a fundamental sense from earlier to later gave us some reason for thinking that the C series had a fundamental sense also, and for thinking that its fundamental sense was the sense in the C series which corresponded to the B series sense from earlier to later. But now that we have discovered independently that the C series has a fundamental sense—the sense from included to inclusive—we have a much stronger reason for supposing that it is the sense which corresponds to the sense from earlier to later.

What we said before was that no element in a perception must be taken as erroneous, unless it has been proved to be so. Now in the C series, which is perceived as the B series, we had seen that its perception as a time-series, connected by the relations of earlier and later, is a misperception. But we had no reason to think that that element in the perception was erroneous which showed the fundamental sense as running from the terms which appear as earlier to the terms which appear as later. And therefore we ought to hold that the fundamental sense of the series does run in this direction.

But now the case is much stronger. For if the fundamental sense, from included to inclusive, which we now know the C series has, is not the sense which appears as the sense from earlier to later, it must be the sense which appears as the sense from later to earlier. And then there would be an additional error in our perception. Not only, as before, should we perceive a sense as fundamental which is not so in reality, but we should also perceive a sense which is fundamental, without perceiving it as fundamental. Nor is this all. We should be perceiving it as *not* being fundamental, for, if one sense of a series is fundamental, it implies that the other is not, and so, in misperceiving the one as fundamental, we should be misperceiving the other as not-fundamental. The very characteristic which really made the fundamental sense run, in some particular case, from P to Q, would be misperceived as making the fundamental sense run from Q to P.

We have no reason to believe that our perception does misperceive reality in such a way as to produce any of these errors, and therefore we must conclude that it does not, and that we are right in perceiving the fundamental sense to run in that direction which we misperceive as being from earlier to later. And, as we know that the fundamental sense runs in the direction from less inclusive to more inclusive, we must conclude that the relation "included in" appears as the relation "earlier than," and that the relation "inclusive of" appears as the relation "later than."

725. It is necessary, as was said in Chap. LIX, p. 353, to scrutinize this conclusion with particular care, because it will lead to desirable practical results. It will lead to the conclusion that, *sub specie temporis*, the future will be for each of us infinitely more good than the past. If, on the other hand, it had been the relation "inclusive of" which appeared as "earlier than," the conclusion would have been that the past had been infinitely more good than the future will be. And, as we saw in the last chapter, the first of these conclusions is much more attractive to us than the second. We must be on our guard, therefore, for fear the attractiveness of the result should induce us to accept it on inadequate grounds. But, after making allowance, as well as I can, for this, it still seems clear to me that we have adequate grounds for holding our conclusion.

726. We have just spoken of our greater interest in the good and evil of the future than in the good and evil of the past. In the last chapter it was said that no reason could be assigned for this. Can we assign one now? If, to a state P, the state Q is future, while the state O is past, then the reality behind the appearance is that P contains the content of O, and something more, while Q contains the content of P and something more. Can our greater interest, from the standpoint of P, in Q than in O, be traced to the fact that Q exceeds P in content, while O falls short of it? I do not think that this has any probability, for it would seem that, on the same principle, I ought, from the standpoint of P, to be more interested in R than in Q, since R, being later than Q, contains the content of Q, and something more. And we have seen in the last chapter that, of two events in the future, the later is not the more interesting to us by reason of its being later.

CHAPTER LXI

THE FUTURITY OF THE WHOLE

727. We have seen in the last chapter that that term in the *C* series which, *sub specie temporis*, appears as the last term, is the term which is the whole of the series—the term which includes all the others, and is included in none of the others. We saw (Chap. XLVII, p. 227) that this last term is a correct perception, and that therefore it does not perceive itself as in time, though it is perceived by all the others as in time. It will therefore never be perceived as present. Nor will it ever be perceived as past, since it is, *sub specie temporis*, the last term. Whenever it is perceived as in time at all, it will always be perceived as future.

728. This result, however, may mislead us if we fail to realize that, while the perception of the whole from its own standpoint is never a perception of it as present, yet it is a perception which has an important similarity with the perception of anything as present. For, since all perception at this stage is correct perception, it follows that at this stage all perception will be perceived as being perception. In other words, all perception at this stage is apparent perception. Now, as we have seen, when perceptions perceive anything *sub specie temporis* it is only perceptions of what appears as present which are apparent perceptions. We never have apparent perceptions of what is perceived as past or future.

Perception of anything as present, then, is the only sort which shares with the perception in the final stage the characteristic of being apparent perception. Our cognition of the past and future, though really, of course, perception, if it appears as cognition at all, appears in the form of judgments, and therefore appears to us as mediated by something else. It is only our perception of anything as present, and the perception in the final stage, which appear as having the directness of perception.

This resemblance to perception as present applies, of course, to the perception from the standpoint of the whole not only of the whole itself, but of any other stage. For all perception at that stage will be apparent perception. But this will include the special case of which we are now speaking—the perception of the whole from its own standpoint.

729. In the second place, when a perception is apparent as a perception, it has, to use Dr Stout's phrase, an "aggressiveness" which does not belong to a perception which appears as a judgment. And this is of considerable practical interest, since it probably accounts for—what certainly exists—the greater importance for our happiness of what we know, by an apparent perception, as present, than of what we know, by an apparent judgment, to be present, past, or future[1].

This aggressiveness, with its practical consequences, will be shared by all perception from the standpoint of the whole—and consequently by the perception of the whole from its own standpoint—since it will be apparent perception. Thus, both in directness and aggressiveness the cognition of the whole from its own standpoint will resemble our cognition of the present rather than our cognition of the past and the future.

730. In consequence of this similarity, presentness will always be an appropriate *metaphor* for the perception of the whole from its own standpoint. If we say that, at the last stage of the *C* series, the whole will appear as present, it will call up a picture far less inaccurate than would have been the case if we had said that at that stage it would appear as past or as future.

But it must be remembered that it is never more than a metaphor. The whole does not appear as present at the last stage of the *C* series, though its appearance does in some respects resemble the appearance of presentness. On the other hand, as we have seen, at all other stages of the *C* series the whole *does* appear as future. And, also, though not really future, it is future in the only way in which anything can be so—as appearing as future. It is just as truly future as all the things which we generally call future—as really future as to-morrow's breakfast. But the whole is not, at any stage in the series, as truly present as those things which we generally call present—never as truly present as my writing these words is now present. *Sub specie*

---

[1] This is, of course, only true *cæteris paribus*.

temporis, the whole *is* now future. It never does more than resemble the present in certain respects.

731. If it had been the case that the relation "included in" had appeared as the relation "later than," and not, as it does, as the relation "earlier than," the positions of the whole and of non-entity would, of course, be reversed. The whole would appear in the *B* series as the earliest term. From the point of view of every other term in the series, it would appear as in the past. Nonentity, on the other hand, would be the limit of the *B* series in the direction of futurity, though, as we saw in Chapter XLIX, it does not itself appear as in time.

732. The whole, then, is as really future as anything can be, and is never present or past. Can we say, then, that the eternal is as future as anything can be, and is never in any sense present and past? We cannot say this. The eternal is what exists and is timeless. And we have seen that all that exists is really timeless. And so other stages of the *C* series are as eternal as the whole, which is the final stage. We cannot say, therefore, that the eternal is never present and past in the sense in which it is future. For all the stages of the *C* series, except the final stage, are present and past in the same sense as that in which they are future— that is, they appear as being so.

But, on the other hand, we must say that the eternal is as really future as anything can be—which is obviously the case, since nothing is really future, and all that appears as future must be really eternal. And further we must say that something which is eternal—namely, the term of each *C* series which is the whole of that series—appears as future from every standpoint but its own, and never appears as past or present.

And, in the second place, the eternal stage which appears temporally only as future, is the only stage which perceives itself as eternal. All the other stages, when they perceive themselves, perceive themselves as in time [1].

733. Again, there are two facts which place this particular eternal stage in a special and unique position. One of these is

---

[1] No doubt it is the case that there can be in any such stage an apparent judgment that it is timeless, and all apparent judgments are really perceptions. But this is compatible with all perception of those stages by themselves being perception of them as temporal. Cf. Chap. LIV, p. 306.

the fact that this stage is, in one sense, all the eternal. It is not every eternal substance, for, as we have seen, all the terms in the C series are eternal substances, and all those terms but one appear also as present and past. But the eternal which appears only as future contains all the content of the universe, since whatever content falls in any of the other terms of the C series, falls also in this term. (It is, of course, for this reason that we have called this term the whole of the series.) And thus there is a sense in which this term contains all the eternal.

734. The second fact is that the term which appears temporally only as future is the term which appears as the latest term of the series. And therefore, *sub specie temporis*, it appears as a term which, though temporal, is not transitory. It appears to begin, but not to end. In the next chapter, and also in Chapter LXVII, we shall see that this fact is of vital importance. At present we only note that it involves that, in spite of appearing as in time, it appears with more likeness to its true eternity than do the other terms of the series. It is not more really eternal, but more obviously so.

735. We have seen, then, that the eternal as such can appear *sub specie temporis* either as present, past, or future, and that no one of these appearances is more adequate to its nature than the other. And we have seen that the eternal substances which are parts of the determining correspondence system—which between them contain all the content of the universe—can, *sub specie temporis*, appear only as future, and not as past or present.

736. It is important to emphasize these results, because they have been denied. It has been asserted that all that is eternal is, at every moment of the time-series, manifested as present. And the tendency is to assert this more emphatically about the universe taken as a whole than about any eternal parts which it may be held to possess.

This might be the case if there were no real eternal series corresponding to the apparent temporal series—if, that is, the serial character as well as the temporal character of the temporal series were due to misperception. But we have found reason to think that this is not the case, and that what appears as the temporal series is really a series, though it only appears to be temporal. And consequently, as we have seen, it follows that some eternal realities appear as being earlier in time, and some as being later. It also follows that every eternal reality, which is a term in the *C* series, with one certain and one possible exception, will appear from some positions in that series to be only past, from some to be only present, and from some to be only future.

The possible exception is that, if the series has a first term, that term will appear from its own position as present, and from other positions as past, but from no position as future. But if there are always terms between any term in the series and its limit of nonentity, there will be no first term, and this exception is therefore only possible.

The certain exception is that, as we have seen, the final term of the series, which is also the whole of the series, appears, *sub specie temporis*, only as future, and never as past or present. And thus it is just the term which, on the view of which we have been speaking, is most emphatically asserted to exist only as present, which in reality never does appear as present. For it is the universe as a whole which is most emphatically asserted to exist as present. And the system of determining correspondence parts is the universe, since it contains all existent content. But the determining correspondence parts are the final terms of the *C* series, the terms which are also the wholes of those series. And it is the final terms of the series which never appear as present.

737. Thus the view which connects the eternal as such, or the universe as a whole, with presentness rather than with pastness or futurity is wrong, and, in the case of the universe as a whole, doubly wrong. And it is an error which is sometimes not without grave ethical consequences. For its supporters are bound to hold that all the good which is in the universe as a whole can be manifested under the conditions of our present life. This obviously restricts to a very great extent the amount and the quality of the good which the universe as a whole can be held to possess. And the restriction is greater than this. For it seems clear that the supporters of such a view are bound to hold, not only that the good must be such as could be manifested under the conditions of our present life, but that it actually is manifested in our present life.

Now our present life falls far short in many respects of the ideals which we form, and hold to represent what is good. If the conclusion drawn is that the universe as a whole has less good in it than our ideals demand, and that consequently the universe is gravely deficient in goodness, the conclusion will be pessimistic, but will have no ethical effect.

But some thinkers who hold that the universe as a whole is adequately manifested in the present, also hold that the universe is very good. Indeed it is sometimes held that the mere fact that it is the universe implies that it is very good. And such thinkers are inevitably led to hold that any ideals which cannot be realized in the present are mistaken and false. This seems to me to destroy ethics altogether. Our present lives contain sin, pain, hatred. And unless we are able to say that these, in spite of their occurrence, are as truly bad, as the virtue, happiness, and love which occur are truly good, the predicate of goodness ceases to have any interest, or, indeed, any meaning.

738. The futurity of the whole invalidates certain criticisms which have been made on such a conception of heaven as is commonly held by Christians. The Christian heaven is usually regarded as future, and not as past or present. We have not been in it before the birth of our present bodies—indeed, most Christians deny that we existed at all before the birth of those bodies. And we are not in it now. We are separated from it by death—not indeed that death alone would place us in it, but that we shall not reach it till we have passed through death. Heaven may be held to be a state of mind, not a place, nor an environment. But it is a state of mind which is still for us in the future. And even if it should be held that it could be reached in this life, still it has to be attained. And thus it was once in the future, and is still for many people in the future.

The Christian heaven, then, is taken as being future. It is sometimes taken as being really in time—as enduring through an endless duration. But it is often taken as timeless. And the criticism has been made that, if heaven is timeless, it cannot be rightly taken as being in the future. It must be conceived, it is said, as standing in the same relation to all stages in the time-process, and, therefore, if we are to use temporal predicates about it at all, it will be best expressed by the metaphor of an eternal present. And this view, I think, is pressed more strongly because a timeless heaven would appear to those in it to be, as it really is, a timeless state, and because, even to us at present, it does not appear as a transitory state.

739. Of course, if heaven is timeless it cannot be really future. But, as we have seen, it may, if certain conditions are fulfilled, be as much future, and as little past or present, as breakfast to-morrow is. And this, I think, would give those who assert a timeless heaven to be future all that they really mean to assert, and something which the critics of whom we have just spoken mean to deny.

The conditions in question were that nothing should be really in time, that whatever appeared to be later than another thing should really have to it a certain non-temporal relation, and that this relation should hold between the events of to-day and heaven, as it holds between the events of to-day and to-morrow's breakfast.

I do not say that these conditions have been seen to be necessary by all those persons who have held that heaven is both timeless and future. And so far the critics have a certain justification. For they have at any rate seen the *primâ facie* difficulties in taking the timeless as future, and many of those who do take a timeless heaven as future have not seen these difficulties. But, when the critics go further and assert that it is impossible that a timeless heaven should be future, and not past or present, in the same way that to-morrow's breakfast is future, and not past or present, then, as I have tried to show, they are wrong. And if—a point to be discussed later (Chapters LXV and LXVII)—the whole which appears as the latest stage of the C series is a state which is infinitely good, then the believers in a timeless heaven in the future have, in point of fact, grasped the truth, though they may not have seen very clearly why it was true.

CHAPTER LXII

IMMORTALITY

740. In Chapter XLIII we discussed immortality, and came to the conclusion that an existent being was to be called immortal if it was a self which had an endless existence in future time. Taking immortality strictly in this sense, no self could be really immortal, since no self is really in time. But selves, though not in time, appear as in time. And we may, I think, fairly say that selves should be called immortal if they appear as having an endless existence in future time. For the use of the word immortality has been determined largely by practical considerations, and, if it is as true to say that my future existence will be endless as it is to say that I have lived through the last twenty-four hours, it would, I think, be in accordance with general usage to say that I am immortal.

741. Do selves appear as having an endless existence in future time? They appear as existing in time, and therefore, unless this apparent existence is endless towards the future, it must have an end towards the future. There are two ways in which it could have an end towards the future. If the self should cease to exist while time continues, so that there are moments of time later than any in which it exists, its existence clearly has an end. And, again, if the self endures through the whole of time, yet it may have an end, if time itself has an end.

742. The first of these alternatives we have already seen reason to reject. We saw (Chap. LI, p. 275) that in the common time-series the latest term of each C series—the term which is the whole of that series—will be simultaneous with the latest term of every other C series. At the other end, each series will have terms, each of which is simultaneous with a term in each other series, till the series stops at the boundary, which is nonentity. It follows that, when they are looked at as terms in a common time-series, each self will appear to have the same duration, and this duration will stretch from the beginning to the end of time.

This is, no doubt, only an appearance, since selves are not in time at all. But it is a *phenomenon bene fundatum*, and it is as true as any other statement as to temporal duration. It is as true that I endure through all time as it is that my repentance came after my crime.

743. There remains, however, the second way in which a self could appear to have an end towards the future. We have seen that the C series has a final term in one direction, and that that direction is the one which appears in the B series as the direction from earlier to later. The B series, therefore, has a latest term. Each self, then, will appear as reaching this term, and will not appear as going any further. Will not this involve that the life of each self will have, *sub specie temporis*, a future boundary, and so cannot be taken as immortal?

But, if we look more closely, this is not the case. The life of the self has a last term. But that term must, *sub specie temporis*, be taken as endless. And therefore the life, which includes that last term, is, *sub specie temporis*, endless also.

For the last term has, of course, no term later than itself. And also the series has, in the direction from earlier to later, no boundary. In the other direction the series has a boundary—namely, nonentity. For the process by which the series passes from a later term to an earlier term is that of diminution of content. And, when this process is carried far enough, it reaches the point where there is no content left. And that is nonentity. Thus, beyond all the terms of the C series in this direction there is one more term in a wider series—the term of nonentity. And this is the boundary of the C series. But, in the other direction, there is no term in a wider series which contains the C series. For the last term includes all the content, and therefore there can be no addition beyond the last term in the C series, as there can be subtraction beyond any term in the C series in the other direction.

Therefore, the last term, *sub specie temporis*, begins but it does not end. A stage in time begins when it passes from future to present (or, if it is the first term, when it becomes present, without having ever been future). And a stage in time ceases when it passes from present to past (and it would do so, if it were bounded, in the direction of earlier to later, by nonentity, when it ceased to be present without becoming past). But the final stage does not become past, because there are no later stages in time, from whose standpoint it would be past. Nor is it succeeded by nonentity, or by any other term in a series which includes the *C* series. And so it begins, but it does not end. It comes, but it does not go.

744. For any ordinary series of stages in time—the life of a man in a particular body, for example, or the duration of a soap-bubble, or of a solar system—the last stages are no more endless than any others. But this is because, although the particular series stops with that stage, the time-series as a whole does not stop there, but has later stages, from whose standpoint the last stage of the particular series is past. And so the last stage of that series ends[1]. Here, however, the position is different. The last stage of the *C* series of any self is at the same point in the *C* series with the last stage of the *C* series of all other selves, and so of the *C* series of the universe as a whole. And therefore, *sub specie temporis*, the last stage of the *B* series of any self is at the last point of time of the *B* series of every other self, and so of the *B* series of the universe as a whole. There is no time which is later than the last stage in the life of any self, and therefore the last term in the life of every self does not end[2].

745. The fact that, *sub specie temporis*, the last stage does not end, and that, when it appears in time, it appears as endless, and not as ended, must not be confounded with the fact that the stage in question is eternal—that is, timeless. It really *is* eternal, whereas it merely appears as in time, and consequently merely appears as endless in time. Nor is it the fact that it is eternal

---

[1] What the timeless reality is which appears as the cessation of a bubble or a solar system will be considered later (p. 377).

[2] The argument in the text depends, of course, on the result, which we reached previously, that there is a common *C* series, and therefore a common *B* series. But even if there were no common *C* series or *B* series, it would still be the case that the life of every self would be, *sub specie temporis*, endless. For it would end in a term—the term which was the whole of the series—which would be the latest term in its own time-series, and which would, therefore, from the point of its own time-series, be endless. And as the time-series of other selves would not, on this hypothesis, be in a common *B* series with it, they would have no temporal relations to it, and none of their terms would be later than its last term, so that it would not cease while they continued.

which makes it appear endless in time. For all the other stages in the *C* series are just as truly eternal, and they do not appear as endless in time. It does not appear as endless because it is eternal, but because it has no stage beyond it.

746. The final stage of which we speak is, in the dimension of the *C* series, one simple and indivisible term. The *C* series, as we have seen, must have simple terms, though it need not have next terms, and these may be called stages. But a group of such terms—the group which appears as an hour or any other period of time—may also be called a stage of the *C* series, though it contains a number, perhaps an infinite number, of simple terms. The final stage is one simple and indivisible term, for it is not reached at all until all the amount of perception in the whole is included, and, when that is done, there is nothing more to include.

The fact that this stage is simple and indivisible may seem to increase the paradoxical aspect of the theory that it is, *sub specie temporis*, endless. For a simple and indivisible stage, occurring elsewhere in the *B* series, might be infinitely short, and would certainly be too short to be perceived separately. But both the appearance of paradox, and the apparent heightening of it by the indivisibility of the stage in question, are alike erroneous. The one characteristic of a stage in the time-series which involves its endlessness is the fact that there is no later stage. No number of parts could make it endless without this, and no absence of parts makes any difficulty if this characteristic is present[1].

747. The paradoxical appearance of the view that the final term is endless arises, it seems to me, entirely from the tendency to suppose that the time-series of a self, like the time-series of a body, a soap-bubble, and a solar system, is a series which ends before the common time-series of the universe, so that there are terms in the time-series later than the last term in the life of the self. In that case, of course, the last stage of the self

[1] It will be remembered that, although the final stage is simple and indivisible in the dimension of the *C* series, it is divided, and divided to infinity, in other dimensions by determining correspondence. The final stage in the *C* series of *G*, for example, which contains the whole content of *G*, is divided into perceptions of the selves who form the differentiating group of *G*, and again divided into perceptions of the perceptions of these selves, and so on without end. This differentiation of the final stage of each self by determining correspondence is, of course, not peculiar to that stage. It occurs also, as we have seen, in all the other stages.

would not appear as endless in time. But the appearance of paradox will vanish if we keep steadily in our minds the result at which we have arrived—that the last term of the time-series of any self is contemporaneous with the last term of the common time-series, and that it is the last term of time, as well as of the self.

748. Thus the life of each ends, *sub specie temporis*, in a state which, *sub specie temporis*, is itself endless. And therefore the life of each self is, *sub specie temporis*, endless. We must therefore say, if we use the word immortality in the sense adopted at the beginning of this chapter, that each self is immortal. It is immortal, that is, *sub specie temporis*. Immortality, as we have defined it, is a term which applies only to time, and therefore the selves are not in reality immortal, since they are not really in time. They are really eternal, but not really immortal. But it is as true that I shall live endlessly in the future, as it is that I have lived a minute since I began this paragraph. And this is what would generally be meant by immortality[1].

It is not only selves which have, in this way, an endless existence in the future. The same is true of all parts of selves which are determined by determining correspondence—the perceptions, that is, which each self contains of selves, and of parts of selves to infinity. And it is true of all groups of selves, including the universe, which is the group which contains all selves. But, although their existence is endless, we should not call them immortal, since we reserved that word, in accordance with common usage, for unendingly existent selves.

749. But is it true, it may be asked, that every group of selves appears as having an endless existence in the future? Is it even true that any of these groups of selves about which we habitually speak appear as having an endless existence? Is it true of a whist party, or of a nation?

To answer this question, we must remember that, by the definition we adopted (Section 121), the identity of a group depends on the individuals which it contains, and not on the relations

[1] As we said in the last chapter, the Christian idea of heaven is often the idea of a timeless state which is still in the future. And the selves which enter this heaven are considered, by the supporters of such a view, to be immortal, although in reaching this timeless heaven they reach the last term of their life in time.

between them. The four selves, *G*, *H*, *K*, and *L*, form a group, since
they are four substances. If they happen to be all brothers, and also
all partners in business, these two sets of relations do not make
them two groups. They are one group—since it is the same four
substances who are brothers and partners—with two sets of relations between them.

If we ask, then, whether a whist party is a group which has,
sub specie temporis, an endless existence in the future, we must
make a distinction. The group is not the whist party as such but
the four selves, *G*, *H*, *K*, and *L*, who do make up the whist party.
That group, like every other group of selves, whatever selves are
contained in it, will appear as having, like the individuals who
compose it, an endless existence in the future. But it will not
appear as being a whist party throughout that endless existence.
At a certain point in the time-series the game, *sub specie temporis*,
is finished, and the whist party breaks up—that is, the relations
between the four selves no longer include the relation which
made them a whist party. The explanation is that the relations,
whatever they are, which appear to us as participation in the same
whist party, do not hold between *G*, *H*, *K*, and *L*, at all stages
of their *C* series, but only at some stages, which are preceded
and followed by stages in which they do not hold. Thus, *sub specie
temporis*, the whist party appears to begin and to cease in time.

750. A similar explanation will hold good of groups of parts
of selves. And we can now answer the question raised on p. 374,
as to the eternal reality which appears as the ending of a living
body, a soap bubble, or a solar system. Whatever appears as a
body, a bubble, or a system, must be, in reality, a self, or a group
of selves, or a group of parts of a self or selves. And thus it has
an existence which is, *sub specie temporis*, endless. But those
relations between members of the group which determine it to
appear to us in this form do not hold between the members at
all stages of their *C* series, but only in some, which are preceded
and followed by others in which they do not hold. And thus the
body, the bubble, and the solar system appear to begin in time,
and to cease in time.

751. We are immortal, then, because we appear as having an
endless existence in the future. We must note—though it has

no bearing on the question whether we are immortal—that we do not appear as having an endless existence in the past. A finite time backwards brings us to the boundary of the series in that direction, in the same way that a finite time forwards brings us to the boundary of the system in the other direction. But the boundary in the direction of the future is, as we have seen, itself a term in the series, and a term which is, *sub specie temporis*, endless, and it therefore makes the series of which it is a term endless in this direction. In the other direction, however, the boundary of the series is nonentity, which is not a term in the series, but a limit. The past consists of those terms which lie between nonentity and the present. And these terms, taken together, are finite in length. Existence, therefore, is not endless towards the past, though it is endless towards the future.

But although our past lives will not appear as endless, they will appear as extending through all past time. For, as we saw in Chap. LI, p. 275, every individual *B* series begins at the beginning of the common *B* series, in the same way as it ends at the end of the common *B* series. The only reason that our past lives do not appear as endless is that past time does not appear as endless.

752. The view that our past lives are of finite length, while our future lives are of infinite length, is the most common view in the Western world. For it follows from a combination of a belief in immortality with a disbelief in pre-existence, and this is the ordinary Western view. But the ordinary view does not hold, as we have done, that each self has existed through the whole of past time—indeed, it is impossible to hold this without accepting pre-existence.

753. And we must further note that our view that past time is only finite, while future time is infinite, does not result from holding, as many people have held, that, while an infinite progress in time is logically possible, an infinite regress in time is a vicious infinite. We have found no reason to make any such distinction. If it had been the case that the relation "included in," and not the relation "inclusive of," appeared as the relation "later than," our lives would have appeared as without a beginning, but not without an end.

CHAPTER LXIII

PRE-EXISTENCE AND POST-EXISTENCE

754. We saw in the last chapter that the life of every self appeared as persisting through the whole of time, and as existing endlessly in the future, and, consequently, as immortal. But the question still remains as to what are the relations of this life as a whole to that particular life in which we are living at present[1]. By a particular life I mean the period which elapses between the birth of any one body and the death of that body[2]. Have we lived before the birth of our present bodies, and shall we live after their death?

755. Existence before the birth of our present bodies may be called pre-existence, and existence after the death of our present bodies may be called post-existence. Pre-existence and post-existence do not involve the existence of the self through all time. It would be possible for a self to have existed before the birth of its present body, and yet not to have existed through all past time, or for it to exist after the death of its present body, and yet not to exist through all previous time.

But, on the other hand, existence through all time does, it is clear, involve pre-existence and post-existence. For I observe the bodies of other people to die, while I myself continue living. And since these other selves must continue to exist while time continues, they must exist after the death of the bodies with which I have known them. And, in the same way, the birth of other people during my life proves that these people, who must have existed during that part of my life which is earlier than their birth, must have existed before the birth of the bodies with which I know them.

756. We have seen that all selves exist, *sub specie temporis,*

---

[1] Pages 383–385 and 389–393 of this chapter are taken, with a few alterations, from Chapter IV of my *Some Dogmas of Religion*. This chapter was reprinted in *Human Immortality and Pre-existence*.

[2] Strictly speaking the period may be rather longer, since it begins with the first connection of the self with the body, which may take place while the latter is still in the womb. But for simplicity we may speak of it as beginning at birth.

through all time, and therefore we must ascribe to all selves both pre-existence and post-existence. The idea of post-existence is one which is very familiar to our thought. For, as we have seen, it is involved in immortality, and the belief in immortality is generally accepted—at any rate in the Western world. But, since the rise of Christianity, the attitude of Western thought to the doctrine of pre-existence has been very curious. Of the many thinkers who regard life after the death of our present bodies as certain or as probable scarcely one regards life before the birth of those bodies as a possibility which deserves discussion. Lotze, for example, treats it as a serious objection to a particular argument for immortality that it would lead to the "strange and improbable" conclusion of pre-existence. Why should men who are so anxious to prove that we shall live after this life is ended regard the hypothesis that we have already survived the end of a life as one which is unworthy even to be considered?

In the Far East, on the other hand, the belief in pre-existence is usually associated with the belief in post-existence. The reason of the difference is, I suppose, that in modern Western thought the great support of the belief in post-existence has been the Christian religion. It followed from this that a form of the belief which was never supported by that religion was not likely to be considered of any importance. And Christians have almost always rejected those theories which place pre-existence by the side of post-existence, though there seems nothing in pre-existence incompatible with any of the dogmas which are generally accepted as fundamental to Christianity.

There are, as it seems to me, various features of our present life which can best be explained on the hypothesis that we existed before the birth of our present bodies. But I do not think that the advantage thus gained is sufficient to afford a proof of pre-existence, while, if we are right in our conclusion that we have existed through all past time, pre-existence, as was said above, follows inevitably. It does not seem necessary, therefore, to consider here the characteristics of our present life which suggest pre-existence[1].

---

[1] I have discussed these in *Some Dogmas of Religion*, Sections 94–99 (*Human Immortality and Pre-existence*, pp. 86–98).

757. Have we any indications of the length of the periods through which we have passed and shall pass, as compared with the length of our present lives? Such indications as there are point to the comparative lengths in both cases being very great. We have been able in Book V to sketch the general nature of a self, as it stands as a primary part in the system of determining correspondence. And this—the whole of the self—appears as the final stage of the *B* series. Now when we compare the nature of the whole, as thus determined, with the nature of that particular state of the *C* series which appears as our present life, we find that the difference is enormous. And when we further take into account the very small advance in any direction which, as experience shows, we are capable of making in the course of any particular life, it seems difficult to resist the conclusion that the length of the future process which will be necessary for us to reach the end of the *B* series will be very great indeed as compared with the length of our present lives.

This, however, is not absolutely certain, because we are not certain that the rate of advance will remain the same, or anything like the same. It is conceivable that, as we pass in the *C* series from nonentity to the whole, the additional amounts of content of the whole taken in at each stage may make a greater and greater difference to the nature of the stage that takes them in. And this rate of increase may itself increase with very great rapidity. Thus it may be the case that a time equal to our present lives may in the future bring our lives far further in the direction of the nature of the whole than our present lives do. But there is no reason to suppose that such a variation in the rate of advance does take place, or that, if it does, it is an acceleration rather than a retardation. And, without a variation and a very great variation, in the direction of acceleration, the future of time and change which lies before each of us must be of very great length as compared with our present lives.

And the past behind each of us, for similar reasons, seems likely to be very long in proportion to our present lives. The amount which we have already gained may not seem so great as what still remains to be gained. But that it is very considerable compared to what we can observe to be gained in the course of a single adult life, cannot be doubted, when we consider the difference between our own development and that which appears to be possessed by the less developed of the lower animals. Here, as in the case of future life, an element of uncertainty is introduced by the possibility that the rate of advance in the past was much greater than the rate of advance in the present. But, in spite of this, we are left with the probability that past life also is of very great length as compared with our present lives.

758. In the last paragraph we spoke of what we can observe to be gained in a single *adult* life. In the period which elapses between birth and adult life the change is much greater, in normal circumstances, than any which takes place during adult life itself. But there is nothing to suggest that the same rapidity of advance took place continuously before birth. On the contrary, the more probable supposition, on the basis of our conclusion that there *has* been previous life, is that the rapid development after birth is due to a recovery from an oscillation which occurs at or before birth, and in which much of what had been previously gained was temporarily lost. And so the rapidity of advance before maturity does not diminish the probability of the great length of previous life in comparison with present life.

This brings us to the consideration of oscillation in general as it bears on our present enquiry. We saw (Chap. XLVI, pp. 218–221) that it was necessary to find a theory of the C series which would allow for the oscillation both of the extent and accuracy of our knowledge. And we saw later (Chap. L, pp. 262–270) that such a theory could be found.

Now this possibility of oscillation gives a possibility of indefinite increase in the length of past and future life. For not only must their length be sufficient to cover the advance from the beginning of the process to our present state, and from our present state to the final state, but, whenever there is oscillation, part of the advance will have to be repeated at least twice over, and may have to be repeated any number of times[1].

---

[1] It may be objected that our theory of the duration of selves is incompatible with the great inequality of their developments at any one time. All selves, we have said, start from nonentity at, *sub specie temporis*, the same time. All selves, *sub specie temporis*, reach the full development, sketched in Book V, at the same time. At any intermediate moment, all selves will contain the same proportion

759. The considerations already mentioned as to the length of past and future life apply, as we have seen, equally to future and to past. But with regard to past life, there is something else which must also be considered. For science tells us that there are convincing reasons for the belief that what appears to us as the material world has existed in the past for a time which is very long in comparison with our present lives. And we have found that every self must have existed, *sub specie temporis*, for the whole of past time, *i.e.* for as long as anything has existed. And this, by itself, gives a great length to past time as compared with our present lives.

760. If, then, past and future life are probably very much longer than our present lives, the most probable hypothesis is that each of them is divided into many lives, each divided from the others by birth and death, in a manner analogous to that in which our present lives are divided from future and past life by birth and death. This doctrine of a plurality of future and past lives may conveniently be called the doctrine of the plurality of lives[1].

Even if we had not found reasons to accept pre-existence, it would still be more probable that our future existence would be divided into a plurality of lives, provided that the period of future change and transition, before the final endless and changeless state is reached, were of considerably greater length than our present lives. We do not know what cause produces the limitation of our present lives by birth and death, but some cause there presumably is, and a cause which produces so

of the full amount of their content. Yet there was a moment when one self was planning *Hamlet*, while other existents, which were really selves, appeared as the bacilli in his blood, or the salt in his salt cellar. It can scarcely be doubted that their developments were really unequal. Is this compatible with our theory?

Such inequalities of development might in some cases be due merely to oscillations. But it is not necessary to account for them in this way. It is quite possible that the original differences in the natures of the selves (which, as we have seen, must have such differences) might be such that equal increments of content might produce more rapid development in the case of some selves at earlier periods, and in the case of others at later periods.

[1] In one sense, of course, a belief in pre-existence and post-existence is itself a belief in a plurality of lives, since it is a belief in three at least. But it will, I think, be more convenient to reserve the name for the belief that for each of us existence on each side of our present lives will be divided into more lives than one.

important an effect is one which plays a great part in our existence, as long as it continues to act.

Now, if we hold that the whole of the very long period of change and development after this life is a single life, not divided off by birth and death, we must hold that the cause, whatever it may be, which operates on each of us so as to determine him to die once, will never operate on any of us for a second time, through the whole of that long period. This is not, of course, impossible. The true nature of death may be such that there is no need, and no possibility, of its recurrence. But this seems improbable.

It is clear that a life which stretched on for a very long period without death, and then ended, not in death, but in an endless and changeless state of consciousness, would differ, very widely and in many important respects, from our present lives. An attempt to imagine how our present lives would be transformed if neither we ourselves nor our fellow men had in future any chance of death, will make this evident. If it is maintained that, on the death of our present bodies, we shall pass at once into such a life, it must be the case that the death which ends our present lives will change profoundly and permanently the conditions of all future life. This is certainly not impossible, but there seems no reason to regard it as probable.

Again, processes begun in this life are sometimes finished in it, and sometimes left incomplete. We continually find that death leaves a fault without a retribution, a retribution without a repentance, a preparation without an achievement, while in other cases, where the life has lasted longer, a similar process is complete between birth and death. And in many such cases the completion seems to proceed, not from the environment, but from the inner nature of the particular self. In such cases, we may expect that these processes, in the cases in which they are not worked out before death, will be worked out in future life. And, if the content of our existence after this death should have so much similarity, in essential features, with the content of our present lives, the presumption is increased that we shall not have changed so far as to have lost the characteristic of periodical death.

761. There seems, therefore, good reason for regarding plurality of lives as the most probable alternative, even if we did not accept pre-existence. But when pre-existence is accepted, the case for plurality of lives becomes stronger. For then the death which ends my present life is at any rate not an unique event in my experience. One life, if no more, came to an end for me before my present life began. Thus any theory would be false which should try to reject plurality of lives on the ground that it was probable that death could only occur once in a self's existence. And plurality of lives would only be regarded as improbable, if there was reason to suppose that an event, which happened twice in a man's existence, could never happen a third time. Now it might be contended—though, as I have said, I do not think that it would be correct—that death presents qualities which make it probable that it could only occur once in a man's existence. But there does not seem to be the slightest excuse for the suggestion that there is anything about death which should make it improbable that it should occur three times, although it was certain that it occurred twice. The rejection of the plurality of lives involves that the causes which break off a life by death, after remaining dormant from the beginning of our existence, act twice within an interval varying from a minute to about a hundred years, and then never act again through all future time.

Thus plurality of lives, the most probable supposition in any case, is still more probable when pre-existence is admitted, though, even in that case, there is no absolute demonstration.

762. If we have existed before our present lives, death and re-birth involve, in this case at least, a loss of memory. For we remember nothing which took place before our present lives— at least it is certain that the great majority of us do not, and there is no evidence that any one does. If, however, there were no plurality of lives, this loss of memory would have much less importance. If the present life were the only one which was limited at both ends by other lives, it would be at any rate possible to regard the loss of memory as a characteristic of this single episodical life, and to suppose that memory of all the past would be regained after death. And, even if this were not so, and each death involved permanent loss of memory, still, if the death which ends our present lives is the last death we shall die, loss of memory will only occur once more, and will never disturb life after this death. But, if all our future life is to be divided into a plurality of lives, we must face the probability that each transition into a fresh life will involve the same loss of memory as the transition into the present life, and that the future life will be accompanied by a continual recurrence of whatever interruption the loss of memory involves.

763. What, then, is involved by the loss of memory? It has sometimes been asserted that the loss of memory is equivalent to the loss of identity, and that, if a man remembers nothing that happened before the birth of his present body, he cannot be the same man with one who lived before the birth of that body. This view, however, seems to be clearly erroneous. We cannot, indeed, rely on the fact that all the C stages which appear, some as the life in one body, and some as the life in the other body, form part of the same substance. For we have seen that any substances in the universe, however slightly connected with each other, form a group (Section 122); and we have also seen that every group is a substance (Section 127). The oldest rabbit in Australia and the last sneeze of Lewis XV form a substance. And, if the different C stages had no more unity with one another than these two, we could not call them a single self, although they would be a single substance.

But the connection between the different C stages which are asserted to belong to the same self is very different from this. In the first place, they all fall within one primary part in the system of determining correspondence. For we have decided that all primary parts, and that nothing but primary parts, are selves. Moreover, the C series comprises the whole content of the primary part. And, as we have seen, the primary parts are the fundamental unities of existence. The differentiation of the universe into primary parts is the ultimate fact from which all the other differentiations of the universe follow. Thus there can be no more vital distinction of two substances from one another than their being within different primary parts, and no more vital union between them than their being in the same primary part. And so the fact that the different C stages of which we are speaking are within the same primary part and compose the whole of it gives them a real and close identity with one another.

764. In the second place, this unity is increased, not only by the fact that they comprise in them the whole content of the primary part, but by the fact that the relation which connects them with each other is a relation of inclusion, so that the content of each stage in the series is part of the content of the next.

765. In the third place, the fact that the rabbit and the sneeze form a single substance gives us no reason to suppose that they share with one another any characteristic which they do not share with all other substances. But here it is different. In the course of a single life we know that there is continuity of character in a self—that the character which it has at one moment it will have at a later moment unless it alters according to definite laws. The loss of memory would not affect this, and therefore there is no reason to doubt that it would also persist from life to life. And it is obvious that this again forms a very vital connection between the different stages.

766. Thus, even if loss of memory were permanent, it would not destroy the identity of the self. But it seems to me that we must conclude that it is not permanent, and that all memory recurs in the final stage of the C series—the stage which is the whole, and which is, *sub specie temporis*, the last and endless stage of the time-process.

767. We cannot argue that the final stage will have perceptions of all other stages of the C series merely from the fact that the other stages do exist, and are parts of the substances perceived in the final stage. For, though we saw that the final stages were correct perceptions, and could contain no error (Chap. XLVII, pp. 228–232), yet there was nothing to show that they must be complete perceptions. They could not perceive anything as having a characteristic which it had not, but they need not perceive a thing as having all the characteristics which it did have. And so a self at the final stage might, so far as this goes, perceive itself and other substances in the final stage, without perceiving them as each containing a C series.

But we saw (Chap. XLVII, p. 227) that each of the states of misperception which form the pre-final stages of the C series perceives not only stages which correspond to itself in the C series, but stages at different parts of the series. To recur to the symbols used in that chapter, $c_xG!H$ will be a perception, not only of c_xH but of c_yH and c_zH, where c_x represents any prefinal stage in the C series, and c_y and c_z any states before or after c_x. That this must be so is evident from the facts of our present experience. For we remember the past, and make judgments about it, and we make judgments about the future. And these apparent memories and judgments are really perceptions. And therefore we must have perceptions (though not apparent perceptions) of the past and the future, as well as of the present.

768. It is true that the amount of the past and future which, in our present experience, is known to us by apparent memories or judgments, is very small, compared with the whole amount to be known. Nor is it always the same part. But we saw in Chap. L, p. 265, that the explanation is that, in our present experience, much that is really differentiated is misperceived as undifferentiated, and much of what is misperceived as undifferentiated is perceived as a vague background to our more definite perception.

But in the final stage this cannot take place. For every perception in the final stage is completely correct. What is differentiated cannot be perceived as undifferentiated, nor can anything be perceived as a vague background, if it is really something quite different.

Now it seems a fair inference that, if the pre-final stages of the C series perceive what is at other stages of the series, the final stage of the C series will perceive them too. And as it cannot perceive them confusedly, which would be to misperceive them, it must perceive them distinctly. And so, when G, in his final stage, perceives either himself or H, it would seem to be the case that he will perceive all the pre-final stages in himself or H, i.e., that he will perceive the whole series which, *sub specie temporis*, is the life throughout time of the self in question.

769. Of course he will not perceive them as being past, or being events in time, because he will perceive them correctly, and they are not really past, or events in time. And, therefore, he cannot strictly be said to remember them. But, on the other hand, he may perceive them as stages which perceived themselves to be present, and which were perceived by various other pre-final stages as past. For these characteristics—to perceive themselves as present, and to be perceived by certain other stages as past—are characteristics which the pre-final stages really have, and which therefore they can be correctly perceived as having. In the final stage, then, each self will be conscious of all that had appeared as its past life, and conscious of it as a series which had appeared as its past life. And such a consciousness only differs from memory in the greater closeness of the link between the consciousness and its object. We may, therefore, when we are speaking, as here, of practical effects, say without impropriety that the final stage remembers all the rest.

770. And thus, even if it should be the case—and I have endeavoured to show that it is not the case—that there is no personal identity unless selves are connected by memory, we should not lose personal identity. For every pre-final stage in the C series would have such identity with the final stage, which remembers them all. And two states of consciousness, which are connected by personal identity with a third state, are connected by personal identity with each other.

771. Since personal identity is not destroyed by loss of memory, we can continue to say that each of us has unending life in the future. But what about the value of that life? It has been maintained that the loss of memory, while not destroying immortality, would destroy its value.

I do not propose to discuss at this point whether any immortality has any value. Some people maintain that all human existence is evil, however favourable its conditions. Others regard existence as of such value that they would be prepared to choose hell rather than annihilation. Among those who differ less violently, some regard the life of the average man on earth at present as having positive value, while others only regard it as worth while if it is the necessary preparation for a better life which is to follow. Such differences as to the value of life must obviously produce great differences as to the value of its unending prolongation. What the value of life is will be discussed in Chapters LXV, LXVI, and LXVII. All that I shall maintain here is that even the permanent loss of memory need not render immortality valueless, if it would not have been valueless without loss of memory. From this it follows, *à fortiori*, that the temporary loss of memory, restored in the final stage, would not make immortality valueless.

772. Let us begin by enquiring what would be the result if the loss were permanent. In the first place, if existence beyond the present life is not expected to improve, and yet immortality is regarded as valuable, it must be because a life no better than this is looked on as possessing value. Now it is certain that in this life we remember no previous lives, whether it be because we have forgotten them, or because there have been none to remember. And, if this life has value without any memory beyond itself, why should not future lives have value without memory beyond themselves? In that case a man will be the better off for his immortality, since it will give him an unlimited amount of valuable existence, instead of a limited amount. And a man who believed that he had this immortality would have a more desirable expectation of the future than if he did not believe it. If, indeed, a man should say that he takes no more interest in his own fate, after the memory of his present life has gone, than he would take in the fate of some unknown person, I do not see how he could be shown to be in the wrong. But I do not believe that most men would agree with him, and to most men, therefore, the prospect of a continuance of valuable existence, even with the periodical loss of memory, would still seem to be desirable.

773. But immortality is not only desired, or chiefly desired, because it would give us more life like our present life. Its attraction is mainly for those who believe that future life will be, at any rate for many of us, a great improvement on the present. Heaven is longed for, not merely because it will be unending, but because it will be heaven.

Now it might be said that our chief ground for hoping for improvement after death would be destroyed if memory ceased periodically. Death, in removing all memory of what we had done in any life, would destroy all the advance we had made in that life. We could no more hope for a permanent improvement than a man on the treadmill could hope to end higher than he started.

If this objection were well founded, it would be serious. In the first place, it would destroy all ground for hoping for improvement till we reached the final stage—a stage which, *sub specie temporis*, may be removed from us at present by any finite period. And, in the second place, it would throw doubt on any conclusion which we may reach later as to the high value of the final stage. For it would be improbable that a great and sudden change of value should arise for the first time in the final stage, if there had been no increase of value at all through the rest of the series. But I think that it can be shown that the objection is not well founded.

774. The chief ways in which memory assists progress are three. In the first place, it may make us wiser. The events which we have seen, and the conclusions at which we have arrived, may be preserved in memory, and so increase our wisdom. In the second place, it may make us more virtuous. The memory, *e.g.*, of a temptation, whether it has been resisted or successful, may help us in resisting present temptation. In the third place, it may tell us that people to whom we are now related are the people whom we have loved in the past, and this may determine, or help to determine, our present love of them.

The value of memory, then, is that by its means the past may serve the wisdom, the virtue, and the love that are present. If the past can help the future in a like manner without the aid of memory, the absence of memory need not destroy the possibility of an improvement spreading over many lives.

775. Let us consider wisdom first. Can we be wiser by reason of experience which we have forgotten? Unquestionably we can. Wisdom is not merely, or chiefly, recorded facts, or even recorded conclusions. It depends primarily on a mind qualified to observe facts and to draw conclusions. Now the acquisition of knowledge and experience, under favourable circumstances, may strengthen the mind. Of that we have sufficient experience in this life. And so a man who died after acquiring knowledge—and all men acquire some—might enter his new life, deprived indeed of his knowledge, but not deprived of the increased strength and delicacy of mind which he had gained in acquiring the knowledge. And, if so, he will be wiser in the second life because of what has happened in the first.

Of course, he loses something in losing the actual knowledge. But it is sufficient if he does not lose all. Most progress is by means of oscillation, and is only progress on the balance. And is not even this loss really a gain? For, while we see things—as we must see them while the time-process goes on—piecemeal and *sub specie temporis,* the mere accumulation of knowledge, if memory never ceased, would soon become overwhelming and worse than useless. Is it not better to leave such accumulations behind us, preserving their greatest value in the faculties which have been improved by their acquisition?

776. With virtue the case is perhaps even clearer. For the memory of moral experiences is of no value to virtue except in so far as it helps to form the moral character, and, if this is done, the loss of the memory would be no loss to virtue. Now we cannot doubt that a character may remain determined by an event which has been forgotten. I have forgotten most of the good and evil acts which I have done in my present life. And yet each has left a trace on my character. And so a man may carry over into his next life the dispositions and tendencies which he has gained by the moral contests of this life, and the value of these experiences will not have been destroyed by the death which has destroyed the memory of them.

777. There remains love. The problem here is more important, if, as I have tried to show, the entire life of each self centres round and depends on his love for other selves, and if, as I believe, it is love which is the supreme value of life. The gain which the memory of the past gives us here is that the memory of past love for any person can strengthen our present love for him. And, if the value of past love is not to be lost when memory ceases, it must still in some way strengthen our present love. The knowledge which we acquire, and the efforts which we make, are directed towards ends which are not themselves, and, if these ends are attained, we need not grieve if the means pass away. But love has no end but itself. If it has gone, it helps us but little that we have kept anything it has brought us.

But past hours of love are past, whether we remember them or not. Yet when they are remembered we do not count their value as lost, since their remembrance makes love in the present stronger and deeper. Now we know that present love can also be stronger and deeper because of past love which has been forgotten. For much has been forgotten in any friendship which has lasted for several years within the limits of a single life— many confidences, many services, many hours of happiness and sorrow. But they have not passed away without leaving their mark on the present. They contribute, though they are forgotten, to the present love which is not forgotten. In the same way, if the whole memory of a life is swept away at death, its value is not lost if the love of the same two people is stronger in a new life because of what passed before.

Thus what is gained in one life may be preserved in another, irrespective of memory, if love can be greater in the second because it was there in the first, and if people who love in one life love the same people in the second. Have we any ground to hope that these two conditions will be fulfilled?

778. As for the first condition; it is clear that the love which two people bear to one another in one life can be greater because they have loved in an earlier life. If by means of love we make, in one life, our relations stronger and finer, then they can be stronger and finer at the next meeting. What more do we want? The past is not preserved separately in memory, but it survives, concentrated and united, in the present. Death is thus the most perfect example of the "collapse into immediacy"—that mysterious phrase of Hegel's—where all that was before a mass of hardearned acquisitions has been merged in the unity of a developed character. If we still think that the past is lost, let us ask ourselves, as I suggested before, whether we regard as lost all those incidents in a friendship which, even before death, are forgotten.

779. Let us pass to the second condition. Is there any reason to think that people who love in one life love the same people in another life? The chance of a love recurring in any future life must depend primarily on the conditions which determine where and how the lovers are born in the future life. For, if memory does not survive death, it will be impossible for any man to love another in any life in which he does not become acquainted with him.

Now our empirical knowledge is sufficient to show us that there are, on this planet alone, a very large number of selves whom we can recognize as selves. This number, however, must, if our theory is correct, be only a part of the number which appear as being on this planet, since, according to that theory, all that exists is really selves. And we have reason to believe that what appears as this planet is only a very small part of the universe, while it may be a much smaller part than it appears to be. The total number of selves must in any case be so great that the chance of lovers meeting again after death would be negligible, if the conditions which determine re-birth, and so juxtaposition in future lives, were not connected with the conditions which determine love. But they are connected.

780. If, in this life, G loves H, this must mean that H is one of his differentiating group. For we have seen that the specially close connection between selves, which is characteristic of love, is found wherever one self perceives another directly, and can scarcely be supposed to exist when one self does not know another directly. It follows that H must be perceived directly by G throughout the whole C series of the latter—for the differentiating group is the same at every stage of that series. This does not, however, imply that H will always be loved by G, even in those states which, *sub specie temporis*, are later than this life, in which he has loved him. For, as we have seen, we must allow for oscillation in the clearness and accuracy of perception. And so it is possible that, in later lives than this, he may not recognize H as a self, or even have a separate perception of H at all. (Cp. Chap. L, p. 265.) And the same possibility of oscillation will prevent it from being certain that, at every later stage in which G does recognize H as a self, he will love him.

But, in spite of these oscillations, the fact that G does love H in this life will involve that he will, in future lives, though not necessarily in all of them, recognize H as a self and love him. For, in the final stage—the whole—he will necessarily recognize him as a self, and love him. Now it is highly improbable that there should be no approximation to this result in the course of the pre-final stages, and therefore it must be expected that the

love for each self who will be loved in the final stage will recur
at intervals throughout the series. There would be no reason to
suppose that this *could* happen, if the same selves were not connected together in each life. And, as we saw (p. 391), if we were
not entitled to believe in the possibility of such a continuous
approximation, it would throw doubts on our view as to the
nature of the final stage. But we have seen that the same selves
are connected together in each life, and that therefore there is
nothing against the likelihood of the recognition and the love
recurring with sufficient frequency to allow of gradual approximation to the final stage.

781. How many such recurrences there would be, and at
what intervals, would depend upon the number of selves in a differentiating group, and upon the number of lives into which a
C series is divided. As to the second of these points, we have seen
that there is a probability that the number is large, but there is,
as far as I can see, nothing more definite to be said. On the first
point I do not see that there are any indications whatever as to
what the number may be.

782. And thus both our conditions are satisfied. Love can be
greater in a later life because it was there in an earlier life, and
the fact that two people love one another in this life is a reason
for holding that they will love one another in various future
lives. The value of love, then, does not cease with the cessation
of its memory.

We have thus seen that, even if death should involve a
permanent loss of memory, the past would not be lost to us, since
it might increase our knowledge, our virtue, and our love. I do
not deny that in each particular life the prospect of the loss of
memory at the end of it will appear as to some extent a breach
of continuity and a loss of value. This will seem greatest, perhaps,
in the case of the values which are not quite the greatest. If
I contemplate the fact that I shall lose at death the memory of
my friend, and of our love in this life, I may well be consoled
by the reflection that, in all the future, I shall never permanently
lose my friend or his love. But I should have no analogous con-
solation when I think of my approaching forgetfulness of my
country and of my school. And, although such loyalties are trivial

by the side of love, yet, by their side, most other things are trivial. Yet it will be the highest that will not be lost.

783. In considering the effects of the loss of memory, we have so far spoken as if that loss were permanent. But we saw earlier (p. 388) that there is reason to think that this is not so, and that, in the final stage, we shall be conscious of all our pre-final stages. And, in this case, the loss of memory, and all losses that it involves, will only be for a finite time, and will be followed by a memory which will be, *sub specie temporis,* endless.

784. Let us turn from the effect of the loss of memory to the more general question of the effect of a plurality of lives. If there is such a plurality extending over a long future, our prospects after leaving our present bodies have possibilities of evil much greater than those generally admitted by theories of immortality which reject—what is generally rejected now—the possibility of an endless hell. Such theories hold, in some cases, that we shall pass immediately at death to a state of complete and endless beatitude. In other cases the view is taken that, before such complete fruition is reached, there must be a period of effort, perhaps of suffering and of strife. But it would be held that this future life, ending in final perfection and unbroken by further deaths, would be altogether on a higher level than our life now. It would be a life from which positive sin would be excluded, and in which effort and suffering would be made more easy by a clear perception of the end to be reached, and by a willing acceptance of it as our own.

We have, however, no right, on the view which we have taken, to share this optimism as to the immediate temporal future. The temporal future will consist of a great number of successive lives. It is true that, in the long run, the later will be better than the earlier. But the rate of improvement may be very slow—so slow that it might be imperceptible for centuries—and it may be broken by periods of oscillation in which a man was actually in a worse condition than he had been previously. With regard to knowledge, to virtue, and to love, we have no ground for supposing that improvement will not be very slow, and that it will not be broken by intervals of deterioration. And with regard to happiness, there is no form of suffering which history records to have happened in the past, which may not lie in the path of any one of us in the future. Some such forms, indeed, seem likely to disappear from this planet. But our future may take us elsewhere.

Such possibilities are not attractive. A universe which excluded them would be better than this universe which admits them. But the universe has evil in it—that is beyond doubt. And the fact that our theory implies that it has this particular evil in it is no reason for rejecting our theory.

785. And the prospect of many such lives as ours has a bright as well as a dark aspect. Not only is there good in all the prefinal stages as well as evil, but those stages, the evil as well as the good, serve for the development of our nature. Such life as ours now, in which sin jostles with virtue, and doubt with confidence, and hatred with love, cannot satisfy us, but it can teach us a great deal—far more than can be learned between a single birth and a single death. Not only because the time is so short, but because there are so many things which are incompatible within a single life. No man can learn fully in one life the lessons of unbroken health and of bodily sickness, of riches and of poverty, of study and action, of comradeship and of isolation, of defiance and of obedience, of virtue and of vice. And yet they are all so good to learn. Is it not worth much to be able to hope that what we have missed in one life may come to us in another?

And though the way is long, it can be no more wearisome than a single life. For with death we leave behind us memory and old age, and fatigue. We may die old, but we shall be born young. And death acquires a deeper and more gracious significance when we regard it as part of the continually recurring rhythm of progress—as inevitable, as natural, and as benevolent as sleep.

Nor will the change, the struggle, and the alternation of life and death endure endlessly. Change changes into the unchanging and then "*Stretched out on the spoils which his own hand spread, As a God self-slain on his own strange altar, Death lies dead*[1]."

---

[1] Swinburne, *A Forsaken Garden.*

CHAPTER LXIV

GOOD AND EVIL

786. We are now in a position to attempt to estimate the relative amounts of good and evil in the universe. For this object it will not be necessary, as it would be necessary for a complete theory of ethics, to determine completely what characteristics would make a thing good, and what characteristics would make it evil. Our enquiry admits of limitation in two ways. In the first place, we need not determine the value of all possible characteristics, but only of those which we have found reason to believe do, or may, exist, leaving out of consideration those characteristics which we have determined cannot exist. In the second place, supposing that we find that the existent possesses a group of such characteristics that its value would be the same according to different ethical theories, we need not enquire which of those theories is correct. If, for example, we find that the existent is more happy than miserable, and more virtuous than vicious, we need not enquire whether nothing is good but pleasure, or whether nothing is good but virtue, or whether they are both good.

I shall use the words good and evil to mean what is good or evil in itself, or intrinsically, excluding what is sometimes called good and evil as means—that which, though not good or evil intrinsically, produces what is good or evil intrinsically. I shall speak of that which *produces* good and evil as possessing utility or disutility. Whatever *is* either good or evil may be said to possess value, or to be valuable.

787. It is generally admitted that it is impossible to define good and evil in terms of anything else. Any assertions that this is possible are due, I think, to a confusion between a definition and a universal synthetic proposition about the thing defined. "The pleasant, and only the pleasant, is good," would not, even if true, give a definition of the good. For this is not an assertion that we mean by the word *good* the same as we mean by the word *pleasant*, but an assertion that either of two characteristics is always found when the other is. And this is not a definition of either characteristic.

788. What is capable of having value? In the first place, I think that it is generally agreed that only the spiritual can have value. This proposition is ultimate and synthetic. It is impossible to prove it. But it is very generally, if not universally, admitted. Even materialists who assert that what other people call spiritual is only a special activity of matter, would allow that these activities of matter have value, and that nothing else has.

We have come to the conclusion that all that exists is spiritual, and, so far then, all that exists is capable of having value. But at this point we must make a distinction which may be expressed by means of the phrases "value *of* anything," and "value *in* anything." When any substance, taken as a unity, possesses value, we shall speak of the value of that substance. But when there is any value either of a substance, or of any of its parts, we shall speak of that value as being in the substance. Thus the value of a part is value in each of the wholes of which it is a part, as well as in itself. But the value of a whole is not in any, or all, of its parts—though, of course a whole, of which there is a value, *may* possess parts of some, or of all, of which there are other values.

789. In the universe, as we have seen, there are parts of selves, selves, and groups of selves, and the universe itself is also a group of selves. Has any group of selves a value? (When there is a value *of* anything, we may speak of it as having a value.)

It seems clear to me that it has not, and that any value which is to be found in a group of selves is to be found in one of the selves which are members of the group. (Whether it is a value of that self, or of parts of that self, will be discussed later.)

790. It follows that the universe, which is a group of selves, has no value. But there is value in the universe, and this value forms a total or aggregate. To say that the universe has value is as incorrect as to say that a town is drunken. A town cannot be drunken, since the inhabitants, as a single substance, cannot drink at all, and therefore cannot drink to excess. But we can speak of the total, or average, drunkenness in a town, by adding together the drunkenness of the drunken inhabitants to get the first, and by comparing this with the total number of inhabitants of the town to get the second. And either of these would often be spoken of as the drunkenness of the town, though, as we have

seen, this would be incorrect. In the same way, people often speak
loosely of the value of the universe, even when they hold that
only selves have value. But they should rather speak of the value
in the universe.

What is true of the universe, in this respect, is also true of all
smaller groups of selves. A nation, a college, a bridge-party, have
no value. But there is value in all of them. For there is value in
the selves which are their parts, and value which is in the parts
is in the whole.

791. The doctrine that value cannot belong to the universe or
to societies, while it does belong to their parts, has often been
condemned as unduly atomistic. This, I think, has been due to the
mistaken supposition that the doctrine involves that the value
of a self is independent of the other selves with which he stands
in relation, or of the relation in which he stands to them. This,
of course, would not be true, but it does not follow from the
doctrine. Drunkenness can only be a quality of a man. It cannot
be a quality of a town. But the drunkenness of a man may be
largely determined, positively or negatively, by the character of
his neighbours and the institutions of his town. In the same way,
the value in each self does very largely depend on his relation to
other selves. But this is not inconsistent with the fact that all
value must be in a particular self.

792. But, it may be asked, should not the atomism of value be
carried still further? Is it not the case that it is only parts of
selves of which there is value, and that there is no value of selves,
though there is value in them? Let us consider this separately
with regard to those parts which appear as simultaneous in a self
(as when a man has a pleasant sensation and a virtuous volition
at the same time), and those parts which appear as successive
stages in his life. (The latter, of course, are really terms in the
C series, and the former are parts of such terms, determined by
determining correspondence.)

793. Taking the first of these, there are some considerations
in favour of the view that the values are of such states, and not
of the selves of which they are parts. We habitually speak of a
feeling of pleasure or of a virtuous volition, as good. And if the
same self has, at the same time, a virtuous volition and a pleasant

sensation, we consider his good as composite—as consisting of a
good referred to his virtue, and of a good referred to his pleasure.
Moreover, if he is at once virtuous and miserable, we regard his
condition as a mixed state of good and evil—good in respect of
the virtue, and bad in respect of the misery.

This, however, could be accounted for otherwise. For we
habitually speak of, *e.g.*, benevolence as good, although a quality
cannot itself have value. What we mean is that the possession of
the characteristic of benevolence makes a self good. And, in the
same way, when we say that a feeling of pleasure or a virtuous
volition is good, we may only mean that a self is good who has
such a feeling or such a volition as one of its parts—in other
words who is happy or who wills virtuously.

794. And there are positive arguments on the other side. It
may be said that we have already found that the selves, being
primary parts, are metaphysically fundamental, in comparison to
all other substances, and that there is therefore a presumption
that they are ethically fundamental, by being the only substances
that have value. I do not think, however, that much weight can
be placed on this argument.

795. There is a stronger argument on the same side. If there
were no value of a self, but only in it, then the value of the feeling
or volition would be no more closely connected with the self than
with any other group of substances, however unimportant and
fantastic, of which the feeling or volition was a member—such
groups, of course, being infinite in number.

A corresponding fact *is* true when the value in the self is taken
as the unit. That value is in the universe, of which the self is a
part. It is also in the nation to which he belongs, and in every
other group of selves of which he is a member, and it does not
belong to any one group more than to any other. But can we say
it would be the same here? Can we say that the value determined
by a virtuous volition is not more closely connected with the self
of whom the volition is a part, than with the universe, or with
any group of selves, or parts of selves, taken at random, of which
the volition in question is a part?

796. In answer to this it might be said that the unity of a self
has been shown in the earlier parts of this work to be closer and

more intimate than the unity of any other substance, since parts
of the same self are in the same fundamental differentiation of
the universe, while parts of different selves are in different fundamental differentiations. And this, it might be contended, would
be sufficient to make the value due to a virtuous volition more
closely connected with the self who had it, than with any other
compound substance in which it was contained, even if the value
was a value of the volition, and was only a value in the self.

I do not regard this answer as satisfactory. For, although no
unity is so close as that of the parts of a self, yet other unities
vary very much in their closeness. The citizens of a nation, or
the members of a cabinet, form groups which have much greater
internal unity than the group of all the red-headed men in
Europe. Yet the value determined by the virtue of a red-headed
cabinet minister is clearly no more closely connected with the
cabinet, than with the group of red-headed men[1]. The value in
the cabinet is as obviously a mere aggregate as the value in the
group of red-headed men.

On the whole, if we confine our attention to the parts which
we have just been considering—those which appear as simultaneous—the balance of the arguments seems to incline to the
view that such parts have not themselves value, but that the fact
that they have certain characteristics affects the value of the
selves of which they are parts.

797. But we have also to consider the case of the parts which
appear as successive, and which are in reality the stages of the
C series. The arguments for holding that these states have not
value, though their nature determines the value of the self, are
the same as before. In the first place, the selves, as primary parts
of the universe, have the same metaphysical predominance over
the parts which appear as successive as they have over the parts
which appear as simultaneous.

In the second place, the value which is determined by one of
the successive parts of a self is much more closely connected with

---

[1] I mean by this that it is no more closely connected with the cabinet than with
the other group by any relation of predication or inclusion. It is, of course, much
more likely to be causally connected, whether as determining or as determined,
with characteristics of the cabinet than with characteristics of the group of red-
headed men.

that self than it is with any of the other wholes of which that part
of the self is also a part. An example of this is the great difference
which is universally admitted to exist between the conduct of a
person who sacrifices his own good at one time to obtain greater
good for himself at another time, and the conduct of a person who
sacrifices his own good to procure greater good for another person.
Both sacrifices would generally be admitted to be right actions,
and to show good qualities in the agents. But while the first
would be held to show only prudence and self-control, the second
would be universally held to manifest a quality which is quite
different—the quality of unselfishness. So different is this held
to be from the others, that, while it is universally allowed that I
can sacrifice my own lesser good for my greater good, it has not
infrequently been maintained that it is impossible for any man to
sacrifice what he believed, at the time when he gave it up, to be
his own greatest good[1]. I believe that this is erroneous, and that
real self-sacrifice is possible and frequent. But the fact that its
possibility has been denied shows how differently we regard it
from the sacrifice of a man's good for his own greater good. This
supports the view that the good is more intimately connected
with the self than with any other whole of which the C stage is
a part. It consequently supports the view that good and evil are
qualities of the self, and not of the C stages.

798. But the objections to this view, in the case of the successive
parts, seem to be much more serious than in the case of the simultaneous parts. When I consider a virtuous volition in the past I
say that the past, and the past alone, is good in respect of that
volition[2]. Or when I anticipate a pleasure in the future, I say that
the future, and the future alone, is good in respect of that pleasure.
Now, if these judgments are true, it cannot be the self which is
good. For my self is not only in the past, or only in the future.
It is the unity which includes all the stages in its C series, and
which therefore, *sub specie temporis*, appears as present, past,

---

[1] This view is not confined to psychological hedonists. It is found, for example,
in Green.

[2] I may, of course, argue that, having made such a volition, I am likely to make
similar volitions, and so argue to my present and future goodness. But this would
not be goodness in respect of my past volition, though it would be inferred from
that volition.

and future. If, therefore, the good were ascribed to the self, it would appear as present, past, and future; and not only as past or only as future.

If, then, the good is to be ascribed to the self, we must say that we are wrong in saying that the good determined by the future pleasure is as future as the pleasure. And the view that we are wrong in this seems very difficult to maintain.

799. It might be possible to say that the simultaneous and successive parts of the self were on different levels in this respect —that all values were values of the successive states of the self, and that the self, on the one hand, had no value of its own, though the values of its parts would be in it, while, on the other hand, the simultaneous parts of each successive state had no value, but would determine the values of the successive states. But the arguments in favour of the view that simultaneous parts have no values are not decisive, and perhaps there is no reason to draw any distinction in this respect between the two sorts of parts.

I am unable to come to any definite opinion on the point. Nor is it necessary for our present purpose. It is important to know whether it is or is not true that every value is in a self—*i.e.*, is a value either of a self or of a part of a self. But, as we have said, there seems no doubt that this question must be answered in the affirmative. And it will not make any difference to the conclusions we shall reach in the rest of this work whether the values are values of selves, or values of their parts.

800. It is to be noticed that, if the true view should be that the value was of parts of selves, and not of selves, this would not involve that in every case in which there was value in a whole, the value was not value of the whole but of its parts. If that were the case, nothing would have any value, since every substance is a whole which has parts. But there might be a value of a part of a self, even although that part had again parts.

801. If, on the other hand, the value is of the self, then, as was said above (p. 401), the relation of a state of the self to the value is that this state determines the value, though it does not possess it. And so a virtuous volition, or a feeling of pleasure, is not good, but the self is good because he has the volition or the feeling. And, whether this explanation is true or not about parts of selves, it is certainly true about the qualities of selves and of parts of selves. Value is clearly only of substances, whether those substances are selves, or parts of selves. But any particular value is only ascribed to a substance because of its possession of certain qualities. If this were not so, there could be no systematic study of ethics, which would be reduced to recording the isolated facts that Smith was good, and Jones bad, and Brown better than Robinson.

It is universally admitted that this is not the case, and that there are general laws by which the qualities of good and evil are connected with other qualities[1]. And when a quality is connected with good or evil by a general law, it is common to say that the quality itself is good or evil. Thus we say that benevolence is good, as well as that a benevolent man is good. Indeed we assert the goodness of the quality more confidently and categorically than we can assert the goodness of the man, or of his state when he is benevolent. For, if misery and cowardice are bad, and the benevolent man was at the same time miserable, or cowardly, or both, he, or his state, might on the whole be not good but bad.

But, in the strict sense, a quality is never good or bad. Benevolence has not goodness for one of its qualities, as it has the qualities of being a quality, and of being a quality of volitions, and of involving a relation to selves other than the benevolent

---

[1] This, of course, leaves the question open whether any of these laws are self-evident *à priori*, or whether they are all obtained by induction from particular judgments of the type that A, who has the characteristic X, is good, that B, who has it also, is good, and so on, leading to the conclusion that whatever has the characteristic X is good.

Again, the assertion that there are general rules in ethics does not involve the assertion that general rules can be found applicable to every ethical problem. It may be the case that there is a general rule that happiness is good, and another that virtue is good, but that there is no general rule as to their comparative goodness. In that case the question whether it would be better that a man should lose a particular amount of happiness rather than commit a particular vicious action could not be solved by general rules, but must be determined, if it is determined at all, by an ultimate particular judgment of value.

We may note in passing that a complex quality is still a quality, and can enter into a general law. Thus, *e.g.*, it does not follow that, because happiness is in some cases better than misery, it is always better than misery. It might be the case that happiness with virtue is better than misery with virtue, but that happiness with vice is worse than misery with vice. This, or something like this, is asserted by believers in vindictive punishment, and, whether it is true or false (it seems to me patently false), it is as much a general law as the proposition that happiness is always better than misery.

person. And, when we say that benevolence is good, what is meant is that, if a man is benevolent, he is, in consequence, better than if he were not benevolent.

802. Another question arises. All value either is of parts of selves, or is of selves and determined by their parts. But, it may be asked, are only conscious parts included in this statement, or are unconscious parts included also?

The phrase "unconscious states of selves" is used ambiguously. It seems clear that it ought to designate those states which have not consciousness as a quality. And it is sometimes used in this sense. But it is often, perhaps generally, used in another sense— to designate those states which are not *objects* of cognitions. And even then it is generally confined to some of such states. In the first place, it is generally confined to those states which are not *perceived* by the self who has them, regardless of the question whether they can be known by inference. And, in the second place, it is generally confined to those states which not only are not perceived, but could not be perceived, excluding such as are not perceived because the self does not happen to be engaged in introspection, or in introspection in that particular direction. It is obvious that there is a great difference between saying that A's state B has not consciousness as a quality, and saying that A's state B is such that A can never have another state C which is a perception of B. But the difference seems to be often ignored[1].

803. I do not think that it is necessary for us to discuss the question whether value can be determined by an unconscious state, in the proper sense of the term. For we are only dealing with the existent, and it seems clear to me that no such state could exist. It is not only that we do not know and cannot imagine what a substance could be like which was part of a self without being conscious, but that the qualities of being a part of a self and being a state of consciousness are positively connected, so that the first implies the second. And it must be remembered that we have come to the conclusion that all parts of selves are

---

[1] Dr Freud distinguishes these ideas clearly. Having first explained why he considers it necessary to hold that there is *ein unbewusstes Seelisches*, he then discusses (and rejects) the view that this can be *ein unbewusstes Bewusstsein*. (*Sammlung kleiner Schriften zur Neurosenlehre*, IV, pp. 294–300.) The distinction, however, is harder to make in English than it is in German.

perceptions or groups of perceptions. How could it be maintained that I could have a perception of anything without that perception having the quality of consciousness?

804. Another question, of course, remains. Can value be determined by a state which is such that the self of which it is a part can never have a perception of it? There seems no reason to deny that such a state could happen. But it could never happen in a self which is self-conscious. For we have seen that any self who perceives a self perceives all the parts of that self. If, therefore, he perceives himself he must perceive all his parts. And therefore the question is of no practical interest to mankind, since we are all self-conscious.

805. It may be objected that, in arriving at this conclusion, we must ignore the whole doctrine of unconscious states of the self which has grown up in recent psychology. To do this, it might be said, would be untenable, in view of the extent to which conclusions, based on the theory of unconscious states, have been verified by subsequent experiment. But our view does not conflict in any way with psychological results. In the first place, there is no such conflict when we assert that there are no parts of the self which have not the quality of consciousness. The position of the psychological advocates of unconscious states is, if I understand it correctly, that the relations to each other of the states which we know by introspection, are such that no simple and harmonious theory can be formed to connect them which does not also connect them with other states of the self, of which we have no direct knowledge, and which we only believe in because their existence is necessary for such a theory. And it is clear that this only requires that there should be states which are not objects of perception, and leaves the question quite open whether they are or are not states of consciousness.

806. But, at any rate, it may be replied, the psychological theory requires that there should be parts of the self which are not objects of perception. In this respect, it would be said, it does conflict with our view that no self who is aware of himself can have any parts which he does not perceive. But even here there is no conflict. For we have seen (Chap. L, p. 265) that our perception of much of what we do perceive may not perceive it as being the separate objects which it really is, but may perceive it only as a vague background to our more definite perceptions. Now all that the psychological theory requires is that there should be states in us whose existence and nature are not known to us by perception, but can only be reached by inference. And, if these states are only perceived by us as a vague undifferentiated background to our more definite perceptions, it is clear that their existence as separate things, and their nature, cannot be known to us by perception, but only by inference.

There is thus no conflict between the psychological theory of unconscious states of the mind, and our assertion that all parts of a self have the quality of consciousness, and that all parts of self-conscious selves are objects of consciousness to that self. Both may be true. Whether the psychological theory is true or not— whether there are such parts of the mind. and whether they do act in such a way—is a question, not for philosophy, but for science.

807. But, although there are no unconscious parts, and although we, who are aware of ourselves, have no parts which are not objects of our consciousness, yet, as we have just said, we may have, and indeed certainly do have, parts which we do not perceive as being separate things, or as having the nature that they do have. And as such parts are states of consciousness, they will determine value. On one of the two alternative theories discussed above they will have values themselves. On the other theory they will determine value in the selves of which they are parts.

It might be thought that this would destroy any possibility of making any judgment as to my good or evil state. If I can have states of happiness or misery, of virtue or vice, of which I can know nothing by introspection, may it not be the case that my state is much better or much worse than I suppose it to be? And, in particular, may I not have moral defects which may render me very wicked, without knowing that I am wicked at all, and, consequently, without any chance that contrition may produce amendment?

808. If such a result were true, it would certainly be unpleasant, but that gives us no reason for not believing it to be true. There are, however, other grounds for disbelieving it. If some of my states cannot be perceived by me as being, as they are, separate states, while others can, there seem to be only two reasons which could account for the difference. The state I misperceive in this manner must either be itself fainter and more confused than the other, or else it must be more separated from the perception of it. As both the percepta and the perceptions are in the same self, the only way in which the perceptum can be separated from the perception is by their position at different points of the C series, and so of the apparent time-series.

A state, then, which I cannot perceive as a separate state must either be faint or confused, or else must be in the future or the past. As to the latter, the fact that our past and future lives contain many elements affecting their value which we do not know in the present, is by no means paradoxical—we know this fact already, independently of any philosophical theory, nor does such ignorance bring any confusion or uncertainty into the moral life of the present. As to those states which are, *sub specie temporis*, present, and yet are not perceived as separate states, they must, as we have said, be faint and confused; and the more faint and confused a perception is, the less will its existence affect the total value in the self who has it. The existence of such perceptions, therefore, will not make our estimates of our present condition valueless, for, although they will affect the value of that condition, they will not affect it greatly. And as to their existence making our estimates not absolutely correct, we knew beforehand that our estimates are often more or less erroneous.

809. Both good and evil are quantitative. Of two good men one may be more good than the other, and of two evil men one may be more evil than the other. Good values then form a series, and so do evil values. And these two constitute together the single series of values, of which the generating relation is "better than" or "worse than." Of any two values, good or evil, one will be better than the other, which will be worse than the first.

This raises the question whether, after all, there are in reality two sorts of values, good and evil, or whether there is no such distinction, and no such positive qualities as good and evil but only relations of better and worse between values not qualitatively unlike. In a series of magnitudes each is larger or smaller than each of the others. But no magnitude is positively large or small. Is the series of values like this?

810. The acceptance of this view would be consistent with maintaining that *qualities* were positively good or positively evil. For, as we have seen, all that is meant by a quality being good or evil is that it makes any substance which possesses it better or worse than it would have been otherwise. And if it is true that selves, or parts of selves, can be better or worse than others, it can be true that a quality will make a substance better or worse than it would have been otherwise. Qualities, consequently, could be positively good or evil in the only sense in which good or evil could ever be attributed to them.

Now for practical purposes the goodness or badness of the qualities is more important than the goodness or badness of the substances which possess the qualities. For it is not so important, for practical purposes, to know whether the state of things is good or bad, as to know how they can be made better or prevented from becoming worse, and this depends on the nature of the qualities. Thus it would make little practical difference if we came to the conclusion that selves or parts of selves were not positively good or bad, but only better or worse than others.

811. But there is a difficulty. We certainly make judgments that the value in a self or an aggregate of selves is positively good or positively evil. Now if there is in reality no positive good or positive evil, we must treat such judgments, not as being objectively true, but as being true only with reference to some arbitrary standpoint. If I say that the value in *A* is a good value, this will have to mean that it is better than the average value of the persons with whom I am acquainted, or that it is a value with which I should not be dissatisfied if it was the value in myself, or something of a similar nature.

Our judgments that anything is large, and not small, or that it is hot, and not cold, can, no doubt, be explained in this way. But can we use the same explanation for our judgments of good and evil? I do not think that we can. For we make such judgments as that, if the values of the universe were of a certain sort, it would be desirable that there should be no universe at all. And, again, we make such judgments as that, if a single indi-

vidual had a nature of a certain description, it would be desirable
that he should not exist at all (considering only the intrinsic
value in that nature, and eliminating any question of his utility
or disutility to others).

Now such judgments could be justified if some values are
positively good, and some positively bad, since it is, I suppose, a
self-evident synthetic proposition that it is desirable that what is
good should exist, and that what is bad should not exist. But I
cannot see how such judgments could be justified if actual and
possible substances do not differ as good and bad, but only as
better and worse.

Either, then, we must hold that values are positively good and
bad, or we must reject all such judgments as invalid. Now though,
no doubt, any particular judgment of this sort may be mistaken,
yet it does not seem permissible to suppose that they are all
necessarily false, owing to the absence of any standard by which
the absolute desirability of anything could be decided. We must
therefore accept the usual conclusion that values are either
positively good or positively evil. Although they are thus qualitatively different, yet, as we have seen, they form a single series
which passes from one to the other through a zero value which
is neither good nor bad[1]. In this respect it resembles various
other series, for example the pleasure-pain series, whose terms
are pleasures on one side of the zero term, and pains on the other.

812. We saw above (p. 398) that good and evil cannot be defined in terms of anything else. But can we define either of them
by means of the other? If we had the idea of value, and the idea
of good, and the idea of variation in the amounts of value and
goodness, could we by means of them define evil? This could not
be done. For, as we have seen, evil is not identical with less good.
Of two things which are less good than some positively good
thing one may itself be positively good, while the other is positively evil. Both good and evil, then, must be pronounced to be
strictly indefinable, though, when both good and evil are known,
it would be possible to define value in terms of good and evil[2].

[1] Although the zero term in the series is neither good nor evil, it is nevertheless
a value. Whatever has zero value is better than anything which is evil, and worse
than anything which is good.

[2] It has sometimes been asserted that, if we did not know evil, we could not

813. What are the qualities which are good or evil—that is, which give a good or evil value to the selves, or parts of selves, which possess them? There are, of course, many different views on this question. But I think the following list will include all which have received any support[1]. Firstly, it has been held that knowledge is good, and that error is bad. Secondly, that virtue is good, and that vice is bad. Thirdly, that the possession of certain emotions is good, and that the possession of others is bad. Fourthly, that pleasure is good, and that pain is bad. Fifthly, that amount and intensity of consciousness which we may call "fullness of life" is good (to this characteristic there is no converse which is held to be positively bad). Sixthly, that harmony in consciousness is good, and disharmony, I suppose, bad.

The third, fourth and sixth of these are sometimes maintained, not only to be good and evil, but to be the only good and evil. I do not think that this is ever maintained of the second, without qualification, nor of the first or the fifth.

The view which I should myself accept is that the characteristics coming under the first five heads are all good or evil respectively, though their importance in respect of value varies. (I should reject the sixth, because I can see no good or evil under this head which does not come under one of the other five.) When good of one sort is incompatible with good of another sort— when, for example, the only alternatives are that a man should have a particular evil volition or submit to a particular unhappiness—I think that we can find no general rules to guide us, but must decide each question by a particular ultimate judgment of value.

It will not, however, be necessary for us to discuss whether this, or any other particular view, as to the qualities which

---

know good. This seems false. No doubt if every reality was good, and equally good, we might have no idea of good. just as we might have no idea of blue if everything was the same shade of blue. But some things which are real, or appear to be so, such as a nation, a chair, and the multiplication table, have no value; and the things which are good have various degrees of goodness. This would, I think, be sufficient to permit us to have a distinct idea of good. And we must remember that, even if a man did not have an idea of good, and so did not know that his state was good, this would no more prevent him from being in a good state than a man is prevented from being humble by his ignorance that he is humble.

[1] Subject to two qualifications which will be discussed later. (Chap. LXV, p. 433, and Chap. LXVI, p. 443.)

determine good or evil, is correct. For we shall see that our judgment as to the relative amounts of good and evil in the universe will be the same, whether we hold that good and evil are determined by qualities under all these heads, or by qualities under any one head, or under any combination of heads.

814. Has the series of values any boundary—whether an extreme term or a limit—in either direction? Have we any reason to think that there is any degree of good so good, or of evil so evil, that no further degree of good or evil is possible? The answer to this question, it seems clear, is in the negative. None of the characteristics enumerated in the last section is such that there is any intrinsic limit to the amount of it possessed by any self. And the amount of good or evil will increase—I do not say increase proportionately[1]—with the amount of the characteristics.

If we consider knowledge, it is clear that there is no amount of knowledge which is the greatest possible. Even if a self should perceive every self in the universe, and every part of every self, and should perceive them as having all the characteristics which they do have—even then he would have a greater amount of knowledge if, in addition to these selves, there were also other selves which he knew. For no number of selves in the universe is the greatest possible number.

So also with emotions. The emotion which is most generally held to be intrinsically good is love. And the amount of love, in the case of any particular self, varies with the number of people he loves, and the intensity with which he loves them. Even if he should love all the other selves in the universe, his love would be greater if there were other selves, whom he also loved. Nor is the intensity of love capable of a maximum. There is always an intensity of love greater than any given intensity[2].

Pleasure varies in amount with the number of sources of pleasure, and with the intensity of the pleasure excited. The number of sources of pleasure is capable of indefinite increase, for, even if I received pleasure from my relations with every other self in the

---

[1] Cp. Chap. LXV, p. 438.

[2] If it should be held that what is good is not love in general, but love of some one person, *e.g.*, of God, such love, though it would necessarily be limited to a single person, would have no intrinsic maximum of intensity.

universe, the pleasure would be greater if there were also other selves, from my relations to which I also derived pleasure. Nor is there a maximum intensity of pleasure any more than a maximum intensity of love.

As to amount and intensity of consciousness, there seems to be no limit to the intensity of my consciousness of any object, any more than to the intensity of pleasure or of love. And, whether this is so or not, it is clear that there can no more be a maximum number of objects of consciousness, than of objects of love, or sources of pleasure, since there is no maximum number of possible selves in the universe.

Harmony might, I suppose, be absolutely complete, and so incapable of any further addition. But, when harmony is taken as being something which is ultimately good, it is, I suppose, considered good not only in respect of the harmony being unbroken, but also in respect of the number and magnitude of the parts of the harmonious whole. And there is no intrinsic limitation of these.

815. The case of virtue is not quite so simple. If by calling a man virtuous we mean no more than that he always desires what he believes to be the good, and that he always carries it out, so far as depends on his will, then virtue obviously has a maximum. If a man never fails to do this, there is no greater degree of virtue than his. Even if we should add that, to be perfectly virtuous, a man must not only desire what he believes to be right, but must have correct beliefs as to what is right, it would be possible that he should always judge rightly about this, and here, too, virtue would admit of no further increase.

This, however, would not enable us to find a maximum good. For it is never maintained that virtue is the only good. Even Kant admits that it is good that the virtuous should be happy. And, as we shall see in the next chapter, the view that virtue is the only good is untenable. Even, therefore, if we could find a maximum of virtue, we should not have found a maximum of good.

816. And, in the second place, it would seem that we must admit more in virtue than has yet been mentioned. If two men should both invariably see and invariably follow the good, yet one must, I think, be accounted more virtuous than the other, if his resolve to pursue the good was so determined that it had resisted, or would have resisted, some temptation to which the other would have yielded. And, if these elements are to be considered part of virtue, it is impossible that there should be any maximum of virtue.

817. There is, then, no maximum good[1]. And, in the same way, we must conclude there is no maximum evil. There is no more an intrinsic limit to the number of false beliefs which a man can have than there is to the number of true beliefs, and error, therefore, is intrinsically as unlimited as knowledge. And there is no intrinsic limit, either to the number of persons I hate, or to the intensity with which I hate them, while pain is intrinsically unlimited in the same way in which pleasure is. Again, the evil of disharmony would, I suppose, increase in proportion as the elements which were inharmonious increased in intensity, and it does not seem that there could be any intrinsic limit to the latter intensity. Nor, with respect to vice, is there any intrinsic limit to the amount of good which a man may be prepared to destroy to attain some particular vicious end[2].

818. Thus we conclude that there is no complete good, and no complete evil. There is no good or evil which is such that it is impossible that there should be a good or an evil greater than it. This does not mean that there is not a greatest possible good or evil, in the sense that the nature of the universe is such as to exclude the occurrence of any greater. Indeed, it is clear that the nature of the universe excludes the occurrence of any good or evil greater than that which does occur, as it does of anything else which does not occur. What is meant is that the value series has not, in either direction, boundaries imposed by its own nature, or, as we have called them above, intrinsic boundaries. Some series have such boundaries. We saw, for example, that the inclusion series had such a boundary. For the terms of that series contained,

---

[1] We shall see, in Chapter LXVII, that, in another dimension, existent good is infinite. But still there is no maximum good. For there would be additional good, if the good, which is infinite in this dimension, should be greater in another.

[2] If the amount and intensity of consciousness is good, there is no evil corresponding to it. There would be some good while there was any consciousness, and when there was no consciousness there could be no evil.

as they passed from less inclusive to more inclusive, more and more of the content of the whole in which they fell, and, when we reached the term which included all that content, it was intrinsically impossible—impossible from the nature of the series itself—that there should be any further terms. So also with the series of closer and closer approximations to the correct solution of a problem, which obviously has its limit in the correct solution itself. But there is nothing in the nature of good and evil which prevents them from increasing as long as certain other characteristics increase. And there is nothing in the nature of those other characteristics which prevents their increasing beyond any point.

We must therefore say that, however great the goodness or the evil in any self, it can never reach complete good or evil, since there is no such completeness. And this is true *à priori* of the value in the universe. For, even if there were a complete good, and it should be attained by all selves, or parts of selves, in the universe, the aggregate of goodness would be greater if the universe contained other selves in addition to these.

819. But, although there cannot be a state of complete good or evil, there can be a state of unmixed good or evil. A man is in an unmixed state of good when he possesses qualities which are good, and none which are evil. Thus, if we assume, to get a simple example, that pleasure and pain were the only good and evil, a man would be in an unmixed state of good if he had pleasure and no pain, and if the only limitation of his good was the fact that there were greater possible amounts of pleasure than the amount which he was enjoying. A state of good which is not unmixed may, of course, be greater than one which is unmixed. A man who does experience pain may experience so much more pleasure that, on the whole, his good, according to a hedonic standard, may be greater than that of a man who experiences no pain, but much less pleasure than the former. And, even if virtue were the only good, a man who committed some sins might be better than a man who committed none, if the devotion of the first, in the cases in which he was virtuous, and the sacrifices he was prepared to make for virtue, greatly exceeded those of the second.

820. There is likewise a justifiable use of the expression that a certain state is very good, or very bad, though the standard must be arbitrary, and must depend on the good and evil experienced by, or known to, the person making the judgment. Thus I may say that our state in heaven will be very good, if I believe, for example, that its superiority in value over the best state known to us on earth, is much greater than the difference in value between the best and the worst states known to us on earth.

821. We have now determined, so far as is necessary for our immediate purpose, the nature of good and evil. In what way can we hope to determine anything as to the relative amounts of good and evil in the universe? (In putting the question in this form, we must remember that we have not yet excluded the possibility that the universe is unmixedly good, and that nothing which exists is in any respect evil. Nor have we excluded the contrary possibility that the universe is unmixedly evil. We shall see, in Chapter LXVI, that both these hypotheses are untenable.) The value in the universe is the aggregate of the values of all its parts which have value—parts which are either selves or parts of selves. Can we find any connection between the characteristic of existence, as it occurs in selves or parts of selves, and the characteristic of value, which will enable us to decide as to the proportion of good and evil in that aggregate?

822. Is there, to begin with, any analytical connection between the characteristic of existence, on the one hand, and either the characteristic of good or the characteristic of evil, on the other? This could only happen in one of three ways. Either existence must be the same thing as good, or as evil; or good or evil must enter into the definition of existence; or existence must enter into the definition of good or of evil. Neither of these is the case. Existence certainly does not mean the same, either as good, or as evil. And we have seen in this chapter that good and evil are indefinable, and we saw previously (Section 5) that existence was so. There is thus no analytic connection.

823. Can there, then, be a synthetic connection between existence and either good or evil, which will enable us to succeed in our object? In the first place, let us enquire if any such connection is immediately evident—if there is any proposition about the relation of good or evil to existence which is immediately and self-evidently true.

Such a proposition might be one of two sorts. It might assert that the quality of existence was itself good or evil, so as to determine the state of all that exists to be good or evil. Or it might assert that whatever had the quality of existence must have also some other quality which made it good or evil.

The connection of existence and goodness sometimes has been asserted by a proposition of the first sort. It has been maintained that existence, as such, is so good that whatever exists, and has value at all, is more good than bad, so that it is always better to exist, under any circumstances, than not to exist. This, of course, is more than to say that existence is a good quality. It is to say that it outweighs all possible bad qualities, however great their intensity, and ensures that whatever has it is, on the whole, good.

Such an assertion as this would be an ultimate judgment of value. All that I can say is that I do not judge it to be true, and that I believe that most persons would not judge it to be true[1].

The contrary belief—that existence as such is so evil that all that exists must be more evil than good, whatever its other qualities—has never, so far as I know, been maintained by any thinkers. Even the most thoroughgoing pessimists regard the existent as necessarily evil, not because the quality of existence is itself evil, but because it is necessarily accompanied by other qualities which are evil in themselves.

824. There remain propositions of the second sort—assertions that it is immediately self-evident that whatever has the quality of existence must have also some other quality which makes it good or evil. What is asserted is not, in most cases, that it is immediately self-evident that what is existent must have certain qualities, and that these qualities are, as a matter of fact, such as to make the existent good. The more usual form is an assertion that it is immediately self-evident that what exists must be

---

[1] Even for those who do accept it as true the belief would have no very comforting effect. A man might hold that it was better to exist unendingly in the hell of mediaeval Christianity than to cease to exist altogether, but the prospect of such an existence in the future would scarcely be compatible with much happiness in the present.

good, together with an assertion that it is immediately self-evident that it cannot be good unless it possesses a certain quality, or one of certain qualities. From these two propositions, accepted as immediately certain, the conclusion is deduced that it must possess that quality, or one of those qualities[1].

As the proposition that what is existent must be good is asserted as self-evident, it is impossible to argue about it. All that I can say is that I see no such self-evidence, and that I see no intrinsic impossibility in the assertion that the universe is evil, or even that it is very evil, although, for reasons which will be given in later chapters, I believe that it is, in point of fact, very good.

825. If there is no immediate connection between existence and value, can we discover any connection which is mediate, and can be demonstrated by reasoning? Various efforts have been made to establish such a connection, several of which rest on the view that evil is the mere absence of good. From this it is argued that a self, or an aggregate of selves, which contains any good, must be good on the whole, since what is called evil is nothing positive which can counterbalance the good, but only the absence of more good. Now there certainly is some good in the universe (cp. Chapter LXVI). And thus the total value in the universe is unmixed good. What is commonly called the presence of evil only means that complete good has not been obtained. And that we have already seen to be impossible.

But the assertion that evil is only the absence of good is quite untenable. It would involve that whatever had no good value would have a value of unmixed evil, and therefore that a piece of matter, if it could exist, or the multiplication table, would have a value—a value of unmixed evil. And this is clearly false.

There are other arguments which attempt to prove this conclusion without falling into this fallacy. Of these, I think, much the most important is Mr Bradley's (*Appearance and Reality*, Chapter XIV). But I do not think that even this can be accepted as valid.

---

[1] Sometimes the first assertion is less ambitious, and it is asserted, not that the existent must be good, but that it cannot be very bad.

826. In the present work, however, we have reached various conclusions as to the nature of the existent. These conclusions, as we have seen, do not admit of absolute demonstration, but we have found reason to accept them as true. Do they afford a basis for determining the proportionate amounts of good and evil in the universe? Our remaining chapters will be devoted to this enquiry.

CHAPTER LXV

VALUE IN THE FINAL STAGE OF THE *C* SERIES

827. In considering the value of the existent, let us begin by considering the value which is in each self in respect of the final term of the *C* series—that term which contains the whole content of the self, and which, as we have seen, appears, *sub specie temporis*, as the latest term in the series, and as itself of unending duration.

We saw, in the last chapter, that various qualities have been held to determine that which has them to be good, while their contraries determine that which has them to be bad. They are knowledge, virtue, the possession of certain emotions, pleasure, the amount and intensity of consciousness, and harmony. It will not be necessary to decide whether all of these qualities are good, or whether only some of them are. For we shall see that, whatever theory we hold on this point, the result will be the same— that the final stage of the *C* series must be pronounced to be a state of great good, and, with one possible exception[1], of unmixed good.

828. It will be necessary to postpone the consideration of virtue until the others have been dealt with. For, in order that a self should be virtuous, his volitions must be directed towards what is good, or at least, to what appears to him to be good. Now, as we saw in Book V, the form which volition must take in absolute reality is that the self acquiesces in all that he perceives, and in nothing else. He will, therefore, not be virtuous unless that which he perceives is either good or useful or appears to him as such. And, as in the final stage there is no error, nothing can appear to him to be good or useful unless it really is so. He will, therefore, not be virtuous unless all that he perceives is good or useful. And therefore we must determine the goodness of the final stage in other respects, before it can be determined whether that stage is virtuous.

---

[1] The exception in question is that which might arise from pain caused by sympathy for the evil in pre-final stages. Cp. Chapters LXVI and LXVII.

829. It follows from what has been said that, if virtue were the only good, there could be no good in the final stage. Nor, if virtue were the only good, could there be any good in any other case where there was no error. For whenever there is no error, nothing can appear to be good unless it really is so, and virtue can be nothing but acquiescence in the good. No doubt I can acquiesce in the existence of a volition, and a volition may be good. But if we attempt to dispense in this way with any goodness but that of volition, we are involved in a vicious infinite. A volition, J, may be an acquiescence in the existence of another volition, K (in the same self or in another). If J is to be virtuous, K must be good. But, since virtue is the only good, K can only be good if it is an acquiescence in something which is good. Should this something be J, we have made a vicious circle in our attempt to establish the goodness of J. For it depends on the goodness of K, which again depends on the goodness of J. It must be established as a condition to its own establishment. If, on the other hand, K is an acquiescence in another volition, L, and so on, we shall have a vicious infinite series. For the goodness of J cannot be established except by establishing that of K, which can only be done by establishing that of L, and so on. Thus the goodness of J cannot be established except by reaching the last term of a series which has no last term.

In a state which allowed of error, on the other hand, it would be possible that something should be good, even though virtue were the only good. For a man might desire something else than virtue—his own pleasure, or other people's pain, or the breaking of the greatest possible amount of crockery—under the mistaken belief that these things were good. And then he might be held to be virtuous, since he was desiring what he believed to be good. He would then be really good. It might, indeed, be doubted whether such virtue would be unmixed, for it might be held that the correctness of judgments of value enters into virtue, and that a man could not be altogether virtuous who devoted himself to torturing his innocent grandmother, with however good a conscience he did it. But, even if a good conscience is not the only thing which is relevant to virtue, it certainly is relevant to it, and a man who tortured his grandmother in the belief that it was good to torture her would certainly be more or less virtuous.

Good, therefore, could be existent, even if nothing were good but virtue. It could exist only so long as its nature was not recognized, and a man who knew what virtue was, could never be virtuous. But such a result is not, so far as I can see, logically impossible.

The view that virtue is the only good, however, is held by few, if any, moralists. Virtue is often asserted to be the highest good, and sometimes asserted to be even incommensurably more important than any other, but it is seldom, if ever, denied that pleasure, at any rate of the virtuous, is good, and that pain, at any rate of the virtuous, is evil.

830. Postponing virtue, then, let us consider knowledge. In the final stage every part of the self is a cognition, and a correct cognition, and therefore we shall have knowledge and no error. We cannot say that there is no ignorance, for it is possible, as we have seen, that there are substances in the universe which we do not know, even indirectly. Nor is it certain that the substances which I do perceive are perceived by me as having all the characteristics which they actually have. But ignorance cannot be held to be a positive evil, even if error is a positive evil. Ignorance could not be more than a limitation of the good. The presence of ignorance thus shows only that, if knowledge is a good, we do not attain complete good. And we have already seen that to reach complete good is impossible. Ignorance does not, as error would, prevent the good from being unmixed. It must also be remembered that ignorance cannot, in the final stage, involve the presence of unsolved problems, which may be a positive evil—though perhaps it is only a disutility, as leading to pain. But the consciousness of a problem, like any other question, is an assumption. And, in the final stage, there is not even the appearance of assumptions—all cognition not only is, but appears as, perceptions.

831. In so far, then, as knowledge is good, and error bad, the value in the self, in respect of the final stage of the *C* series, is unmixed good. It will also be a state of very great good, as compared to our present experience. In our present experience the perception caused in us by various substances is so confused that it does not give us separate perceptions of them (cp. Chap. L, p. 266). And therefore in our present experience we do not know these substances. But, in the final stage, all perception is clear and distinct, and we shall know all these substances. And the number of these is infinitely greater than the number known in present experience. For the number known in present experience is finite. But, in the final stage, we shall know the parts of parts to infinity of all substances that we know—that is to say, we shall know an infinite number of substances.

It is true that, in the final stage, there will be no knowledge by judgments. But, as we have seen, this will not prevent us from knowing characteristics, since we perceive the substances as having certain characteristics, and as having characteristics which have again characteristics. And in this way we shall know, and know without any error, all characteristics of which we have any knowledge at any stage in the C series.

832. Let us now consider emotions. In the final stage, as was pointed out in Chapter XLI, I shall feel emotion towards every substance which I perceive. For other selves whom I perceive directly I shall feel love, for myself I shall feel self-reverence, for other selves whom I perceive indirectly I shall feel affection, and for the parts of all these selves I shall feel complacency. Thus towards all these substances I shall feel emotions which would be admitted to be good by all persons who allow that emotions can be good at all.

There are certain other emotions which we saw (Chap. XLI, pp. 167–168) are not impossible in the final stage, though we cannot be sure that they will be there. These are sympathy, approval, disapproval, pride, humility, gladness, and sadness. Now none of these emotions is intrinsically bad. They may be condemned in certain cases, as based on mistaken grounds. But this cannot be the case in the final stage, since in that stage there is no error. And sympathy, disapproval, humility, and sadness sometimes give rise to pain, and so may have disutility, if pain is an evil. But they are not evil emotions.

On the other hand, there are emotions which would be admitted to be evil, if any emotions are good and evil (though, like other evil things, they may have utility if they promote other goods).

But I think no emotion would be pronounced evil which does not occur among the following—hatred, repugnance, malignancy, anger, regret, remorse, jealousy, envy, and fear. And we saw (Chap. XLI, p. 166) that none of these could occur in the final stage. Thus we are led to the conclusion, that, if emotion has value, the value in the self in respect of the emotional qualities of the final stage, must be unmixedly good.

833. It will also be a state of very great good. To begin with, we have seen (Chap. XLI, pp. 158–161) that love will be more intense, in the final stage, than it ever is in our present experience. And it will also be more extensive—a love, that is, of more persons. For it will be a love of every person who is perceived directly by the self at all. Now all knowledge, as we have seen, is really a perception, though generally a misperception, of selves or parts of selves. And the parts cannot be perceived without perceiving the selves of which they are parts. It is, however, possible that many of these selves may only be perceived indirectly—by perceiving other selves' perceptions of them. It is therefore impossible to decide how much of a man's present experience is direct perception of selves, not explicitly recognized as such. And, even if that could be decided, it could not be decided of how many selves it was a direct perception. Again, of those selves which are explicitly recognized as such, we cannot tell to what extent the same selves will be grouped together in different lives, and to what extent each life will involve meeting fresh people.

But still, when we consider how little of any man's present experience is of selves explicitly recognized as such, and when we consider how, even in a single life, the persons with whom we associate change from time to time, it seems in the highest degree probable that the number of selves whom he perceives directly is much larger than the number of those which, in this life, he explicitly recognizes as selves. And, in the final stage, he will recognize all of these as selves, and he will love them.

Even the selves which are met and explicitly recognized as selves in present experience are not all loved by the self who recognizes them, and, when love does occur, it is never unbroken, since no one thinks of any other person without intermission. Both these limitations will be removed in the final stage.

It seems certain then that, in the final stage, love will be more intense than at present, and almost certain that it will be more extensive. And it must be remembered that for a self perceived indirectly we shall feel affection, though not love. Self-reverence, again, will increase with the increase of love. If, therefore, the possession of emotions is good, the final stage will, in respect of emotions, be a very great good.

834. Our conclusion that the final stage will, in respect of emotion, be one of great and unmixed good may perhaps be thought to be overthrown by the possibility that love should not be reciprocated. (Chap. XLI, p. 158.) It is possible that B might perceive C directly, while C did not perceive B directly. In that case B would love C, while C would not love B. It is possible that C should perceive B indirectly. He might perceive D directly, and so perceive B through D's perception of B. And then he would regard B with affection, since he would love D who loved B. But he might not perceive B, even indirectly. And then he would not know anything of a person who loved him.

I think, however, that this possibility will not affect our conclusion that, in the final stage, love is a great and unmixed good. To begin with, it is clear that the fact that love is unreciprocated, should it be so, would not prevent love from occurring, or from being very great love. There are cases, no doubt, in our present experience, when the fact that love was unreciprocated has prevented it from becoming great, or has destroyed it altogether. But there is nothing in the nature of love which makes reciprocation necessary to it. It could exist without it, and it is higher in proportion as it can exist without it. Our argument, therefore, that there will be great love in the final stage, is not destroyed by the possibility that such love should not be reciprocated.

Nor will unreciprocated love introduce evil by introducing pain. In our present experience, no doubt, it can cause intense pain, but that is because we desire that our love should be returned, and because that desire is ungratified. In the final stage there can be no ungratified volitions, and so no desire that love should be reciprocated when it is not. And thus the want of reciprocation can cause no pain.

It does not seem, therefore, that the absence of reciprocation could introduce any positive evil in respect of the final stage. If indeed, *C* did love *B*, *C*'s state would be better, since he would love one more person than he does now. And it is possible that, if *C* loved *B*, this might increase *B*'s love of *C*. And it would certainly increase *B*'s happiness. But this does not show that, if the final stage contained unreciprocated love, it would have any evil in it. It only shows that it would not have complete good. And we have seen that complete good is in any case impossible.

835. The relations which would hold, in the final stage, between two persons, one of whom loved the other with a love which was not returned, would be different from those which would hold in present experience. For in that the normal case is that *C* knows *B* in the same way as *B* knows *C*, but that his cognition of *B* has not the quality of love, as *B*'s cognition of him has. Knowing *B*, he does not love him, and perhaps hates him. But, in the final stage, each self loves every self whom he perceives directly. And so, if *C* does not love *B*, he does not perceive him directly. And, if he neither loves him nor regards him with affection, he perceives him neither directly nor indirectly—*i.e.*, he does not know him at all.

836. The next point to be considered is pleasure. There will certainly be pleasure in the final stage. For there will be love, and, as was said earlier (Chap. XLI, p. 169), even if all love does not give pleasure, it seems certain that love must give pleasure when, as is the case in the final stage, it is not combined with ungratified volition. And the same can be said of self-reverence, affection, and complacency.

There is pleasure, therefore, but can there also be pain? It is certain that at this stage we acquiesce in all that exists. But this does not exclude the existence of pain. For it is possible to acquiesce in a painful state, even when, as here, there can be no malignancy joined to the acquiescence. A man may acquiesce in his own pain, *e.g.*, either because he believes that he deserves it, or because he believes that the pain is pleasing to God. But this acquiescence does not prevent the state from being painful.

No doubt acquiescence in a painful state diminishes the amount of pain which we should otherwise have, since it spares us the secondary pain of protest and revolt[1]. But the original pain remains pain, and, if all pain is bad, remains bad.

837. Thus pain is not excluded by acquiescence. But, apart from this, is there any room for pain in the final stage? It is clear, to begin with, that in that stage there can be nothing of what is commonly called physical pain. (The expression is rather unfortunate, since all pain, however caused, is a state of the mind.) Physical pain is a quality of certain perceptions which perceive their percepta as sensa. But the percepta are not really sensa, because there are no sensa. All such perceptions, therefore, are misperceptions, and, in the final stage, where there are no misperceptions, they cannot occur. In that stage, therefore, there will be no physical pain.

838. Is it possible, again, that the final stage should contain pain of any other sort? Our perceptions in this stage are all of percepta as they really are—as selves, or as the perceptions which are the parts of selves. Can any of our perceptions of these have the quality of pain? All our perceptions have as qualities either love, self-reverence, affection, or complacency; they can none of them have the qualities of hatred, repugnance, malignancy, anger, regret, remorse, jealousy, envy, or fear; and they can none of them be ungratified volitions. How can knowledge of selves—one's own or others—which had such a nature, positive or negative, be in any way painful?

The only way in which such a state could be painful, as far as I can see, is that it might have the pain of sympathy for some evil. Such sympathetic pain, however, cannot be the only evil. For with what would it sympathize? Since the final stage excludes error, it cannot sympathize with imaginary evil which did not exist. And, if sympathetic pain had nothing but sympathetic pain with which to sympathize, there would be a circle or an infinite series, either of which would be vicious. It must sympathize, then, with some other evil. And, as the final conclusion of this chapter will be that there is no other evil in the final stage,

---

[1] It does not necessarily follow that it diminishes the amount of evil, unless pain is the only evil. If the painful state in question is something in which we ought not to acquiesce, then the volition in question would be vicious, and, if vice is evil, then the evil of this might overweigh the good produced by the absence of secondary pain.

the only possibility left is that there might be in the final stage
the pain of sympathy for evil in the pre-final stages. To this we
shall return in Chapter LXVII.

839. With this reservation, the final stage must be one of
unmixed pleasure. And it would seem that it must be one of very
great pleasure, as compared with any pleasure which we experience at present. For it depends on the emotions of love, self-
reverence, affection, and complacency, which we feel towards the
objects we know, and we found reason to hold that these emotions
would be much greater in the final stage than they are in our
present experience. And, even in our present experience, we know
that love may yield a pleasure equal to any pleasure which we
now enjoy. We may therefore conclude that a pleasure in love
which was much greater than this would much exceed in intensity any other pleasure which we now enjoy.

840. We now come to amount and intensity of consciousness.
It has been held that this is in itself a good, independently of the
quality of the consciousness. Take, for example, two men, in each
of whom there was the same excess of virtue over vice, and of
pleasure over pain. It would be held that the value in one of
them was greater than that in the other, if the first led a life of
the fullness and intensity possessed by the average able man,
while the other's life was that of a man who was feeble-minded,
or always partially under the influence of drugs. And it would
also be held that there might be greater value in the first self,
even if it were, within certain limits, inferior in respect of the
balance of virtue or happiness.

To this positive good there is, as was mentioned above, no
opposite in the shape of a positive evil. If the amount and intensity of consciousness were continually diminished, we should
never pass through a zero point of value into evil, but should
arrive finally at the complete extinction of the self, which would
of course be the limit beyond which there could be no further
diminution. This would not be at a zero point in the scale of value,
but would have no place in the scale of value at all, since it
would not be a self, nor a part of a self.

If amount and intensity of consciousness is a good, it is a good
which every self must have in some degree, since every self is

more or less conscious. And, as there is no correlative evil, it must be unmixed so far as this particular sort of value goes, though, if it is not the only good, the self which has it may have some other characteristic which is evil. All, then, that we have to enquire is whether, in respect of the final stage, this will be a very great good.

841. The answer is that it will be a very great good. In the final stage, as we have seen, we shall know clearly and distinctly a part of the ultimate content of the universe which in our present experience we do not know clearly and distinctly. And, as of every substance that we know we shall know clearly and distinctly all its parts to infinity, we shall know clearly and distinctly an infinitely greater number of substances than at present. Our knowledge, also, of all these substances will have the intensity and directness of apparent perception, while at present much of it is in the less vivid form of apparent judgment. There will be emotion towards every substance which we perceive, and the sum of that emotion will be much greater than any in our present experience. Also we shall acquiesce in all that we know—there will be nothing which is merely neutral in respect of volition, as so much of our present experience is. And, as will be pointed out later, we shall acquiesce much more intensely than at present. Thus the amount of our consciousness which is clear and distinct will be infinitely greater than at present, and its intensity will be greater than at present, so that the good arising from it will be a very great good as compared with our present experience.

842. There remains the view that harmony, or, as it is sometimes called, harmonious self-development, is intrinsically good. (It is sometimes, indeed, said to be the only good.) I find it difficult to gather from the works of the advocates of this view exactly what is meant by such a harmony, nor am I able to see for myself what sort of internal harmony could exist in the self which could be called a distinct sort of goodness, differing from, and possibly replacing, the other sorts of goodness of which we have spoken.

I do not find it easy, therefore, to determine whether, according to this standard, our state would be good in respect of the final stage. But it seems to me that the place of the self in the final stage is one which ought to be called internally harmonious. In that state there will be no conflict, no sacrifice of one of two incompatible goods to assure the other[1]. Again, all good would depend on a single characteristic of the self—its perception of selves and their parts. And the perceptions of selves and their parts constitute the entire content of each self. Thus we get an absence of conflict, and a positive unity in the internal nature of the self, which perhaps have a fair claim to the title of harmonious.

Thus, with regard to every quality which has been asserted as good, except virtue, we have found that the value in every self, in respect of its final stage, must be held to be a good which is very great, and which, with a possible exception due to sympathetic pain, is unmixed.

843. We can now return to virtue. In the final stage we shall acquiesce in all that we perceive. It is certain that some of the five characteristics, other than virtue, which we have enumerated, must be such as to make what possesses them good. For, except these five and virtue, nothing has ever been put forward as making what possesses it good. And we have seen that virtue cannot be the *only* characteristic which does this. Some of the other five then must make what possesses them good. And we have seen that, whichever of them we take, we come to the conclusion that the value in the final stage is a great good, and, except for sympathetic pain, an unmixed good.

In so far, then, as our perceptions in the final stage have as their objects anything which is itself in the final stage, they will perceive nothing which is evil, and much which is good. It does not follow that all they perceive will be good. For we shall perceive parts of parts to infinity, and it is not certain that the parts of what is good are also good. (If, *e.g.*, nothing were good but selves, no parts of selves would be good. And if, *e.g.*, nothing were good except states of love, the parts of states of love

---

[1] I do not say that there are not other goods, which are incompatible with those enjoyed in the final stage, and which therefore cannot be there. But there is no conflict and no sacrifice. For that would involve a desire for the good which is sacrificed. Such a desire would be, as a cogitation, not a perception but an assumption, and, in the final stage, an assumption can neither exist nor appear to exist.

would not be good, since they are not states of love.) But all that we perceive will either be good, or else something which is not evil, and whose existence is involved in the existence of that which is good. And our acquiescence in all these will be virtuous.

844. But this is not all we shall perceive in the final stage. We shall perceive also the sympathetic pain which exists in that stage. And we shall also perceive in that stage the contents of the other stages—the pre-final stages of the C series (cp. Chap. LXIII, p 388). We shall perceive these as being, in many respects, evil. And, since we shall have no ungratified volitions, we shall not wish them to be otherwise. Will not this be vicious, and prevent our virtue from being unmixed?

The answer to this question will be discussed more conveniently in Chap. LXVII, pp. 465–468. I shall endeavour to show there that our actual acquiescence in these objects will be virtuous. If this conclusion is justified, then all our volitions in the final stage are virtuous, and that stage is a state of unmixed virtue.

It is also a state of very great virtue. For the number of separate perceptions in the final stage is infinitely greater than the number in any part of our present experience. And, while at present many of our cognitions are not volitions at all, in the final stage all our cognitions will be acquiescences in what is good, or in what is essential to the good. All our cognitions, therefore, will be virtuous volitions. Moreover, since our emotions towards what we perceive will, as we have seen, be more intense in the final stage, and since our acquiescence will not in any case be hindered by any ungratified volitions, we may conclude that our virtuous volitions will not only be more numerous than in present experience, but more intense, and therefore more virtuous.

Our conclusion, then, is that, whichever of the six qualities we take as our criterion, the value in the final stage will be a good which is very great, and which is unmixed, except in so far as this result may be modified in respect of the sympathetic pain we may feel for the evil of pre-final stages, or for our acquiescence in such of their content as is evil.

845. The question may be asked whether the good which is in each self in respect of its final stage includes all kinds of good. This question is ambiguous. It does include all kinds in the sense that it includes good determined by each of the six characteristics which we have been discussing. But, if we take "kinds of good" in more detail, it does not include them all. To have the pleasure of swimming or of admiring beautiful scenery would be good. To show virtue in a virtuous act would be good. To love an existent God would be good. But, if we have been right in our theory of the nature of reality, none of these kinds of good can be found in the final stage. We cannot swim, or see beautiful scenery, if there is no matter. We cannot act virtuously, or act at all, if there is no time. We cannot love an existent God, if no God exists.

All these statements are equally true of all other stages in the *C* series, as well as of the final stage. But in other stages it is possible to appear to swim, or to perceive what appears to be beautiful scenery. And this gives real pleasure. It is possible for real virtue to show itself under the appearance of a virtuous act. And it is possible to believe erroneously that God exists, and to feel love towards this supposed being. Whether this would properly be called love might be disputed. But I think it is really love, and, whether it is really love or not, it is certainly really good. And therefore there are kinds of good, in this narrower sense of kinds, which can be found in the pre-final stages, and not in the final stage. For, in the final stage, there is no error, and nothing can appear as matter, or as temporal, or as God, if in reality it is not so.

846. The fact that there are, in this sense, some kinds of good which cannot be in selves in respect of the final stage does not, of course, invalidate our conclusion that the condition of selves in respect of that stage is good. It only shows that it is not a state of complete good, which we already knew, on other grounds, that it could not be. Nor does the fact that it cannot contain certain kinds of good which we can have in present experience prove that it is not a state of much greater good than anything in present experience. For the particular pleasure, virtue, or love which could not exist in the final stage might be outweighed, and much more than outweighed, by other pleasure, virtue, or love which is found in the final stage, and not in present experience. And we have found reason to believe that the pleasure, virtue, and love, which are found in the final stage, do greatly outweigh any that can be found in present experience.

847. It might, however, be further asserted of some particular form of virtue, pleasure, or love, that its absence not only excluded a particular form of good, but was itself a positive evil. This would, of course, be an ultimate judgment of value. If this were asserted about any particular form of good which could not be enjoyed in respect of the final stage, it would involve that our state in respect of that stage could not be a state of unmixed good, since it would contain some positive evil. If, further, it were asserted about any such form of good that its absence was so great an evil that it would outweigh all possible good, it would involve that the final stage was on the whole evil.

What effect this would have on the comparative value of the final stage and the earlier stages would depend on whether the good which could not be possessed in respect of the final stage could be possessed in respect of any of the others. If, for example, the good in question should be the contemplation of a natural beauty which really existed, or the love of a God who really existed, the evil of its absence, whatever that evil might be, would extend to all stages. In no stage would it be possible to contemplate natural beauty which really existed, or to love a God who really existed, if neither matter nor God did exist.

But if the good in question were the contemplation of apparent natural beauty, or the love of a God who was believed to exist, whether the one or the other really existed or not, then the result would be different. For it is only the final stage in which error is impossible, and in any other stage it would be possible that matter or God might appear to exist. And thus we might be led to the conclusion that our state in respect of the final stage was not much better than our state in respect of the pre-final stages in our present experience, or even that it was worse.

848. I believe that, when it is asserted that the absence of some particular good of this sort is a positive evil, the assertion is often due to one of two mistakes. It is sometimes due to a confusion between the proposition that the absence of this particular sort of good is the absence of good—which of course it is—and the proposition that its absence is a positive evil. But these are quite different propositions, and the second does not follow from the first. And it is sometimes due to a confusion between the simple absence of this particular good, and its absence combined with a desire that it should be present. In the latter case, no doubt, a positive evil is involved—the pain of ungratified desire. But this evil cannot be present in the final stage, in which, as we have seen, there can be no ungratified volition.

849. The same thing seems to me to be true about the assertion that a final stage which lacked certain goods which we can enjoy in our present experience would necessarily have, or produce, less good and more evil than our present experience does. I believe that this further assertion is often due to one of two confusions. The first of these is the same as the second of those mentioned in the last paragraph. It is assumed that, in the final stage, we can have, and shall have, the same desire for that particular good which we have now, and that this desire, which in our present experience is sometimes gratified, will, in the final stage, be always ungratified, and always painful. But, in the final stage, there can be no ungratified volition. The second confusion is one between the essence of the quality which makes a state good, and the form in which that quality manifests itself. For example, in our present experience, which is an experience of change and conflict, virtue would be worth little or nothing if it did not manifest itself in action and in the resistance to temptation. But it does not follow that, in a timeless experience, there cannot be great virtue without either action or resistance. Now it is sometimes supposed, I think, that, when we leave behind us the form which, in our present experience, is essential to virtue, we abandon virtue itself. And this, as we have seen, would be an error, and an error which would make us think too unfavourably of the final stage.

But, when all these confusions have been cleared away, it is possible that it would still be maintained by some thinkers that the absence of particular qualities from the final stage would produce positive evil, or even that it would produce so much positive evil as to outweigh all the good produced by the other qualities of that stage. Such judgments would, I suppose, be

ultimate, and so would not admit of refutation. But I do not
think that they are true, nor do I believe that most people would
think them true[1].

850. Can we say any more about the value in selves in respect
of their final stage? We have seen that it will be one of very great
good, if good is to be judged by any one, or any combination, of
the qualities of knowledge, virtue, emotion, pleasure, or fullness
of life—under which last name we may sum up both amount
and intensity of consciousness. Is there anything more?

I hold, as I have said, that the possession of any one of these
qualities makes that good which possesses it. But it seems to me
also that those thinkers are right who attribute an unique and
supreme goodness to love.

In what way can love be supremely and uniquely good? It
cannot, I think, be taken as the only good. It seems to me as
indubitable that certain other emotions are good, and that knowledge, virtue, pleasure and fullness of life are good, as that love is
good.

Nor can we say that all other goods are dependent on love. The
other good emotions, if our conclusions are correct, do depend on
it. Our only certainty that we shall reverence ourselves comes from
our certainty that we shall love others. We are certain that we
shall regard with affection selves which we know indirectly only
because they will be selves which are loved by those whom we
love. We are certain that we shall regard parts of selves with
complacency, only because they will be the parts of selves whom
we love, or of other selves whom they love, or of ourselves who
love them.

And our only certainty that the final stage will be a state of
pleasure depends on our certainty of the emotions which we shall

---

[1] The two characteristics with regard to which this would most often and most
seriously be maintained are, I should say, firstly, change, and, secondly, the
knowledge and love of God. The view of change as intrinsically good is, no
doubt, sometimes an ultimate judgment. But I think that, in addition to the two
confusions mentioned in the text, it is sometimes forgotten that the changelessness in question is not the comparative changelessness of monotonous duration,
but the absolute changelessness of the eternal—the changelessness which means,
not limited experience, but completed experience. I have discussed the importance
of the other characteristic in *Some Dogmas of Religion*, Chap. VIII.

feel towards what we perceive. But knowledge and fullness of life are independent of love. And so, consequently, will virtue be independent of love. It cannot, as we have seen, be independent of all other good, but, if knowledge and fullness of life are good, it would be virtuous to acquiesce in them.

Again, the supremacy of love cannot lie in the fact that it is really eternal, for everything is really eternal. Nor can it lie in the fact that it appears as eternal. For everything in the final stage appears as eternal—the knowledge, the virtue, and the pleasure as much as the love.

Can we say that love is incommensurably better than any other good? This seems attractive, but I cannot think it is correct. If it were so, it would follow that, starting from *any* standpoint—my own at present, for example—the smallest conceivable increase in love would be better than the greatest possible increase in knowledge, virtue, pleasure, or fullness of life. And it does not seem to me that this is true.

851. Is there any other way in which love could hold a supreme and unique position? I think that there is. It would hold such a position if it were true that love is capable of being so good, that no possible goodness arising from knowledge, virtue, pleasure, or fullness of life could equal it. And it is this view—a view which has been held by many people, mystics and non-mystics—which I believe to be true. It seems to me that, when love reached or passed a certain point, it would be more good than any possible amount of knowledge, virtue, pleasure, or fullness of life could be. This does not, so far as I am concerned, spring from any belief that I have reached such a point. It is a conclusion which seems to me to follow from contemplating the nature of love, on the one hand, and of the other qualities on the other hand.

852. But is not this inconsistent with the results we reached in the last chapter? For we saw there (pp. 413–415) that there could be no complete good in respect of any of the five qualities. And the reason was that there was no intrinsic limit to the amount of any of them. And, as the amount of good increases with the amount of the quality, the good could never be so great as not to admit of further increase. Now, it might be thought, this involves that the good in respect of any of these qualities has no limit. And, if this were so, the conclusion reached in the last paragraph must be false. For then, whatever the good belonging to an amount of love, there would be some amount of each of the other qualities which would be better.

But this would be mistaken. For, as was pointed out in the last chapter (p. 413), it does not follow that, because the good increased with the increase of each quality, it increased proportionately to it. If, in the case of the other qualities, the good, after a certain point, should only increase asymptotically—each successive increment of the quality yielding a smaller increment of good—then, in the case of those other qualities, there would be a limit to the good it yielded. The good would never be complete, for another increment would always be possible. But these increments, continually diminishing as they would be, would never raise the amount beyond the limit. And if, on the other hand, the goodness of love did not increase asymptotically, but directly in proportion to the love, then a certain amount of love would be more good than any amount of the other qualities could be.

The only reason for believing this theory to be true is that it seems the only possible way of reconciling two conclusions—the first being that good increases with the increases of each of the five qualities, and that the qualities have no limit to their increase; while the second is that a certain amount of love would be more good than anything except a greater amount of love. If, therefore, the second of these conclusions is not accepted, there is no reason to adopt the theory. But, as I have said, I think that the second is true, though its truth is not so certain as that of the first.

853. Is there any reason to think that there is a similar limitation at the other end of the scale of value? Is there a certain amount of love which is not only better than any possible amount of knowledge, virtue, or pleasure, but outweighs any possible amount of error, vice, or pain? There seems to me no difficulty in accepting this about error, but more in accepting it about vice, and still more about pain. As to pain, however, it must be remembered that, if we are taking the value in a single self, we can scarcely imagine love co-existing with very great pain, because the latter is so dominant and absorbing. But this incompatibility may be due, not to the intrinsic nature of love and pain, but to something incident to our present state. And it *might* be that, under other circumstances, love would be compatible with any amount of pain. And then, again, it *might* be the case that a certain amount of love would be more good than any possible amount of pain would be bad. But all this seems very doubtful. And, if our conclusions about the general nature of the universe are justified, the nature of great evil has less practical importance for us than the nature of great good.

In any case, there seems no reason to suppose that hatred has a supreme and unique position among evils, even if love has a supreme and unique position among goods.

Thus of all goods we may say that the greatest is love. It does not follow that, even in the final stage, love will be so great that it will be better than all possible other goods. We have seen that it will be very great, but there is nothing which makes it certain that it will be as great as that. But the fact that love has this potentiality of unlimited goodness makes it, even where it is limited, express more perfectly than all else the nature of goodness.

CHAPTER LXVI

VALUE IN THE PRE-FINAL STAGES
OF THE C SERIES

854. We have now to consider what can be said about the value which is possessed by the self in respect of stages of the C series other than the final stage. About this we can tell much less than we could about the value in respect of the final stage. For, in the final stage, we perceive everything as it really is, and thus, from the conclusions we had reached as to what the nature of the existent is, we were able to deduce what would be the nature of our perceptions, and so to conclude how that nature affects the value in the selves which possess them. But in the other stages of the C series there is always misperception, and thus we cannot conclude from the nature of what is perceived to the nature of the perception. Nor is there any way of determining the nature of the perception, except empirically. We cannot lay down any general laws as to what that nature must in all cases be. We can only observe what it is in those cases which fall within the scope of our observation.

855. The amount of all the perception in the universe in any pre-final stage is so enormous, compared with the very small amount which each one of us can observe in himself, or, by inference, in others, that the conclusions which we could reach by such observations would be of no philosophical importance, except in one respect. If I find reason to believe that there is any positive good in what I can observe, then the value in the pre-final stages of the universe is not unmixed evil. If I find reason to believe that there is any positive evil in what I can observe, that proves that the value in the pre-final stages of the universe is not unmixed good. And, if I find reason to believe that there are both, this proves that the value in the pre-final stages of the universe is mixed.

This last conclusion is clearly true, and can be sufficiently proved by the contemplation of any person's experience, without any consideration of the experience of others. My present experience—taking that to include the near past which I know by memory—certainly does contain what is positively good, whichever criterion of good be adopted. It certainly contains some knowledge, since, as we have seen (Chap. XLIV, p. 197), the supposition that all that I believe is, or might be, erroneous, is incompatible with its own assertion. In the second place, it cannot be doubted that it contains some virtue. For it is certain that I do sometimes desire what I believe to be good, and that I sometimes act in order to promote what I believe to be good. And, whenever this occurs, I am virtuous. In the third place, I do sometimes feel love. Nor would it affect this, if it should be the case—which is not strictly impossible, though it is extremely improbable—that my present love is always directed towards some person whom I believe to exist, but who does not really exist. For the emotion of love would be there all the same. In the fourth place, I certainly sometimes feel pleasure. And, in the fifth place, since I exist, it is certain that I have consciousness of a certain amount and of a certain intensity.

My present experience, then, does possess good characteristics, and in respect of those characteristics there is more or less good in me. But it is equally certain that my present experience also possesses evil characteristics. We saw (Chap. XLIV, pp. 197–198) that it is certain that some of the cognitions in my present experience are erroneous. And it is beyond doubt that I do sometimes desire that which I believe to be evil, and that I do sometimes act in order to promote what I believe to be evil. And, whenever this occurs, I am vicious. Again, I do sometimes feel hatred and malignancy. And, although they may sometimes have utility as means, it seems as certain that these emotions are evil as that love is good. Finally, it is as certain that I do sometimes feel pain as that I do sometimes feel pleasure. And thus my present experience is partly evil, as well as partly good. And this is sufficient to prove that the value in the pre-final stages of the universe—and, therefore, of course, in the final and pre-final stages taken together—is partly good and partly evil.

856. The presence of some good in our present experience is almost universally admitted. There are pessimists who assert that the universe as a whole is much more bad than good, or even that

every part of the universe is much more bad than good. But they
seldom, if ever, deny that parts of the universe have qualities,
because of which they are really good, however much this good
may be overweighed by the evil which arises in other respects.

But, with regard to evil, the position is different. Here we do
find many attempts to deny the occurrence of any of those characteristics which would make things evil, and to assert that the
values in the universe are completely and unmixedly good. There
are, I think, two reasons for this. One is that certain systems of
philosophy which have found considerable acceptance have denied
on general grounds the existence of evil. The other is our natural
tendency to believe what it is pleasant to believe. It is very
pleasant to believe that good is predominant over evil; and the
most effective predominance would be one which excluded evil
altogether. But few people would deny that there was something
evil in sin, hatred, and pain. The only alternative is to deny that
sin, hatred, and pain ever exist.

It may be argued that, after all, each of us may be mistaken
in thinking that they do exist within his own experience. For
why is it that I am so certain that I have sinned, hated, and
suffered? It is that I have perceived myself, or perhaps am now
perceiving myself, as having these qualities. But our previous
results have shown us that perception can be misperception, and,
indeed, that all perception in our present experience is more or
less misperception. May it not be the case, then, that when I
perceive myself as having any evil characteristic, I am misperceiving myself, and that I, and all other selves in the universe,
are perfectly good?

857. But this view is untenable, because, if the evil which
is *primâ facie* existent is explained away as an error, then the
error in question will be evil, and so evil will remain, while any
attempt to avoid this by maintaining that it is an error to hold
that the first error exists will involve a vicious infinite series. Let
us consider this in detail.

In the first place, it is clear that, if error is an evil, we cannot
get rid of evil by saying of what appears as error that it is not
really so, and that the appearance of error is erroneous. For then
the appearance of the unreal error is itself a real error.

With regard to pain, I think that the case is exactly similar, and that an appearance of being in a painful state is itself a painful state. In that case the appearance of pain proves the reality of pain, in the same way as the appearance of error proves the reality of error. But, even if the appearance of being in pain were not painful, it would certainly be evil. A world in which there was no pain, but in which everyone was under the illusion that he was in intense pain, would certainly be a world with a good deal of positive evil in it.

The illusion that I was vicious, or that I hated someone, would certainly not be itself a state of vice, or of hatred, but, if vice and hatred are themselves evil, it would be a state of evil. Imagine a universe of virtuous men, each of whom was under the illusion that he was a very wicked man. Would not this illusion introduce positive evil into such a universe?

Thus all such illusions involve evil. It must be noted that they do not merely involve the evil of error, if error is an evil. Suppose, for example, that error as such was not an evil, but that hatred was, then the illusion of hatred would necessarily be also an evil.

There are, then, evils different from, though connected with, those evils enumerated in Chap. LXIV, p. 412. If vice is an evil quality, then it is an evil quality to have an illusion that I am vicious, although the illusion is not itself vicious. In the same way, if hatred is an evil quality, it is an evil quality to have an illusion that I hate anybody, although this illusion is not itself hatred. And, if pain is an evil quality, then an illusion that I was in pain would be an evil quality even if the illusion were not itself painful.

It is impossible, then, to deny the existence of evil on the ground that our perceptions of anything as evil may be all misperceptions. For such misperceptions, as we have seen, would themselves be evil. The existence of evil is therefore certain.

858. Is the existence of good equally certain, and for the same reasons? There is, as we have said, less tendency to deny the existence of good, than to deny the existence of evil. But it will be worth while to consider what reply could be made, if any one did deny it. Let us take each of the qualities which have been maintained to be good, and consider the possibility that it does not exist, and that the perception of anything as having it is a misperception.

If knowledge is good, we cannot logically doubt the existence of this good on the ground that all perception of anything as knowledge may be misperception. For, in doing this, we are asserting that such a misperception is possible. If this is not true, our contention falls to the ground. But if it is true then we know the possibility we are asserting, and there is therefore some knowledge.

If amount and intensity of consciousness is a good, it is impossible to get rid of good from existence by saying that my perception of myself as having amount and intensity of consciousness may be in that respect a misperception. For a misperception is an act of consciousness. There must be existent consciousness, and all consciousness must have some amount and intensity.

And, if pleasure is good, it is impossible to get rid of good from existence, for reasons analogous to those which we saw were valid in the case of pain. An appearance of being in a pleasurable state is itself a pleasurable state, and therefore a misperception of myself as having pleasure would involve that I really had pleasure· And, even if the appearance of being in a pleasurable state were not pleasurable, it would certainly be good.

859. But, while pleasure is in this respect analogous to pain, it does not seem that virtue is analogous to vice, or the possession of a good emotion to the possession of an evil emotion. I do not think that we can say that, if virtue is a good, an illusion that I am virtuous must be good, or that, if love is a good, an illusion that I love must be good[1]. And thus, if we were to hold that, of the six qualities which we have considered, none was really good except virtue and the possession of certain emotions, there might be nothing contradictory in the hypothesis that no good exists. On the other hand, as we have seen, there would always be a contradiction in the hypothesis that no evil exists.

But, although there is no contradiction in the hypothesis that no good exists, it is a hypothesis which must be rejected—even

---

[1] An illusion that I love means, of course, not a real love for a non-existent person, but an illusion that I am loving, when I am not loving.

if goodness were limited to virtue and emotions. For I perceive myself as sometimes virtuous and sometimes loving. And we have no right to doubt the correctness of a perception in any point, except one in which we have reason to believe that it cannot be correct. If we did, we should be reduced to that complete scepticism which is incompatible with its own assertion. We came to the conclusion that the perception of anything as temporal, or as a sensum, must be, in that respect, a misperception, because we had previously found reason to believe that nothing could really be in time or be a sensum. But we have found no reason to suppose that nothing in present experience can be virtuous or loving. On the contrary, since all selves in the final stage are really both, there is some sort of presumption that they will sometimes be so in the pre-final stages. At any rate, there is no reason at all why I should distrust my perception of myself as at present virtuous and loving, and therefore it must be accepted as correct.

We must, then, retain our conclusion that the value in myself, in respect of the pre-final stages of the *C* series, is not unmixed, but contains both good and evil. And so the value in respect of all the stages, including the final one, will be both good and evil, since, whatever value is added to a mixed value, the total will be mixed.

860. But a difficulty may seem to arise. We found that the only evil which could exist in the final stage was that of sympathetic pain[1]. In the pre-final stages, on the other hand, there are other evils besides pain, and other pain besides sympathetic pain. But the final stage contains all the others as parts—there is no ultimate content in any of the other stages which is not also contained in the final stage. If, therefore, the final stage contains all the others, and contains nothing evil except sympathetic pain, how can the other stages contain any other evil? (The denial that any evil can exist at all is often based on an argument of this type, though the precise form would vary with the system in which the argument is found.)

[1] Cp. Chap. LXV, p. 428; though, as remarked there, the correctness of the statement depends on the proof, postponed to Chap. LXVII, pp. 466–469, that the acquiescence in evil things, which is found in the final stage, is not itself evil.

861. But the objection is invalid. The pre-final stages of the
C series are all parts of the final stage, but are, of course, separate
substances. Now, as has been pointed out previously, a substance
which is a part has often very different qualities from the substance of which it is a part. And therefore there is nothing
surprising in the fact that each of the stages in the *C* series
which are parts of the self should have qualities which are absent
from that stage in the *C* series which is the whole self—the
qualities in respect of which the pre-final stages have in them
evil, other than sympathetic pain.

An analogy may make the matter clearer. Let us suppose a
community of a hundred members, which is engaged in trade.
Let us suppose that among the members of that community there
are ninety-eight partnerships for trade, the first consisting of two
members only, *A* and *B*, the second consisting of those two and
a third, *C*, and so on until the last contains all the members of
the community but one. And let us suppose that, on some
particular day, the trading of the community as a whole has a
balance-sheet showing only assets, without liabilities, while each
of the smaller partnerships has both liabilities and assets. It is
evident that the wealth in the community would be mixed, since
it would comprise both the positive element of assets and the
negative element of liabilities, and that this would be quite consistent with the facts that the community as a whole had assets
without liabilities, and that nothing was contributed except by
partnerships, each of which was part of the community. Here
the community will correspond to the final stage in the *C* series
of a self, and each partnership to a pre-final stage in that *C* series.

862. The value in myself, then, is a mixed value of good and
evil. And this by itself would be sufficient to prove that the total
value in the universe is a mixed value of good and evil. But,
besides this, I cannot reasonably doubt that the values in other
selves are also mixed. I cannot, indeed, be as sure of the accuracy
of my conclusions with regard to another man as I can with regard to myself. It is quite possible—indeed, it often happens—
that *A* should judge *B* to be very happy when he is actually
very miserable, or should judge him to be acting very viciously
when he is in fact acting very virtuously. But, while such mistakes

happen in individual cases, it is wildly improbable that all my judgments about other people are as wrong as they would be if the state of all other people was unmixedly good or unmixedly bad. Of course there may be some selves who are unmixedly good or bad in the pre-final stages. But this does not seem probable. And, if fullness of life were good, the result could not be unmixedly bad; nor unmixedly good if error is evil. But the result in all of us together is mixed.

863. It follows from what we have said about value in this chapter, and in the two which preceded it, that the aggregate value in the universe is a quantity which depends on other quantities through five stages. In the first place, it is an aggregate of the values in all the selves in the universe. In the second place, the value in each self is a quantity which depends on the nature of the *C* stages of that self. In the third place, the extent to which any *C* stage of a self influences the value in that self is a quantity which depends on the nature of the parts appearing as simultaneous in that *C* stage[1]. In the fourth place, the extent of the influence thus exerted by one of the parts appearing as simultaneous is a quantity which depends on the nature of the good or evil qualities which that state possesses. And, in the fifth place, the extent of the influence possessed by any quality is a quantity which depends on the intensity of that quality.

864. Such a view of value would be rejected by many thinkers as too quantitative. The main objection, if I understand it rightly, is that such a view involves that the aggregate value in a collection of valuable things consists of the good values of those things, less their evil values, and, again, that the effect, which the valuable qualities of any substance have on value, depends on the amount of the good characteristic, less the amount of the corresponding evil characteristic—the amount, for example, of pleasure, less the amount of pain. This, it is said, and I think rightly, involves that the quantities in question can—at any rate ideally—be reduced to units, and that it is possible—at any rate ideally—

---

[1] The second and third statements will be true whether the value *in* the self is a value *of* the self, or whether it is an aggregate of values of its stages in the *C* series, or of values of parts of those stages.

to form a calculus of values and of valuable characteristics. And this, it is said, is impossible.

But why should it be regarded as impossible to reduce such quantities to units, or to form a calculus of values, or, for example, of pleasures? So far as I know, this asserted impossibility is based on the nature of intensive quantity. The quantity of value in anything in which there is value is an intensive quantity. It is not made up of parts, each of which is also a separate value[1]. The intensity of qualities is also an intensive quantity. And it is maintained that it is impossible to reduce intensive quantity to units, or to form any calculus of quantities into which intensive quantity should enter.

But, as we have seen (Chap. XLVIII, p. 243), it is theoretically possible to measure by a common unit the differences between two intensive quantities which are qualitatively similar. And, if the series of such states has a first term or an initial boundary, it is possible to measure, not the intensive states themselves directly, but the total amount of increment in each of them, which is an indirect way of measuring the intensive states themselves. Now all the series of which we have been speaking start from a zero point, and so, in this direction, they are not unbounded. It is, therefore, theoretically possible to measure them in terms of a common unit, and to assert that one quality varies proportionately to another, and to establish a calculus of values.

It is, of course, quite a different question whether this is practically possible, or whether the difficulties of observation and calculation are too great for the human intellect. But, if this were so, it would only prove that our decisions as to the right course of conduct would be in many respects unsystematic and untrustworthy. It would not affect our conclusion that the aggregate value in the universe had a quantity which did depend on other quantities, although we were unable to determine it with accuracy[2].

[1] I am speaking here, of course, of those things *of* which there are values. When there is a value *in* a thing which is not a value *of* it, the value in the thing is made up of parts, each of which is a separate value *of* one of the parts of the thing.

[2] It seems clear, however, that some of the intensive quantities which determine value can be measured with sufficient accuracy to form a guide for conduct which

865. A second objection to such a quantitative conception of value as we have proposed appears to lie in the belief that of the qualities which have been held by different thinkers to be good or evil, none admits of measurement except pleasure and pain, and that, therefore, if it is possible to measure amounts of value, nothing can be good but pleasure, and nothing evil but pain. And this limitation of good and evil is held—as I think, rightly—to be indefensible. This objection is in some cases reinforced by the curious confusion of thought which induces some thinkers to believe that Ethical Hedonism involves Psychological Hedonism—that is, that the proposition that nothing is good but the pleasure of some self implies the proposition that no man can act for any motive but the expectation of pleasure for his own self.

But both these beliefs are erroneous. Ethical Hedonism does not involve Psychological Hedonism, and therefore, even if our conception of value did involve Ethical Hedonism, this would not compel us to adopt Psychological Hedonism. Nor does a quantitative conception of value involve Ethical Hedonism. For, as we have seen, all the other characteristics which are ever taken as being good or bad are also quantitative, and are therefore ideally as capable of measurement as pleasure and pain are, even if it should happen that the practical difficulties in measurement may be greater than in the case of pleasure and pain.

866. A third objection which is sometimes raised depends on the assertion that it is impossible to sum the quantities in question, because those quantities will vary according to the manner in which they are combined. No doubt it may be the case that two qualities, or two parts of selves, or two selves, may

has practical value. This is perhaps especially obvious in the case of pleasure. Take the case of a man choosing between two glasses of wine, equally expensive and equally wholesome. The pleasure he anticipates from each will have extensive quantity varying with its length, but will also have intensive quantity in respect of the different pleasure at each instant which is due to the character of the wine. If the pleasure to be derived from each glass is taken as a whole, it must be done by a comparison of these intensive quantities throughout each extensive quantity. If we are not able to do this in practice, even with approximate accuracy, it follows that a man is likely to get as great pleasure if the question whether he shall drink port or madeira is settled by chance, as he would if it were settled by his own choice. And I suppose it would be generally admitted that this is not the case.

have different effects on value, or different values, if they are combined in one way from what they would have if they were combined in another. But this will not affect the fact that, in whatever combination they do occur, they will have definite quantities, and that these quantities can be summed and compared.

The only ground on which it could be asserted that the alteration of the quantities by different combinations rendered it impossible to sum and compare them would be that the summing or the comparison recombined the quantities in such a way as to alter the amounts which they had before they were summed or compared. And this is certainly not the case. L, M, and N may have a different effect on value, or different values, if L and M are combined, and N isolated, from what would be the case if L were isolated, and M and N combined. But in either case they will have a definite value, which will be no more altered because an observer sums it with others, or compares it with others, than the magnitude of the attraction of the sun for the earth is altered when a fresh schoolboy learns of its existence.

CHAPTER LXVII

TOTAL VALUE IN THE UNIVERSE

867. The total value in the universe consists of the value in the final stage, together with the value in the pre-final stages. Of the latter we saw in the last chapter that we only know that they are partly good and partly evil. We could tell nothing about the relative amounts of good and evil in them. And we saw in Chapter LXV that the relative amount of good and evil in the final stage depended in part on the relative amounts of good and evil in the pre-final stages. For we saw that the final stage was a state of unmixed good, except in so far as the contemplation, in that stage, of the evil in the pre-final stages, might introduce the evil of vice or of sympathetic pain.

It would seem at first sight, therefore, as if we could determine nothing about the relative amount of good and evil in the total value of the universe. But we shall find that we can determine a good deal about it by determining, in the first place, the relative magnitude of the value, whether good or evil, in the final stage, as compared with that in the pre-final stages.

868. We have seen (Chap. LXII, p. 373) that the final stage occupies, *sub specie temporis*, an infinite time. Now it seems to me beyond doubt that when two states which have value appear as being in time, their values, *cæteris paribus*, vary with the length of time which they appear to occupy[1].

We saw in Chap. LI (pp. 276–279) that "the length of time which a state appears to occupy" is an ambiguous term. It may mean what a believer in the reality of time would call the length of time which it did occupy, and what on our theory is a *phenomenon bene fundatum*, in which the only element of error is that it appears as being time at all. Or it may take into account also the way in which states appear to be longer or shorter because

---

[1] It is, of course, irrelevant to remark that, if the time is prolonged, pleasure may become wearisome, or pain, under different circumstances, more or less bearable. For in that case the other factors would not be equal.

of their qualities—because they are exceptionally wearisome, or exceptionally interesting, or the like. In which sense do we use the term here?

I think it is clear that we ought to use it in the first sense. In the first place, we saw (Chap. LI, p. 277) that there is reason to hold that, when states appear longer or shorter in the second sense, the appearance is only retrospective—they appear, from the point of view of a later stage, to have been longer or shorter because they were wearisome, or interesting, and so on, but they do not appear to be longer or shorter for such reasons while they are actually taking place. And their value cannot be greater or less because of misperceptions of them which are found in other stages of the series. And while a state may certainly have a different value if it is, for example, wearisome, than it would have had if it had not been wearisome, this is due to a difference of quality, and so the cases are not *cœteris paribus*. For, as we saw (Chap. LI, p. 277), it is not wearisome because it appears longer, but it appears longer because it is wearisome. It is, therefore, in the first sense of "appear" that we must say that two stages which have value, have that value, *cœteris paribus*, in proportion to the length of time which they appear to occupy. And in that sense, as we saw (Chap. LI, p. 278), stretches of the C series which have any given proportion to one another will appear as occupying periods of time which have the same proportion to one another. As this appearance as periods of time, although only an appearance, is a *phenomenon bene fundatum*, no confusion will arise if, for the sake of shortness, we state our principle in the form that the values of states in time vary, *cœteris paribus*, with the time they occupy.

869. From this principle it follows that any value, which has only a finite intensity, and which only lasts a finite time, may be surpassed by a value of much less intensity which lasts for a longer time. Take a life which in respect of knowledge, virtue, love, pleasure, and intensity of consciousness, was unmixedly good, and possessed any finite degree of goodness you choose. Suppose this life prolonged—for a million years if you like—without its value in any way diminishing. Take a second life which had very little consciousness, and had a very little excess of pleasure over pain, and which was incapable of virtue or love. The value in each hour of its existence, though very small, would be good and not bad. And there would be some finite period of time in which its value would be greater than that of the first life, and another period in which it would be a million times greater[1].

870. This conclusion would, I believe, be repugnant to certain moralists. But, in the first place, a conclusion may be rightly repugnant to us, and yet it may be true, since the universe is not completely good. And, in the second place, I can see no reason for supposing that repugnance in this case would be right. So far as I can see, it rests on a conviction that quality is something which is inherently and immeasurably more important than quantity. And this seems to me neither self-evident nor capable of demonstration.

Nor is there any reason to doubt our conclusion because it is highly probable that many people, if offered the million years of brilliant life, followed by annihilation or by a state whose value should be neither good nor evil, but zero, would prefer it to any length of an oyster-like life which had a slight excess of good. For it must be remembered that men's choice in such cases is very much affected by their imagination. Now it is much easier to imagine the difference between the two sorts of life which we have considered, than it is to imagine the difference between an enormously long time and another time which is enormously longer. And, again, we are generally affected more than is reasonable by the present or the near future in comparison with the far future. And a change which will only happen after the end of a million years is in a very far future.

871. If our conclusion is right, it follows that a portion of the C series, which is bounded on each side by a further portion of the C series, or on one side by such a portion and on the other side by nonentity, will have a value which is finite in that dimension. And, as we saw in Chapter LXIV, all values in any one

---

[1] This, of course, is only true on the assumption, made in the text, that the pleasure of the second life had some value. If it were held that nothing was good, except, or without, virtue or love, or if it were held that pleasure which was below a certain intensity had no value, then the situation described in the text would not arise.

self are finite in their other dimensions, so that the value as a whole will be finite. We will speak of this as Case *Z*.

872. How about the *C* series as a whole? The *C* series, by definition, is that series which appears as a time-series. Now suppose the *C* series was a series bounded on both sides by nonentity. That is, let us suppose that the first term in the series stands to nonentity in the same relation as the second stands to the first, but also that the last term in the series stands to nonentity in the same relation in which the last term but one stands to the last term. We have seen (Chap. LXII, pp. 373–374) that the first of these suppositions is true, and the second is not. For the first term is reached from the second by the subtraction of content, and a further subtraction will take us to nonentity. But the last term is reached from the last but one by the addition of content. And it contains all the content, so that there can be no further content to be added—nor, if there were, would it produce nonentity. The *C* series, therefore, is not bounded on each side by nonentity. But, if it were, then the whole value of the *C* series would be finite in this dimension and, therefore, finite altogether. We may call this Case *Y*.

I think that there can be no doubt about these two cases. For in each of them what we were considering would have, *sub specie temporis*, a finite duration, and therefore its value could, in that dimension, be only finite.

873. Now suppose a *C* series which differs from the actual *C* series in another way. Suppose that, either in one direction or in both, there was a term of the *C* series beyond every term. In that case the *C* series would appear, *sub specie temporis*, as being a time which was infinite, either in the past, or in the future, or in both. And thus its value would be infinite in this dimension, and so altogether. Let us call this Case *X*.

874. We now come to a case, which we will call *W*, which is not so simple as *Z*, *Y*, or *X*. And it is much more important than *Y* or *X*, because it is the case of the *C* series as it really is, and not of the *C* series as it is not. For it is the case in which the *C* series is bounded at one end by nonentity, and has no boundary at the other end. And we saw (Chap. LXII, p. 373) that the *C* series is really bounded by nonentity at the end which appears.

sub specie temporis, as the earlier, while it has no boundary at the end which appears as the later.

And therefore, as we also saw in Chap. LXII, p. 373, the last term of the C series appears, *sub specie temporis*, as endless. It resembles in this respect the whole C series in case X, since this also appears, *sub specie temporis*, as endless. Does it resemble it as having a value which is infinite in this dimension?

875. The transition from the appearance of infinite time to the reality of infinite value is not as clear here as it is in case X. For, in case X, the C series would appear *to itself* as occupying infinite time. That is, each term would perceive itself as present, and would perceive the terms on either side of it as past and future respectively, with a term later than each future term[1]. In W the final term will appear to other terms as being endless in time, but it will not appear so to itself. For in that term, as we have seen (Chap. XLVII, p. 232), there can be no error, and therefore it cannot perceive itself, or anything else, as in time, since nothing is really in time.

And there is another difference between the two cases. In case X the infinite value of the whole series is made up of the values of the different terms of the series. Each of these is finite in value, and the infinity of the value of the whole is due to the fact that the terms are infinite in number. But, in W, that, if anything, which has infinite value is not made up of an infinity of terms, or even of a plurality of terms. There is only one term—the last of the C series. And the infinite value, if there is one, must belong to this undivided term.

876. Now what is it which gives the infinite value in case X? There are two characteristics of the series in this case. It has no boundary towards the apparent future. And it is made up of an infinite number of terms, each of which appears, *sub specie temporis*, to be of finite duration, and each of which has a finite value. This second characteristic depends upon the first. For, if the series had a boundary towards the apparent future, then, since it has a boundary towards the apparent past, it would be made

---

[1] In order to get a supposition more closely parallel to W, I have taken the form of X in which the series is unbounded towards the apparent future but not towards the apparent past. But what is said would apply equally to the form of X in which the series is unbounded in the other direction, or in both directions.

up of a finite number of the terms which have finite values, and its own value would be only finite.

In case W, on the other hand, the final term is unbounded towards the apparent future, but it is not made up of an infinite number of terms, each having a finite value.

877. The question is whether the infinity of value depends simply on the unboundedness, or whether it depends on an unboundedness which determines the infinite number of terms. Either supposition will account for the first three cases. It is admitted that in Z and Y the values are finite, and in X the value is infinite. And these results will follow equally from the first hypothesis—that the unboundedness is sufficient for infinite value—and from the second hypothesis—that there must be an infinite number of valuable parts. But for W the two hypotheses give different results. On the first hypothesis the final term of the series will have infinite value, since it will have no boundary in one direction. On the second hypothesis it will not have infinite value.

878. It seems to me that the first hypothesis is right, and that the value of whatever is unbounded in the series in at least one direction is infinitely greater than the value of anything in the same series which is bounded in both directions[1]. If so, the final term of the actual C series will have infinite value.

In the first place, I am disposed to think that I can to some extent realize what a state, of love, $e.g.$, would be like, which appeared to itself as being not temporal but eternal. And it seems to me that the value of such a state would be a value which was not bounded in the way in which the value of a state of love which appeared to last an hour was bounded.

I do not, however, attach much importance to this. It is very difficult to get anything like a clear idea of what a timeless and eternal state of consciousness would be, since we always in our present stage misperceive states of consciousness as being in time[2].

---

[1] Strictly speaking, we should have said "infinitely greater in respect of that dimension." But since, as we have seen, values are finite in respect of their other dimensions, it follows that a value which is infinitely greater in respect of this dimension will be infinitely greater altogether.

[2] The question how it is possible that we should know anything of the nature of characteristics which we do not perceive anything as possessing was discussed in Chap. LIV, pp. 309–313.

In particular, it is difficult to exclude the possibility that we are considering it, not as a timeless state, but as a state lasting through an infinite time. And this, of course, would be as grave an error, on the one side, as would be, on the other, the consideration of it as an instantaneous present.

Moreover, as will be seen later in the chapter, if we are entitled to accept the view that the final stage has an infinite value, we shall be able to arrive at conclusions as to the proportions of good and evil in the universe, and as to the amount of sympathetic pain in the final stage, which are of an optimistic nature. But, if we cannot attribute infinite value to the final stage, there seem no grounds for optimistic conclusions on these points. And as every man is always under a strong temptation to believe what he wishes to believe, I dare not attach much importance to the results of such contemplation.

Some importance, however, I am disposed to attach to it. And, when we consider certain consequences which would follow if the second hypothesis were true, we do get, I think, good grounds for believing that that hypothesis is false, and, consequently, that the first is true.

879. If the second hypothesis were true, what value would the final stage have? It would, of course, be finite, but how would its amount be determined? We saw above (p. 451) that the values of things which appear as occupying finite times vary, as to that dimension, with the time which they appear to occupy, and, consequently, in proportion to the stretches in the C series which they really do occupy. If a state of happiness in Smith occupies a stretch of the C series which is one-millionth of the whole, and a state of happiness in Jones occupies a stretch of the C series which is one ten-millionth of the whole, then, in respect of this dimension, the value of Smith's happiness is ten times as great as the value of Jones' happiness.

If, therefore, the second hypothesis is true, and the value of the final term is finite, it seems inevitable that its value should be proportionate to the amount of the C series which it occupies. This is the consideration which determines the amount of all other finite values, and there seems no other possible consideration which would determine this particular finite value.

Now we have seen (Chap. LXII, p. 375) that the final stage is one simple and indivisible term of the C series. It therefore occupies a smaller proportion of the C series than any stage which appears as a divisible time, for such a stage must contain more than one simple term of the C series. And whatever appears as lasting a second appears as lasting a divisible time, since we can distinguish shorter intervals than seconds. It follows that the final stage occupies a smaller proportion of the C series than anything which, *sub specie temporis*, lasts a second. How much less we do not know, as we do not know how many simple terms would appear as a second. But the number of terms which appear as a second might be any finite number. It might even be infinite, if in the series of simple terms there were no next terms.

The value of the final stage, then, would, in respect of this dimension, be less than the value of a state which lasted a second, might be less in any finite proportion, and might even be infinitely less.

880. Now this result, I think, must be condemned as absurd. A state which appears to itself as transitory, and as lasting two hours, is, in respect of that dimension, twice as valuable as a state which appears to itself as transitory and as lasting only one hour. But a stage which appears to itself as eternal and not transitory at all, would be, in respect of this same dimension, several thousand times, at the least, less valuable than a stage which appears as lasting an hour.

This becomes clearer when we take two particular states which are similar in other respects. Take, for example, two states of love, equally intense and equally pure and unselfish. Suppose that one lover perceives his love as a state which is present, and which is passing, and that after he has perceived it as present for an hour, he perceives it as past. The other perceives his own love as eternal and unending. Is the second several thousand times less valuable than the first? It seems to me, at any rate, absolutely certain that it is not.

881. It may be said on the other side that, since the final stage is a part of the C series, the presumption is that its value will be determined in the same way as the values of other parts of that series—proportionately to the amounts of the C series which they occupy. But when it is said that the final stage is part of the C series, it must be remembered that this requires qualification. The C series is that which appears as the time-series. And so a series is only a C series relatively to certain selves at certain points. The final stage is in the C series for all the pre-final stages, for, as we have seen, it appears to all of them as being future. But it is not in the C series for itself, because it is free from error. It sees everything, including itself, as it really is, and, therefore, not as being in time. Now the value of anything may be dependent on the way in which it appears to itself, because that appearance is part of itself. But it cannot be dependent on the way in which it is misperceived by something else[1]. And therefore the fact that the final stage is in the C series—in the sense in which it is in that series—gives us no reason to suppose that the amount of its value is determined, as that of the other stages is, by the amount of that series which it occupies.

882. And, indeed, if we could have settled the matter by taking the final stage as it stands in the C series, the answer must have been that the value of the stage was infinite. For, as was said earlier in this chapter, when the final stage appears as future, it appears as an endless future, and the value of a stage which lasted through endless time would be infinite. The reason we could not settle the question in this way was, as we saw there, that the final stage did not appear in time to itself—*i.e.*, that its position in the C series, in the sense in which it was in that series, was not decisive as to its value.

The final stage is a term of the inclusion series in the same sense in which the pre-final stages are[2]. But its position in the inclusion series differs very much from that of all the other terms. For it is the only term in the series which includes without being itself included. And it is just this difference which, on the first hypothesis, makes the value of the stage infinite. For it is the fact that it is not, like the other stages, both included and

---

[1] I have used neuter, and not masculine, pronouns, because the question is not only of a perception by a self, but of perception by a self in one of its stages, and the stages of a self would be spoken of, not as "he," but as "it."

[2] The final stage is, of course, not a term of the third series of which the pre-final stages are terms—the misperception series. For in the final stage there is no error.

inclusive, which makes it unbounded in one direction. And it is this which, on the first hypothesis, gives it infinite value.

883. We saw earlier (p. 455) that, if the value of the final term in case W was infinite, it would differ from the infinite value of the whole series in case X, because, in case X, the infinite value was made up of an aggregation of an infinite number of finite values, each of which belonged to one part of the series, while in case W the infinite value would be the value of a single simple term.

In this, again, there seems to be no difficulty. The value of the final term, it is true, is not made up of separate values the sum of which is infinite. But, if we take any finite value we know, or any finite multiple of that value, the value of the final term will be greater. And, if so, it is infinite.

884. The value of the final term will be an intensive magnitude. For it will not be made up of parts, each of which is a lesser value. And the values with which we compare it will not be intensive, but extensive. For they will be values of stretches of the C series, appearing as durations of time. Now the value of a duration of two hours is made up of parts, each of which is a lesser value. It is made up, for example, of the value for the first hour and the value for the second hour, or, again, of the value for the first hundred minutes and the value for the last twenty minutes. And the intensive and extensive magnitudes must be comparable in respect of their magnitude, since we are able to assert, in each case, that the first is greater than the second.

To compare intensive and extensive magnitudes with one another is, however, quite possible. The wealth of a rich man is an intensive magnitude. For the wealth of one rich man is not made up of the wealth of two men who are not so rich. The wealth, on the other hand, contained in a collection of sovereigns is extensive. For the collection is made up of smaller collections, and the wealth contained in the whole is made up of the wealth contained in the parts. Yet we can say of the wealth of a rich man that it is equal to the wealth contained in a certain collection of sovereigns, and greater than the wealth contained in another collection of sovereigns. In the same way, we can say that the value of the final stage is greater than any value in the pre-final stages, or than any finite multiple of that value.

It will be true of the value of the final stage, as of any other intensive magnitude (cp. Chap. XLVIII, p. 243), that, although it is not made up of separate values, it has increments which are amounts of value. Thus a certain amount of the value in it will be equal to the amount of value in any extensive value. And the infinity of the intensive value can be expressed by saying that the amount of its value always exceeds the amount which is equal to the amount of any finite value.

885. It seems, then, to me at least, quite clear that we must reject the second hypothesis, and accept the first. That is, we must say that, since the final stage is unbounded in one direction, the value of that stage is infinite. The value of a state, then, may be said to be infinite, when it appears to itself, not as transitory, but as eternal. But, if we say this, we must remember that what makes it infinite is the fact that it is unbounded. The connection with the appearance as eternal is only through the fact that the unbounded state is in point of fact one in which there is no error, and that therefore it must perceive itself as being—what all the stages really are—eternal.

886. If the whole—the all-inclusive term in the inclusion series—had been, *sub specie temporis*, the first term, instead of the last, it would make no difference as to its perceiving itself as eternal. For it would still be a term in which there would be no error, and it would therefore have to perceive itself as eternal which it was, and not as temporal which it was not. And its value would still be infinite, since it would be unbounded in one direction. But, *sub specie temporis*, it would appear to us, in our present stage, as an unbeginning past, instead of an unending future. And therefore it would not excite as much interest in us.

887. It should be noticed that the infinite value of the final stage is not due to an infinite intensity of the qualities which give value—knowledge, virtue, happiness, love, and so on. Nothing has been said which affects our previous conclusion on this subject (Chapter LXV). The intensity of these qualities in the final stage will be much greater than it is at present, but it will always be such that there could be a greater intensity. Nor have we any reason to suppose that it will be much greater than it is in those pre-final stages which are nearest to the final stage. It may, no doubt, be much greater, but we have no reason to think that it will be. The infinite quantity of value is in another dimension. It depends entirely on the unboundedness, which is not itself quantitative at all, although, *sub specie temporis*, it appears as infinite time.

There is thus no reason to suppose that there is any sudden change of content in passing from the pre-final stages to the final stage, although it is possible that there may be such a change. The only difference of which we are certain is that the value is infinite in the final stage. And this difference, as we have seen, is due, not to a change of content, but to the final stage being unbounded.

888. It should also be noticed that, although the final stage contains the content of all the other stages, yet the value of the final stage does not contain the values of the other stages. Each stage has its separate value, which is not part of the value of the final stage. And, of course, *à fortiori*, the value of the final stage is not the sum of the values of the pre-final stages. Indeed, if it were, it would not be infinite, but finite.

889. This raises another point. It is possible that there are selves in which there are no inclusion series of perceptions. There is no contradiction in the hypothesis that somewhere in the universe there may be self-conscious selves, who have no inclusion series of perceptions, but only the single stage which contains all the content, and so corresponds to the final stage in the selves in which there is such a series.

What would the value of such a stage in a self-conscious self be? It seems clear to me that it would be infinite. To begin with, it must have some value, for its nature will be the same as that of a final stage in the cases where there is a series. And we saw in Chapter LXV that such a final stage had value.

890. As compared with stages in an inclusion series, this stage is simple and undivided. For they are differentiated by including a plurality of the simple terms in the inclusion series, while the stage which we are considering has no plurality at all in this dimension. On the other hand, it is not bounded in the dimension of the inclusion series, since it does not form part of such a series.

If it were maintained that its unboundedness in respect of this series does not make its value infinite, because it is not a member of such a series, there is nothing else that is relevant except the fact that it is not differentiated in respect of that series. And then its value can only be equal to that of a simple and undivided term in the inclusion series. That is to say, a state of love in such a person, which appears to him to be, as it really is, eternal and non-transitory, will be, at the least, several thousand times less valuable than a state of love, equally pure and equally intense, in another person, which appears to itself as having the duration of an hour. And this seems to me to be clearly false, for the reasons given above (p. 458).

891. We must then hold that the value of such a stage is infinite. And therefore we must hold that infinity of value depends, not on the comparatively positive quality of being an unbounded stage in an inclusion series, but on the comparatively negative quality of not being a bounded stage in an inclusion series[1].

From this it follows that infinity of value in this dimension (and therefore infinity of value as a whole, since one infinite factor makes the whole infinite) is—if I may so express it—the normal characteristic of anything which has value at all. Finitude of value is only introduced in the case of stages in the inclusion series which are bounded on both sides by other stages, or on one side by another stage and on the other side by nonentity.

There seems no reason why we should distrust this conclusion. Nor is it of any practical importance for us. We knew before that the final stages of inclusion series had infinite value; it still remains the case that the pre-final stages of inclusion series have only finite value; and, even if there are any selves without states of consciousness forming an inclusion series, they are not among selves with which we are at present acquainted. An emotional importance, however, I think that it does have. It seems to me to increase the significance of the universe if we realize that any of the qualities which give value—virtue, pleasure, love, and the rest—do, normally and intrinsically, give infinite value to whatever possesses them, and that the value can only be reduced to finitude by the intervention of special circumstances. Nor does it seem unimportant from this point of view that, although it is not an error that there are finite values—they are as truly existent as the infinite values—yet nothing has finite value

---

[1] The difference, of course, is that anything which is not a stage in the inclusion series has the second quality, but not the first.

except errors. For nothing has finite value except the pre-final stages of the inclusion series. And we have seen that all such stages are misperceptions.

892. The value of the final stage, then, is infinitely greater than the aggregate value of all the other stages. For the value of the final stage is infinite, while the aggregate value of all the others is finite, since they all of them together occupy a bounded portion of the series[1], which appears, *sub specie temporis*, as a finite time.

And, apart from its greater value, the final stage has, from our present position, a greater importance than any other stage. For, *sub specie temporis*, it is now in the future, it will be in the present, but it will never be in the past. It will, as we have said, begin, but it will never end. And, since the present and the future are of more interest to us than the past, the final stage will be of special importance to us, apart from the infinitely greater magnitude of its value[2].

893. What, then, is the quality of this value which is infinite in quantity? Is it good or evil? We saw, in Chapter LXV, that, judged by any criterion of goodness which has been put forward, the value of the final stage would, subject to two reservations, be good, would be a very great good as compared with any that we experience at present, and would be unmixedly good. The reservations in question arose from the probability that the self in the final stage perceived, not only the final stages of himself and of other selves, but also their pre-final stages. In these pre-final stages there is much evil, and in the final stage, in which there is no error, this must be perceived as being evil. Now in the final stage we can have no ungratified volitions. We shall therefore not wish that these evils should be otherwise. And is not the knowledge of evil, without a wish that it should be otherwise, vicious? And, again, the contemplation of such evil may cause us sympathetic pain, and so prevent our good being unmixed, if pleasure

---

[1] Bounded in one direction by the final stage, and in the other direction by nonentity.

[2] As we saw on p. 461, if the all-inclusive term in the inclusion series had been, *sub specie temporis*, the first term instead of the last, it would still be a term which would perceive itself as eternal. And its value would still be infinite, since it would be unbounded in one direction. But it would not have the special interest for us mentioned in the text, since, *sub specie temporis*, it would appear, not as a future which will never end, but as a past which never began.

is a good. These are the two reservations. I shall endeavour to show, in the first place, that, though we shall not wish the evil states to be otherwise—indeed we may very possibly acquiesce in their existence—yet this will not prevent our virtue from being unmixed. With regard to the evil of sympathetic pain, the results reached in the earlier part of this chapter will enable us to show that such pain, while it really exists in the final stage, will be infinitely less in amount than the pleasure which exists in the final stage.

The question only arises with regard to virtue and to pleasure. For it is quite clear that the four other qualities which have been held to determine goodness can belong just as well to a perception which perceives something as being evil, as to a perception which perceives something as being good. If knowledge, to begin with, is good, and error is bad, the perception of something as being evil may itself in this respect be unmixedly good. For that which is evil (whether as being erroneous or for any other reason) can be known as fully and perfectly as that which is good.

The same is the case about the emotions which have value, either as good or as evil. We saw in Chap. LXV (p. 424) that in this respect we should be good in the final stage, because each self would regard other selves with love or with affection, and himself with self-reverence, and because he could not feel hatred, repugnance, malignancy, anger, regret, remorse, jealousy, envy, or fear. And, if we consider the reasons which were given in Chapter XLI why we must feel the one set of emotions and cannot feel the others, we shall see that they do not depend on the selves which are perceived being in an unmixedly good state, and cannot be invalidated by the fact that they are partially in an evil state.

It is true that a perception of anything as evil may very probably have the quality of being an emotion of sadness. But this does not make the person who has it evil in respect of his emotions, for it is not evil to have the emotion of sadness in perceiving what is evil. Any evil which could arise could only be from the fact that sadness is painful to the person who is sad. And the question of pain is a distinct one, which is to be discussed later.

Again, so far as we place good in the amount or intensity of consciousness, it is clear that this is not diminished by the fact that some of the substances of which we are conscious are in a partially evil state. And, finally, if good is to be found in harmony, or in harmonious self-development, I do not see that it would be affected. There might be want of harmony in the pre-final stages which were perceived, but that would not involve any want of harmony in the final stage which perceived them. And the reasons for thinking that there would be such a harmony in the final stage, given in Chap. LXV (p. 431), remain unaffected.

894. We now come to virtue. In the final stage we shall perceive evil which we shall not wish to be otherwise, since we can wish nothing to be otherwise. And, further, it is possible that we may acquiesce in the existence of substances which are, and which we perceive as being, evil. But this would not prevent our virtue from being unmixed, for such acquiescence would be virtuous.

Our perceptions are of substances, and when a perception is also an acquiescence, it will be an acquiescence in the existence of that substance. But substances are perceived as having various qualities, and acquiescence in the existence of a substance may be determined by one or by another of the qualities which it is perceived as having. And whether the acquiescence is virtuous or vicious depends on the quality which determines the acquiescence.

Now, if I do acquiesce in all that I perceive in the pre-final stages, why should I do so? There is only one characteristic common to all of them which could determine such an acquiescence. And that is that they are all parts of one or another self whom I love, or for whom I feel affection or self-reverence.

Now this characteristic has the quality which we called utility (Chap. LXIV, p. 398). It has utility because its possession by a substance is a means to another substance being good. It is good that selves should love—though they do not love because it is good to love, but because they love. And, if they are to love, there must be selves to be loved. And those selves cannot exist unless their parts exist. Thus the parts of the selves which are loved have utility as means to other selves having the good quality of loving. And, if our acquiescence in a substance is determined by its possession of a characteristic which has utility, then that acquiescence is virtuous, regardless of what the other qualities of the substance, which have not determined our acquiescence in it, may be[1].

895. In our present experience, indeed, the case is different. Here, if we realize that anything is partly desirable, and partly undesirable[2], we shall not, if we are virtuous, acquiesce in it. We shall form an assumption that the thing has its desirable qualities without its undesirable qualities, and this assumption will be, from the point of view of volition, an acquiescence in the content of the assumption. And, since the assumption does not agree with the fact which we perceive, we shall not acquiesce in that fact.

But in the final stage nothing appears as an assumption. All our cogitation appears there, as it really is, as perception. And so in the final stage it is, as we have seen earlier, impossible to desire anything except that which does exist. The only alternatives are to desire what does exist, or to remain volitionally neutral about it.

If, therefore, in the final stage we acquiesce in the pre-final stages on account of desirable qualities which they possess, the virtuous character of that acquiescence is not destroyed by the fact that we do not desire the absence of the undesirable qualities of those stages. We are not indifferent to better possibilities— we just do not contemplate them, or anything else except the existent. The only thing which could make our acquiescence vicious would be if it were determined by some undesirable quality of the existent—if we acquiesced in it on account of an undesirable quality. And, as we have seen, we do not acquiesce in it on account of an undesirable quality, but on account of a quality which is desirable.

896. It might, however, be objected—and I think rightly— that it would not be virtuous to acquiesce in the existence of anything because of some slight and trivial goodness or utility

---

[1] There will be an analogous argument in the cases where the parts in which I acquiesce are parts of a self for whom I feel affection, or of my own self for whom I feel self-reverence.

[2] I call a thing desirable if it is either good or useful, and undesirable if it is evil, or if it possesses the quality of disutility.

which it possessed, while it was in all important respects undesirable. If a man were really virtuous, the presence of great
and predominant evil or disutility would prevent him from
acquiescing in such things, though he could not desire that they
should be different.

This, as I have said, I think is true. But then the utility of the
quality which determines our acquiescence is not in this case
slight or trivial. If love is the supreme good in the universe, then
it is not a slight or trivial quality to be a part of a loved being,
and so a means to the existence of love. If so, we shall be virtuous
if we acquiesce in states of sin and pain, in others and in ourselves, because they are parts of selves who are beloved, and of
selves who love.

We are not without examples of such acquiescence as this even
in our present experience. For many men—if not most men—
know what it is to regard even the sins and the miseries of their
friends as something precious—even supremely precious—just
because they are parts of a life of one who is beloved. The
experience does not come always, and may not come often, but
it does come. This, it is true, is not, in our present experience,
a state of acquiescence. Acquiescence would be reserved for the
assumption that this part of my friend's life, while still a part of
his life, should be free from sin and misery. But the recognition
that it is precious just because it is part of my friend's life is a
recognition of that connection which forms the ground of the
final acquiescence. And, even here and now, we can see how far
this is from being slight and trivial.

897. I do not think it possible to determine whether, in the
final stage, we shall in fact acquiesce in everything in the pre-
final stages for the reasons given above, or whether the intrinsically bad qualities of some of them—their sinfulness or
misery, for example—may prevent our doing this, and leave our
perceptions of them without any volitional quality. But it has
been shown that, if we did acquiesce in them, our acquiescence
would be virtuous. And it follows, *à fortiori*, that our virtue will
not be mingled with vice merely because we shall not, in the
final stage, be able to condemn anything in the pre-final stages.

898. There remains the question of pain. And here, I think.

we must come to a different conclusion. The perfection of knowledge, of emotion, of virtue, of amount and intensity of consciousness, and of harmony, are not, as we have seen, affected in the final stage because it perceives the pre-final stages. But its perception of the pre-final stages does introduce pain into the final stage—the pain of sympathy.

Experience shows us that we often feel pleasure in contemplating the pleasure of others, and pain in contemplating their pain. It is this which is primarily called sympathetic pleasure and pain. But experience also shows us that we often feel pleasure in contemplating good in others, which is not itself pleasure, and pain in contemplating evil in others, which is not itself pain. It is convenient to include the pleasure and pain of these latter contemplations under the title of sympathetic pleasure and pain. I think that it is clear, also, that in some cases a man feels sympathetic pleasure and pain for himself (generally in cases where the pleasure or pain sympathized with is past before the sympathetic feeling begins).

At present, then, we feel sympathetic pleasure and pain. Nor is there anything in the difference between our present experience and the final stage that could make us suppose that sympathetic pleasure and pain do not occur in the final stage. It is true, of course, that when, in the final stage, we are aware of the pain of others, we shall not desire that it should be otherwise, because, in that stage, we can desire nothing but what actually exists. But, while this will spare us the further pain of fruitless protest and revolt, it will not affect the occurrence of sympathetic pain, which has no dependence on volition.

899. It does not always happen that we feel sympathetic pain in contemplating the pain of others. But I think that it always happens when we contemplate the pain of people whom we regard either with love or with affection. In the final stage, then, sympathetic pain will exist. It will be felt by every person who regards with love or affection any person who suffers any evil in the pre-final stages. And as, in the final stage, everyone will regard with love or affection any person whom he knows, it will be felt by everyone who knows any person who has suffered any evil in the pre-final stages. And all of us know such persons— indeed, none of us knows anyone except such persons.

There is, then, pain in the final stage. And this will prevent the final stage from being one of complete goodness, if pain is an evil. It seems to me to be perfectly clear that pain is an evil, and, indeed, the contrary view is seldom, if ever, maintained. Even those thinkers—and they are few—who would maintain that pain was not intrinsically an evil, would always, or almost always, admit that undeserved pain was an evil. And sympathetic pain cannot be said to be deserved, since it does not arise from any moral defect in the person who feels it.

900. Can we arrive at any conclusion as to the proportion which the evil in the final stage bears to the good in that stage? The only evil in the final stage will be pain. If knowledge, or emotion, or virtue, or amount and intensity of consciousness, or harmony, are any of them good, then in respect of them the final stage is a state of unmixed good. And it seems clear to me that all the first four of these *are* good. The existence of this good, with no corresponding evil to balance it, is a fact of great importance, but it will not help us to any quantitative estimate. For we cannot, I think, lay down any general rules as to the relative value of goodness of two different sorts—pleasure and virtue, for example—although we are often able to decide, by an ultimate particular judgment of value, whether a particular instance of good of one sort has more or less value than a particular instance of good of another sort. If we are to get any quantitative result in our present enquiry, it must be as to the relative amounts of pleasure and pain in the final stage.

901. Let us give the name of Original to that good or evil which is not sympathetic pleasure or pain. We shall have, in the first place, sympathetic pleasure and pain in sympathy with original good and evil, and, again, we may have sympathetic pleasure and pain in sympathy with sympathetic pleasure and pain.

When one self, *G*, sympathizes with another self, *H*, it may be accepted (subject to one possible qualification, which will be discussed later) that the amount of pleasure and pain in his sympathy will be proportionate to the amount of original good and evil which he perceives *H* as possessing. In the final stage there is no error, and, therefore, when *G* perceives *H* as having, in respect of any particular stage, good or evil, or both good and evil in certain proportions, then it will be true that *H*'s state is, in this respect, what G perceives it as being. Thus the amounts, in the final stage, of G's pleasure and pain from sympathy with H will be in proportion to the amounts of good and evil in H.

Now the original value of H's final stage will, as we have seen, be unmixedly good, since the only evil in the final stage is sympathetic pain. The original value of H's pre-final stages will probably be partly good and partly evil. But, since the value of the final stage is infinitely greater than that of all the pre-final stages, it follows that the original value of all the stages, final and pre-final together, will be infinitely more good than evil, though the evil—what there is of it—is just as real as the good.

G's sympathy, then, in the final stage, with the whole of H's original good and evil, will be infinitely more pleasurable than painful, though it will contain real pain. The same will be the case with G's sympathy with any other self whom he perceives. Thus G's sympathy with original value will be infinitely more pleasurable than painful. And, as this will be the case with other people's sympathetic pleasure and pain, G's sympathy with sympathetic pleasure and pain will also be infinitely more pleasurable than painful.

Thus the value of the final stage of any self will be infinite and unmixed non-hedonic good, together with hedonic good and evil, where the hedonic good (consisting of original pleasure and sympathetic pleasure) will be infinitely greater than the hedonic evil (consisting exclusively of sympathetic pain).

As the value of the final stage in each self is infinitely greater than the value of the pre-final stages, it follows that, whatever the value of the pre-final stages, the value of both taken together in all selves—that is, the total value in the universe—is infinitely more good than evil.

902. I said above that there was one possible qualification to the proposition that the amount of pleasure and pain in G's sympathy for H will be in proportion to the amount of original good and evil which he perceives H as possessing. We find by experience that we sympathize with present good and evil much more than with past good and evil. Now that which appears, *sub specie temporis*, as present, is really that which is at the same stage of the C series—the inclusion series—as the perception of it as present. And that which appears, *sub specie temporis*, as past, is really that which is at a stage of the inclusion series which is included in the stage at which it is perceived as past. Now, of the original value which is sympathized with in the final stage, all which is itself in that stage is unmixed good. The evil is all in the stages which are *included* members in the inclusion series. If, then, the greater sympathy with the apparent present than with the apparent past comes from their real positions in the inclusion series, we shall, in the final stage, sympathize with good more completely than with evil. But I do not think that there is any reason to hold that the greater importance which we attach to the present is due to the real relations of what appear as present and past. It is just as possible that it depends on the fact that they do appear as present and past. And this does not happen in the final stage, where nothing appears as temporal. But, however this may be, it does not affect our conclusion that the amount of sympathetic pleasure in the final stage is infinitely greater than the amount of sympathetic pain in it, and that, consequently, the total amount of good in that stage, and so the total amount of good in the universe, is infinitely greater than the corresponding amount of evil.

903. This conclusion, as we have seen, depends on our earlier conclusion that the value of the final stage was infinitely greater than that of the pre-final stages. If we had not accepted that earlier conclusion, what should we still know about the final stage? We should know that it was a state which, in respect of knowledge, emotion, virtue, and amount and intensity of consciousness, was a state of unmixed goodness, and of goodness which was very great in comparison with anything we now experience. We should know that it was a state which contained pleasure which was very great in comparison with anything which we now experience. We should know that it contained no evil but pain, and no pain but sympathetic pain. And we should know that it was a state which, *sub specie temporis*, we should reach in a finite time, and which, *sub specie temporis*, would never end. But we should have no definite knowledge as to the proportion in it of good and evil, and we should have no knowledge at all as to the proportion of good and evil in all the stages taken together. Our knowledge on these points is dependent on our acceptance of the infinite value of the final stage.

CHAPTER LXVIII

CONCLUSION

904. If we know that we shall, in a finite time, reach an endless state which is infinitely more good than bad, we know what is doubtless a very important fact—the most important fact—about the future. But there are other questions which are by no means unimportant. For the time which separates us from this final and endless state, while it must be finite, may be of any finite length. How long it may be we cannot tell. And we have seen (Chap. LXIII, pp. 381–382) that there is, not indeed a certainty, but a strong presumption, that that length will be very great compared with the length of a human life, or, indeed, with the entire time known to history.

It concerns us very greatly, therefore, to know what is the value of that part of our lives which falls between the present and the final stage. Can we tell anything about this?

905. It might seem at first sight as if we could count on a steady improvement during this period. The final stage is one of very great goodness as compared with the present. The difference between the two depends on the fact that the final stage is one of complete perception of substances of which the present stage is one of only incomplete perception. And, as we pass, *sub specie temporis*, from earlier to later, and approach the final stage, the amount of perception increases steadily, since each term in the series contains the amount in the previous term and also an addition to it. We might be tempted to think that this would involve a steady increase in the amount of goodness of each successive stage—though what the rate of increase would be must remain uncertain, since we know neither how great the increase would be, nor over how long a future it would be spread.

906. But this conclusion would be unwarranted. We have seen already (Chap. XLVI, pp. 218–220) that two of the qualities which have been held to determine goodness—amount and intensity of consciousness, and amount and accuracy of knowledge—oscillate in such a manner that, in going from earlier to later, while they sometimes increase after they have diminished, they sometimes diminish after they have increased. It follows from this that a uniform increase in the amount of perception does not involve a uniform increase in those characteristics. They can diminish while it increases, and therefore it is also possible that, while it increases, they can increase at a slower rate, or remain unchanged.

And similar oscillations may be observed in the other qualities which have been held to determine goodness. With regard to happiness it is obvious that, as time progresses, I am often first more happy, then less happy, and then more happy again. With regard to virtue the matter is not quite so clear. So many different circumstances may come into account when we have to consider how virtuous a man is, that we cannot be as certain that amounts of virtue change as that amounts of happiness change. Still there can be no practical doubt that a man is often first more virtuous, then less virtuous, and then more virtuous again. And there seems the same certainty about those emotional qualities which have been held to determine goodness. In so far again as any definite meaning can be assigned to the criterion of harmony, it would seem that the amount of it possessed by any self oscillates in the same manner.

Thus, with regard to each of the qualities which have been held to determine goodness, we find that it need not increase regularly as time goes on, and, indeed, that it may even for a certain time diminish. If this is the case with each of these qualities, it must be held to be the case with goodness itself. It is, of course, not absolutely impossible that, whenever one characteristic which determines goodness temporarily diminished, another should always so increase that the net result was a steady increase of goodness. But there is nothing whatever to make us believe that this coincidence—for it could scarcely be anything more—does take place. The state of each self, then, does not always get steadily better as the increase of perception increases, and as, *sub specie temporis*, time goes from earlier to later. And not only does it not always get better, but, owing to oscillations, it may for a period get worse.

907. What can we determine about these oscillations? (It is only those of them which are, *sub specie temporis*, in the future, which have any practical interest for us, but our information about them is neither greater nor less than our information about those in the past.) Since time is finite, and since every oscillation must have a duration in time, it follows that the number of such oscillations must be finite, and that the duration of each must be finite.

The length of such oscillations is more important to us than their number. If, that is, there is oscillation over a given time, it is less serious if there are many short oscillations than if there is one long one. For this there are two reasons. If there is one long oscillation, one side of it will be a long and unbroken period of deterioration, and the prospect of this, I think, would always be more depressing than that of many short periods of deterioration, alternating with periods of recovery. In the second place, the longer the period of deterioration, the lower, *cæteris paribus*, will the self fall below the position which he had previously attained. If, *e.g.*, no period of deterioration could be longer than a life in a single human body, the limits of deterioration for any self would, *cæteris paribus*, be narrower than if it could last a thousand such lives.

908. As to the length of the oscillations—and consequently of the periods of deterioration—we know, so far as I can see, only two things. We have seen that they must all be finite in length. And we see by observation that they can in some cases last for the whole of an adult's life in one body. It is, I suppose, impossible to be absolutely certain in the case of any particular man that his state is on the whole worse just before his death than it was when he was twenty. But there are many cases in which it would be agreed that this conclusion was almost certain, and it is impossible to suppose that we are mistaken in every one of them. And, if one such judgment is correct, it proves that a period of deterioration may last as long as this—broken, no doubt, by minor oscillations, but still being, as a whole, a period of deterioration.

Beyond these two points, I see nothing that we know. The nature of the series in which the oscillations occur will tell us nothing more—indeed it could not tell us that there were oscillations at all. We only know that fact by observation of the oscillations which fall within our experience. And by observation we can only know that they do occur, and that they may endure for the greater part of a single life. For it is impossible, at any rate at present, to carry our observations in any case beyond the limits of a single life. Even if I did happen to know, personally or in history, two lives which were, in fact, successive lives of the same self, I should not know that they were lives of the same self.

909. It might seem as if we could observe, in a nation or a race, periods of deterioration which extend for much longer than a single life. But this would be a mistake. Nothing is intrinsically good or bad, as we have seen, except either selves or parts of selves, and the deterioration of which we are speaking must in every case be deterioration within some particular self. Nations and races, indeed, are made up of selves. But when a society is such that individuals enter it and leave it, replacing and being replaced by others, it is not permissible to argue from change or want of change in the state of the society to change or want of change in the individuals who pass through it.

If, for example, we annually visited a class-room in a school, and no changes took place in the assignment of class-rooms, we should find in that room each year boys who were learning the same lessons, making similar mistakes in them, and showing on the average the same amount of knowledge. But it would be wrong to infer from this that no progress was made in the class-room. The unchanged standard of knowledge would be due to the fact that, as the boys made progress, they left the form, and were replaced by others who had that progress still to make.

And, further, if the assignment of the class-rooms was changed every year, so that the room occupied in one year by the Upper Fourth was occupied in the next by the Lower Fourth, and in the next by the Upper Third, we should at each visit find a lower standard of knowledge in that room, which would leave it quite possible that every boy who had been in the room at any of the three visits knew more at the date of the third visit than he did at the date of the first.

Thus, if we could be certain that the state of a nation, of a race, or of the inhabitants of a planet was worse at the end of a thousand years than it was at the beginning, this would show nothing as to the length of periods of deterioration, unless we had reason to believe that, during this period, all selves, or most selves, who died in the society in question, were re-born as members of the same society. And I do not see on what grounds this view could be maintained.

910. There is thus no evidence for asserting that any deterioration does last longer than a single life. It is true that it sometimes happens that a man dies while he is undergoing an active process of deterioration, and it might be asked whether this does not give a presumption that the same factors which produced the deterioration before the death of this particular body will continue to produce it afterwards. On the other hand, death is certainly an important event in the life of the person who dies, and it is conceivable that its effect should in all cases be to check any process of deterioration which may be in progress when it occurs. But, while it is conceivable that it should be so, I do not see that there is the slightest reason why it should be so[1]. And thus, while we have no right to assert that any single deterioration does last longer than a single life, there is no impossibility, or even improbability, in its lasting through many lives.

911. We must realize that, here as elsewhere, we get our knowledge about the universe in two ways, and that it is impossible for us to ascertain—if the expression is permissible—the common scale of their magnitude. On the one hand, we can arrive at conclusions which are valid about the whole universe, or about everything in the universe which has a certain nature. On the other hand, we each of us know—partly by our own observation, and partly by what we learn from others—certain facts about that part of the universe which is open to such observation by us and by our fellow men. But we do not know what proportion this field of observation bears, either in extent or duration, to the whole universe. As to extent, the universe may be infinitely

---

[1] We must not forget that, since deterioration has been taken to mean passing from a better state to a worse, it would include passing from a state of happiness to a state of misery, or of less happiness.

greater than that field, and there seems good reason to regard it as very much greater. In duration, indeed, it cannot be infinitely greater, but it may exceed it in any finite proportion.

This insignificance of the field of our observation, as compared to the whole universe, is a result which is in some ways unattractive. It is disappointing when we realize that what can be observed is so small in comparison with the whole that it is impossible to obtain any information about the whole by induction from what is observed—to argue, for example, that the whole development of the universe will go in a certain direction, because it has done so in this planet since the dawn of history. And—passing from theoretical to practical interests—this greatness of the universe reduces to insignificance within it, not only the importance of a single self, but the importance of all those groups of selves which we are accustomed to regard with sympathy or loyalty—nations, races, and the human race itself. Our hopes that our aspirations, for ourselves or for others, can be realized, must depend, not on any importance which we or they can have relatively to the rest of the universe, but on general considerations—such as those brought forward in this work—which indicate that the nature of the universe is such that the eventual good of each self is secured by it. And even those Idealists who accept the view that the nature of the universe does secure the good of each self, seem generally unwilling to adopt a view which makes the selves that we know numerically insignificant in the universe. Finally, the conclusion that the time to be passed through before the goodness of the final stage is reached may have any finite length, cannot be altogether attractive to those who feel how far our present life is from that great good.

912. Hegel is perhaps the strongest example of this unwillingness to accept the largeness of the universe. The suggestion that conscious beings might be found in other planets besides this seems to have roused in him that special irritation which is caused by anything which is felt to be unpleasant, and which cannot be proved impossible. And, while he did not explicitly place any limits to the development of the universe in time, he seems to have regarded its significance—and Hegel would scarcely have held that it could have continued without developing fresh

ignificance—as pretty well exhausted when it had produced the
urope of 1820.

913. But the universe *is* large, whether we like its largeness
r not. And, if the conclusions which we have reached as to the
oodness of the universe are true, the greater the extent of the
niverse the greater balance of good will it contain. Duration is
 a different position, since increase of duration increases the
mportance of the only part of the universe in which original
vil is to be found—that is, the pre-final stages. The shorter the
pparent time which separated us from the final stage, the better
ould it be for us. And the only limitation we have found for
hat time is that it must be finite.

Nor can we limit the evils which may meet us in this future
ny more than we can limit its duration. There may await each
f us, and perhaps await each of us in many different lives,
delusions, crimes, suffering, hatreds, as great as or greater than
ny which we now know. All that we can say is that this evil,
however great it may be, is only passing; that our lives are, with
however much oscillation, gradually approximating to a final
stage which they will some day reach; and that the final stage
is one in which the good infinitely exceeds, not only any evil
co-existent with it, but all the evil in the series by which it is
attained. And thus the very greatness of evil which we endure
gives us some slight anticipation of the greatness of the good
which outweighs it infinitely.

Of the nature of that good we know something. We know that
it is a timeless and endless state of love—love so direct, so
intimate, and so powerful that even the deepest mystic rapture
gives us but the slightest foretaste of its perfection. We know
that we shall know nothing but our beloved, and those they love,
and ourselves as loving them, and that only in this shall we seek
and find satisfaction. Between the present and that fruition there
stretches a future which may well need courage. For, while there
will be in it much good, and increasing good, there may await
us evils which we can now measure only by their infinite insignificance as compared with the final reward.

INDEX

OF TERMS DEFINED OR TREATED AS INDEFINABLE

For terms not included in this index see index to Volume I

References are to Sections